T0358634

Strategic and Sustainable Management of Workplace Facilities

This book introduces the reader to contemporary issues in the management of facilities in the African context and includes case studies from across the continent and internationally. It consolidates theory and practical information useful for managers and researchers across Africa and other developing countries. It covers a cross section of the key elements of facilities management, including customer relations management, emergency preparedness, development of a facilities strategy, sustainable buildings management, and management of intelligent buildings in developing countries.

The book has been developed from a combination of degree level and professional course materials and academic resources and is therefore suitable for use by students on undergraduate and postgraduate degree programmes, professional short courses, and for practicing facilities managers and members of IFMA, SAFMA, IWFM, HEFMA, RICS, Association of Facilities Management Practitioners of Nigeria, Egypt Facility Management Association, Botswana Facilities Management Association, and the Africa Facilities Management Association as a key reference. This book is key reading for anyone:

- Studying for a degree in Facilities Management, Real Estate, Estate Management, Workplace Management, or Surveying
- Wanting to be more strategic in their facilities management and operations
- Transitioning into a facilities management role from another profession
- Benchmarking and analysing the performance of their built assets
- Training themselves or their staff in emerging areas such as workplace productivity, sustainability, and intelligent buildings systems
- Interested in researching emerging areas of facilities management in developing countries

This is the essential guide to the growing field of facilities management in some of the world's fastest developing countries.

Yewande Adewunmi-Abolarinwa is an Associate Professor at the University of Witwatersrand, South Africa, where she lectures undergraduate and graduate students in Property Studies and Facilities Management. She pioneered and established the facilities management program at the University. She was also part of the team that established the facilities management program at the University of Lagos. Yewande has also published in reputable research outlets and consulted in FM trainings and projects in Africa and the UK.

Strategic and Sustainable Management of Workplace Facilities

Yewande Adewunmi-Abolarinwa

LONDON AND NEW YORK

Designed cover image: © Nitat Termmee/Getty Images

First published 2025
by Routledge
4 Park Square, Milton Park, Abingdon, Oxon OX14 4RN

and by Routledge
605 Third Avenue, New York, NY 10158

Routledge is an imprint of the Taylor & Francis Group, an informa business

British Library Cataloguing-in-Publication Data
A catalogue record for this book is available from the British Library

ISBN: 978-1-032-65667-0 (hbk)
ISBN: 978-1-032-63872-0 (pbk)
ISBN: 978-1-032-65666-3 (ebk)

DOI: 10.1201/9781032656663

Typeset in Times New Roman
by SPi Technologies India Pvt Ltd (Straive)

Contents

Acknowledgements

Appreciation goes to God first for making the project possible; without him, I can do nothing. Many thanks also go to my family, especially my husband - Prof Joshua Abolarinwa, for his patience and support and for helping with the building energy management systems and the process mapping. Also, many thanks to my parents (late Professor and Mrs Adewunmi) for their mentoring and prayers. Thanks to my siblings and my in-laws, who provided support, assistance and beneficial resource information for the book. Thanks to my nephew - David Akintunde, and his friend Malvin Maparutsa, for helping with some of the diagrams.

I want to thank the organisations that contributed case studies to the book: AfricaWorks, South Africa and Nigeria, PAWA254, Kenya, Makanak, Egypt, BaseCamp Initiative, Ghana, Officephase Rwanda, Workville, New York, Absa Bank, Kenya, American University of Cairo, Schneider Electric, South Africa, Vetasi, South Africa, Eskom, South Africa, Discovery, South Africa, Euromonitor International, International Facilities and Property Information Limited, United Kingdom, and Ayo Abolarinwa and Associates (AAA), Abuja, Nigeria.

Special thanks are due to the following professional bodies for their significant contributions in the form of case studies and documents – the International Facility Management Association, Royal Institution of Chartered Surveyors, Higher Education Facilities Management Association, South Africa, Africa Facilities Management Association, the Green Building Council of South Africa, British Standards Institute, the US Green Building Council and the Continental Automated Building Association. Your contributions have been instrumental in shaping the content and quality of this book.

I would also like to thank all my past and current students at the University of the Witwatersrand, the University of Lagos, and the Namibia University of Science and Technology. I am particularly grateful for the contributions of Mr Steven Molloy in environmental management systems, Mr Ashley Naidoo, Mrs Nontando Bukaza and Ms Neo Mosebo in intelligent building systems, Mr Tebogo Motlung and Mr Michael Mafuze, Mr David Pierre Eugene, Mr Justin Brandon, Mr Matla Selamolela and Mr Patrick Katabua for assisting with information regarding FM strategy. Thanks to Mr Oluwaseun Alabi for helping with the work area recovery strategy. Thanks to Mr Pieter Jobert for helping with some help desk information. Thanks to Mr Hendrix Wannenburg for helping with information on environmental systems. Ms Lungile Mashiyane, for the benchmarking information. Mrs Lilias Makashini, thank you for providing some background information for Zambia. Your dedication and hard work have greatly enriched this book. Thanks to Panduleni Ntinda for your assistance.

I also thank my professional colleagues and mentors, Prof Modupe Omirin and Prof David Root, for encouraging me to write a book. I appreciate the beneficial suggestions of

Dr Margaret Nelson and Prof Kathy Mitchell in making the book better. Many thanks to my former colleagues at the University of Lagos, especially Prof Hikmot Koleoso, for your support and the late Mr Chika Udechukwu for encouraging me to write a book. Many thanks to the encouragement and support provided here by colleagues at the University of the Witwatersrand - Prof Kola Akinsomi and Dr Prisca Simbanegavi, and the Head of School, Prof Laryea. Also, thanks to my outside work mentors, Prof Bernard Williams of IFPI, UK and Mr Ayo Abolarinwa from AAA, Nigeria, for the benchmarking resources. Thanks to Mr Sherif Maged of AFMA for the net zero carbon information and Mrs Irene Thomas Johnson of IFMA foundation and JLL, USA, for assisting with getting some useful international information.

Preface

The concept and discipline of facilities management (FM) is not widely accepted in many developing countries but has witnessed a significant shift in recent years. Facilities management is still confused with other real estate management functions, each having its focus. A noticeable trend of outsourcing services to companies has emerged, offering a transformative potential for the sector's expansion. This shift in dynamics allows organisations to concentrate on their core business and presents numerous opportunities for the sector's growth. Facilities contribute to many organisations' budgets, underscoring the importance of considering the life-cycle cost of operating these facilities. With managing the shifting economic landscape in developing countries, FM helps organisations save costs through facilities and infrastructure problems; the workplace managers help manage issues and ensure business continuity. The FM market also plays a significant role in African employment since many are employed in the sector. Since the worldwide COVID-19 pandemic, FM has contributed to managing the workplace by helping to put the right measures in place for a safe return to work.

With up to 20 years of experience doing research in the FM space and having published in journals in facilities management, benchmarking, corporate real estate, and sustainability, the author has been able to disseminate her research in the FM space and related fields. Also, having collaborated with other colleagues and worked collaboratively in establishing programs and teaching FM in Nigeria, South Africa, and Namibia, namely at the University of Lagos, the University of the Witwatersrand, and the Namibia University of Science and Technology (on sabbatical). There is a need to improve the industry's knowledge base regarding the field's strategic content so that FM can be better positioned. There is a call for decolonisation of the curriculum, and students get local and international knowledge since they are mainly local change agents. The experience from my research and teaching will be useful to researchers, students, practitioners and training providers who need more knowledge in the field. In line with RICS requirements, the strategic aspect of FM includes strategy, planning, service delivery and review. Also, only a few research studies in developing countries have focused on FM's strategic content, especially in customer relationship management.

There is also a pressing shift towards sustainable facilities management as building managers are custodians of facilities where activities that damage the environment occur; they are positioned to develop strategies to help limit adverse activities. Some of the trending topics from IFMA in the last 15 years have been strategy, emergency preparedness, sustainability, globalisation, use of technology, intelligent buildings and well-being. Few studies have focused on emergency preparedness and intelligent buildings in the workplace in facilities management in developing countries.

This book covers the following subject areas:

- The meaning, nature, and scope of facilities management in Africa and other parts of the world
- Work productivity
- Strategy
- Customer relationship management
- Emergency preparedness and business continuity
- Benchmarking
- Sustainability
- Intelligent buildings

These topics are some of the essential parts of sustainable FM and FM strategy and the areas where I have been researching and teaching since we launched the FM programme at the University of the Witwatersrand and formed part of the curriculum. Also, my doctoral research at the University of Lagos focused on benchmarking, where I acquired a knowledge base in benchmarking. During my doctoral training at the International Facilities and Property Information Ltd in the UK, I understood the rudiments of benchmarking in practice, and I have been involved in collaborative projects in benchmarking in Nigeria and South Africa. The book discusses existing knowledge in the subject areas and conceptualises useful guidelines and frameworks for managing workplace facilities in developing countries. It combines theory with practice by bringing in case studies from different countries in Africa and internationally.

List of Acronyms

A1-A5	Upfront of Embodied Carbon
AAA	Ayo Abolarinwa and Associates
AC	Alternating Current
AEPB	Abuja Environmental Protection Board
AFM	African Facilities Management Association
AHU	Air Handling Unit
AI	Artificial Intelligence
AM	Asset Management
ANSI	American National Standards Institute
ATS	Automatic Transfer Switch
AUC	American University of Cairo
AutoCAD	Computer-aided Design Software
AWS	Alternative Workplace Strategies
B2B	Business-to-Business
BACnet	Building Automation and Control Networks
BAS	Building Automation System
BASE	Building Assessment Survey and Evaluation
BBC	Buy Back Centres
B-BBEE	Broad-Based Black Economic Empowerment
B-C	Embodied Carbon
BCI	Business Continuity Institute
BCM	Business Continuity Management
BCP	Business Continuity Planning
BEEG	Building Energy Efficiency Guideline
BEMS	Building Energy Management Systems
BIA	Business Impact Analysis
BIFM	British Institute of Facilities Management
BIM	Building Information Modelling
BMS	Building Management System
BPR	Business Process Engineering
BREEAM	BRE Environmental Assessment Method
BREEM	Building Research Establishment Environmental Assessment Method
BSC	Balanced Scorecard
BSI	British Standard Institute
BUS	Building Use Studies Occupant Survey
BWA	Bernard Williams and Associates

CABA	Continental Automated Buildings Association
CA-CP	Clean Air-Cool Planet Carbon Calculator
CAFM	Computer-aided Facilities Management
CAPEX	Capital Expenditures
CAT	Climate Action Tracker
CCTV	Closed Circuit Television
CDC	Centres for Disease Control
CFM	Corporate Facilities Management
CFP	Carbon Footprint
CIB	Chartered Institute of Building
CMMS	Computer Maintenance Management Software
CO2	Carbon Dioxide
COO	Chief Operating Officer
COP26	Conference of the Parties
CPM	Corporate Property Manager
CREM	Corporate Real Estate Management
CRM	Customer Relationship Management
CSFs	Critical Success Factors
CSIR	Council for Scientific and Industrial Research
DC	Direct Current
DR	Disaster Recovery
DRC	Democratic Republic of Congo
DRI	Disaster Recovery Institute
DRM	Disaster Regulatory Management
DVRs	Digital Video Recorders
ECM	Energy Conservation Measures
EcoStruxure	Schneider Electric's open, interoperable, IoT-enabled system architecture and platform
EDGE	Excellence in Design for Greater Efficiencies
EEAA	Egyptian Environmental Affairs Agency
EIA	Environmental Impact Assessment
EIB	European Installation Bus
EIBG	European Intelligent Building Group
EMAS	Eco-Management and Audit Scheme
EMS	Environmental Management System
EN	European Standard
EPA	Environmental Protection Agency
EPBD	Energy Performance of Building Directive
EPR	Extended Producer Responsibility
ESG	Environmental, Social, and Governance
EU	European Union
FCP	Foreign Corrupt Practices Act
FCUs	Fan Coil Units
FEPA	Federal Environmental Protection Agency
FEWS NET	Famine and Early Warning Systems Network
FM	Facilities Management
FMS	Facilities Management System
FOM	Facilities Operation and Management
FSOP	Facilities Strategic Optimisation Project

FSS	Facilities Support Service
FT	Full time
GBC	Green Building Councils
GBCSA	Green Building Council of South Africa
GCT	Global Consumer Trend
GDP	Gross Domestic Product
GFA	Gross Floor Area
GHG	Green House Gases
GHO	Green Health Observatory Indicator
GIA	Gross Internal Area
GPRS	Green Pyramid Rating System
GUI	Graphical User Interface
HEFMA	Higher Education Facilities Management Association
HFC	Hydrofluorocarbon
HMI	Human Machine Interface
HOD	Head of Department
HR	Human Resources
HRM	Human Resource Management
HSE	Health, Safety, and Environmental
HVAC	Heating, Ventilation and Air Conditioning
IAQ	Indoor Air Quality
IB	Intelligent Buildings
IBI	Intelligent Building Institute
IBS	Interdisciplinary Business Systems
ICS	Integrated Communication Systems
ICT	Information Communication and Technology
ID	Identification
IDIC	Identify, Differentiate, Interact, and Customise
IEC	International Electrotechnical Commission (Fieldbus Standards)
IEEE	Institute of Electrical and Electronics Engineers
IEQ	Indoor Environment Quality
IFMA	International Facility Management Association
IFPI	International Facilities and Property Information Ltd
IFRC	International Disaster Assistance in Southern Africa
IIWTMP	Integrated Industrial Waste Tyre Management Plan
IoT	Internet of Things
IP	Internet Protocols
IPD	Invest Property Databank Limited Company
ISO	International Standards Organisation
IT	Information Technology
IWFM	Institute of Workplace and Facilities Management
IWMS	Integrated Waste Management System
JIBI	Japanese Intelligent Building Institute
JLL	Jones Lang LaSalle Incorporated
KL	Kilolitre
KPI	Key Performance Indicator
kVA	Kilo-volt-amperes
LAN	Local area network
LASEPA	Lagos State Environmental Protection Agency

LAWMA	Lagos Waste Management Authority
LEED	Leadership in Energy and Environmental Design
LMI	Local Management Interface
LON	Local Operating Network
LonTalk	Communication protocol for networking devices over media
LonWorks	Local operating network for networking platforms
LTS	Long-Term Strategy
MBP	McDonough Bolyard Peck
MD	Maximum Demand meter
MIF	Macro Infrastructure Framework
MIMO	Multiple-Input Multiple-Output
MP-C	Multimedia Personal Computer
MP-V	Music Photo Video
MSW	Municipal Solid Waste
MT	Metric Tonnes
MTN	Mobile Telecommunication Network
MWh	Megawatt Hour
NCCRS	National Climate Change Response Strategy
NDC	Nationally Determined Contributions
NDMA	National Disaster Management Agency
NDMO	National Disaster Management Organisation
NEM	National Environmental Management
NEMA	National Emergency Management Agency
NFA	Non-assignable Floor Area
NGO	Non-Governmental Organisation
NIESV	Nigerian Institution of Estate Surveyors and Valuers
NOSA	National Occupational Safety Association
NWMS	National Waste Management Strategy
NWU	North-West University
OHS	Occupational Health and Safety
PABX	Private Automatic Branch Exchange
PDCA	Plan-Do-Check-Act
PERT	Programme Evaluation Review Technique
PESTEL	Political, Economic, Sociological, Technological, Legal, and Environmental
PF	Power Factor
PHEIC	Public Health Emergency of International Concern
PLC	Programmable Logic Controller
PM	Property Management
POE	Post Occupancy Evaluation
POM	Portfolio Management
PowerTag	Multi-socket plug
PowerTap	Energy sensor for monitoring and measurement of power
PPE	Personal Protective Equipment
PROFIBUS	Process Fieldbus
PV	Photovoltaic
QCI	Quality Competitive Index
QEM	Quality Environment Modules
QR	Quick Response
REC	Renewable Energy Certificate

RFI	Request for Information
RFP	Requests for Proposal
RICS	Royal Institution of Chartered Surveyors
ROI	Return on Investment
RS232	Recommended Standard
RSESA	Rivers State Environmental Sanitation Authority
RTO	Recovery Time Objective
SA	South Africa
SABS	South Africa Bureau of Standards
SACU	Southern African Customs Union
SAFMA	South African Facilities Management Association
SANS	South African National Standards
SARS	Severe Acute Respiratory Syndrome
SARS-CoV-2	Severe Acute Respiratory Syndrome Coronavirus 2
SBP	Sustainable Building Project
SCDI	Southern Corridor Development Initiative
SD	Sustainable Development
SDG	Sustainable Development Goals
SERVPERF	Service Performance
SERVQUAL	Service Quality
SFP	Strategic Facility Planning
SGS	Société Générale de Surveillance
SL	High-level structure for modern ISO Standards
SLA	Service-Level Agreement
SME	Small and Medium Enterprises
SMU	Singapore Management University
SOPs	Standard Operating Procedures
SPM	Strategic Property Management
SWM	Solid Waste Management
SWOT	Strengths, Weaknesses, Opportunities, Threats Analysis
TFM	Total Facilities Management
TV	Television
UI	User Interface
UK	United Kingdom
UN	United Nations
UNEP	United Nations Environment Programme
UNFCCC	United Nations Framework Convention on Climate Change
UN-Habitat	United Nations Human Settlements Programme
UPS	Uninterrupted Power Supply
USA	United States of America
VAT	Value-added-tax
VAVs	Variable Air Volume
VOCs	Volatile Organic Compounds
VPN	Virtual Private Network
WBCSD	World Business Council for Sustainable Development
WEHAB	Water, Energy, Human Health, Agriculture, and Biodiversity
WHO	World Health Organisation
WLCA	Whole Life Cycle Assessment
WM	Waste Management

1 Introduction

1.0 Introduction

Facilities management is "an organisational function which integrates people, places and processes within the built environment to improve the quality of life of people and the productivity of the core business" (International Facility Management Association (IFMA), 2003). FM is practiced at three levels in any organisation – the operational, tactical, and strategic levels. The facilities manager should aim to practice FM strategically, which involves establishing goals in response to the intended use of the FM functions and conducting long-term planning while also considering the outside world's needs. The contributions of FM at the strategic level are covered in Chapter 3 of this book.

Facilities management and other building support practices are not yet properly conducted in developing countries, as compared to developed countries. Previous research efforts in Africa showed that the role of facilities managers is yet to be recognised at the design stage. As such, buildings are designed without fully considering the operational requirements of maintaining those buildings and meeting users' needs. In many instances, maintenance of buildings is not a priority, which results in problems with the safety, comfort, and usability of the buildings.[1] Other problems include disruptions in the organisation's activities, loss of revenue, loss of productivity levels, and inability to comply with some local and international regulations. FM can add value to an organisation as it helps to manage the relationship between the different stakeholders in an organisation, thus improving the satisfaction of building users regarding services; it is also of benefit with regard to providing value for money, which assists with the performance and quality of services and not just cost savings in the management of such services. FM also contributes to society in that it is at the forefront of sustainability in many institutions.[2] The contributions that FM makes to sustainability are discussed in Chapter 7. The focus of this chapter is to introduce the concept of facilities management.

1.1 Facility

Before we move to what facilities management is, let us begin by looking at the meaning of "facility", since facilities management is about the management of facilities. A facility can be defined as a physical building or installation. It might be the entire building, a section of a building, or a building together with its location and surroundings. A facility can also encompass both physical buildings and the services provided therein.[3] Lifts, sewage treatment plants, security devices, water tanks, laundry equipment, air conditioning, and catering equipment are some examples of facilities. This shows we can have school, hotel,

DOI: 10.1201/9781032656663-1

Figure 1.1 Word cloud of the meaning of a facility.

Source: Author

medical, and residential facilities (see Figure 1.1). The use or type of organisation will determine the facility type.

1.2 Facilities Management

The phrase “facilities management” still needs clarification in many countries. For instance, upon enrolling in a course in FM in Africa, some students were questioned about their grasp of the idea of FM. The following are some of their preconceived notions about FM:

> I thought it would be an addition to a subject called building services but in more depth. In a way, it is similar to some aspects of building services related to facilities management.
>
> I thought it was about repairing toilets and fixing door locks that would not open. How misinformed I was! Facilities Management is a broad and vast sector. It can range from elements of human resources all the way to security and cleaning services. All the elements then work together to provide the optimum user experience for the building users.
>
> I thought FM was just a branch of property management. I thought FM was only the maintenance and cleaning of a building. I just found out that FM is a discipline on its own. I was unaware of how the facilities manager is important to the core business and how he can help the business achieve profit maximisation and reduce expenses.
>
> I understood that facilities management was only based on security, health and property cleaning. Facilities managers had the sole role of ensuring that residential properties were suitable for residents and that commercial properties were effective working spaces that could enhance employee productivity. They only focus on what meets the eye and only manage so much. In addition, facilities managers are similar to property managers with fewer responsibilities.
>
> Initially, I did not have as much knowledge of the facilities management industry, and I believed the work was narrowed down to just having the technical role of assessing building components and ensuring that strategies are in place to do with how you

Figure 1.2 Word cloud of some of the thoughts about the concept of FM from interviews with FM students at a university in South Africa.

Source: Author

should go about repairing them or running maintenance on them. Facilities management for me, as mentioned above, was more to ensure the elevator doors operate and the HVAC system functions optimally while carrying a toolbox around and running inspections due to the job being demanded on physical labour, and nothing more outside of that.

The first time I came across the term "Facilities Management," I took the most literal definition where I assumed it is the management of facilities, which would mainly focus on the cleaning, security, and maintenance of a facility.

The interviews above show that FM still needs clarification with regard to other aspects of real estate management, such as building maintenance, property management, and asset management. Figure 1.2 shows that the words mostly associated with FM include "building," "property," "management," and "FM managers,"

A benefit of facilities management is that most buildings or facilities are large investments for organisations, and the cost savings and quality obtained from FM can affect an organisation's net profit. It helps organisations to focus on their core business by providing support services; since the running of an organisation is a complex, coordinated process, it is good to take an integrated view. FM enhances an organisation's culture and image, enables change in space use, and provides a competitive advantage to its core business.

1.3 Real Estate Management

People usually need clarification on the different aspects of real estate management, including building maintenance, property management, and asset management. Hence, there is a need to examine the meaning of and differences between these aspects. One of the focuses of real estate management is facilities management. As explained by Tuomela and Puhto (2001), citing Leväinen (1997), real estate management may be seen from three distinct perspectives – asset management, property management, and facilities management.

1.3.1 Facilities Management

According to IFMA, FM covers multiple disciplines to ensure the built environment functions by integrating people, place, process, and technology. FM provides the necessary tools and services for effective buildings and work environments, which results in the best operation of an organisation's assets and the realisation of its primary goals. In other words, facilities management may help an organisation realise its goals by improving the performance of a facility.

1.3.2 Property Management

This is the general term; it stands for managing and carrying out the daily responsibilities of operating real estate assets. It encompasses administrative management tasks such as collections, record keeping, and reporting; marketing management tasks such as tenant selection and rent schedules; physical management tasks such as maintenance, rehabilitation and renovation, space management, acquisition and disposal, and security management may also be performed.

1.3.3 Asset Management

This area includes responsibilities concerning the real estate organisation's purchasing, selling, and portfolio management. Real estate income is the goal of a well-executed asset management strategy. Asset management is the general process of managing all aspects of real estate assets, including acquisition and disposition, developing management strategies, overseeing building and real estate operations, financial management, and all facets of accounting and reporting on real estate held, according to Lapides and Frank (1991), cited in Tuomela and Puhto (2001).

1.3.4 Maintenance Management

Responsibilities include site supervision, root cause analysis, condition monitoring, plant shutdown, materials and components management, and other engineering operations.

In facilities management, the emphasis is on the worker at a workplace who is interested in the space and services that support his or her work or the firm's output, as opposed to property management, where the technical manager is focused on the building and its equipment. The item of interest is also different: facilities management (FM) is focused on space and services, while asset management (AM) is focused on capital, and property management (PM) is on buildings. However, a primary real estate management component is portfolio management (POM).

1.4 Differences Between Property Management, Asset Management, and Facilities Management

Maxi Gold Consulting explained the differences between asset, facilities, and property management (2019). Lease agreements, rents, and occasionally service contract management – i.e., the administration of the numerous agreements with facilities service providers – are part of property management. The building operations and asset value are the property manager's main concerns. It is encouraged by conventional standards and service delivery methods.

The main goal of facilities management is to fulfil the organisation's demands while initially helping to determine those needs. It emphasises the selection and use of resources, the efficient enabling, communication, and management of people, and the creation of true value for money by reducing costs and raising standards. The demands of the organisation influence it. Larger commercial sites with more complicated building management and operation are usually where facilities management is used. In asset management, the owner and investor prioritise company profitability, while the asset manager stresses cost-effectiveness.

1.4.1 The Property Manager and the Facilities Manager

The facilities manager's top priority is increasing the user's primary business effectiveness and productivity. To do this, the facilities manager must consider the entire life cycle of an asset. The facilities manager also focuses on the needs and demands of the end user and occupier workplace. Facilities managers team up with the building user or occupier.

The property manager aims to increase the building's net operating income and value. The property manager considers the planned business cycle of an asset, which is lease-determined, and "owner/tenant relations management" is focused on achieving a balance of interests.

1.5 The Meaning of Facilities Management

There is no standard definition of the term "facilities management." It has been defined in various ways by individual writers, professional bodies, and individual organisations in different parts of the world.

Globally, the definition and scope of facilities management are still up for debate, and different meanings are used depending on the local culture, organisational needs, and individual interests. Even though meanings vary, the effects of globalisation are causing definitions to become increasingly unified. Researchers and practitioners globally have developed definitions that outline the goals and boundaries of FM. These definitions have given FM a unified foundation for continuous advancement.

FM plays a strategically significant role in the company structure. FM addresses various problematic areas, including people, processes, the environment, and health and safety. FM includes the whole property life cycle, including planning, constructing, financing, and operating.

In the UK, Barrett (1995) defined facilities management as "an integrated approach to operating, maintaining, improving and adapting the buildings and infrastructure of an organisation to create an environment that strongly supports the primary objectives of that organisation." With the help of this description, one can see why some individuals started referring to facilities management as business infrastructure management.

Since organisations embrace additional "working environment" components, including IT, finance, and human resources, to create a comprehensive infrastructure for an organisation, FM should be responsible for all services that support the core business, including human resource management (HRM). Instead of using the phrase "facilities management," the terms "service management" or "business infrastructure management" (BIM) can be used.

A closer look at some available FM definitions suggests that repetitive themes have helped FM to forge its identity. It can be summarily defined as: "The integrated management of the workplace to enhance the performance of the organisation" (Tay & Ooi, 2001, p. 359). The first theme is the workplace, where the work is carried out. Thus, it is not limited to commercial office buildings but also includes other types of workplaces, such as medical, educational, and industrial workplaces. Also, FM relates to all organisations since they all occupy space for their work. Furthermore, FM plays a supporting role in enhancing the organisation's performance; it is a strategic tool that enhances the performance of organisations and must be supported in their core activities (Dettwiler, 2008). Finally, an integrated approach is required to practice FM. FM can also be applied not only in workplaces but in residential and even community areas. FM adds value and is not just a means of coordinating an organisation's non-core and property-related activities.

In Africa, professional bodies such as SAFMA have defined FM as an enabler of sustainable enterprise performance through the whole-life management of productive workplaces and effective business support services.

Various authors have also defined FM. For example, from Egypt, Waly and Helal (2010) defined FM as "the comprehensive tasks necessary for the efficient operation of the Facility to allow its occupants to satisfactorily and effectively pursue their daily activities while sustaining the value of the building". In Nigeria, Adewunmi (2014) defined FM as "integrating non-core services with the organisation's people, process, workplace, and technology using a multi-disciplinary approach to deliver the appropriate services needed to add value and achieve the organisation's core objectives." In Kenya, Kavinya (2019) defined FM as the professional management of effective and efficient service delivery of support services in an organisation that integrates people, processes, systems, places, and technology to achieve its goals. These definitions from Africa are emblematic of definitions of FM in other parts of the world. The reason could be the influence of globalisation on FM practice.

Overall, as shown by the many definitions, the concern is that the facilities management (FM) is facing an identity crisis as it operates in "an ever-widening and poorly defined scope of activity." Based on developments in the field, FM meets several criteria that qualify it as a recognised academic discipline; the profession needs to keep working as a multi-disciplinary discipline (Junghans & Olsson, 2014).

1.6 Scope of FM

The scope of FM has evolved in different parts of the world. Property management (real estate), property operations and maintenance, and office administration formed the foundation of FM. The scope of FM is more comprehensive than building operations and maintenance. In the UK, facilities management was believed to cover a wide range of services, including real estate management, financial management, change management, human resources management, health and safety, and contract management, in addition to building management, domestic services (such as cleaning and security), and utility supplies. FM encompasses workplace, facility, support services, property, corporate real estate, and infrastructure. FM's scope covers strategic facilities planning, strategic asset management, asset maintenance management, and facilities service management.

The following nine core competencies are the globally recognised competencies of FM, based on IFMA's position as the most recognised professional institution in the global FM space.

- Preparedness for emergencies and business continuity
- Sustainability and environmental responsibility
- Business and finance
- Human elements

- Strategy and execution
- Operations and upkeep
- Project administration
- Quality\Communication
- Property management, real estate, and technology

In addition, the work of a facilities manager potentially covers the following duties and functions:[4]

- Facility strategic and tactical planning
- Facility financial forecasting and budgeting
- Real estate procurement, leasing, and disposal
- Procurement of furnishings, equipment, and outside facility services
- Facility construction, renovation, and relocation
- Health, safety, and security
- Environmental issues
- Development of corporate facility policies and procedures
- Quality management, including benchmarking and best practices
- Architecture and engineering planning and design
- Space planning and management
- Building operations, maintenance, and engineering
- Supervision of business services such as reprographics, transportation, and
- catering
- Code compliance
- Telecommunications

Generally, support services concerning FM range from building operational services to construction management and real estate services. Also, the scope of FM services comprises nine groups and 61 services.[5] The responsibilities of FM can be grouped into nine significant functions captured under the disciplines of strategic management, maintenance, support services, IT, environment, property management, occupational health and safety, space management, and project management (Figure 1.3). It is not easy to perform all FM roles because it is multi-disciplinary.

FM includes building and service-related tasks and may be divided into hard and soft FM. Environmental management is also included in FM.[6] It is a crucial organisational area because more organisations are linking how they manage their buildings and other workplace assets to how well their daily operations run. FM may consider sustainability to be the most important factor when it comes to demographic changes, a mix of cultures, labour shortage problems, and workplace facilities. The breadth of FM is undervalued.[7]

Within the context of Africa, Amos et al. (2019) provided an example of the classification of the scope of FM in Ghana's hospital sector, which included cleaning, waste management, and estate management, and was based on local hospital sector practice. Based on SAFMA's classification in South Africa, the scope of FM can be categorised into hard services, soft services, business support services (Table 1.1).

In Kenya, Kavinya (2019) believed that the role of FM included resource management, time management, customer relationship management, contract management, project management, budget management, health and safety management, human resources management, and strategic management. In Nigeria, the scope of FM covers strategic facilities

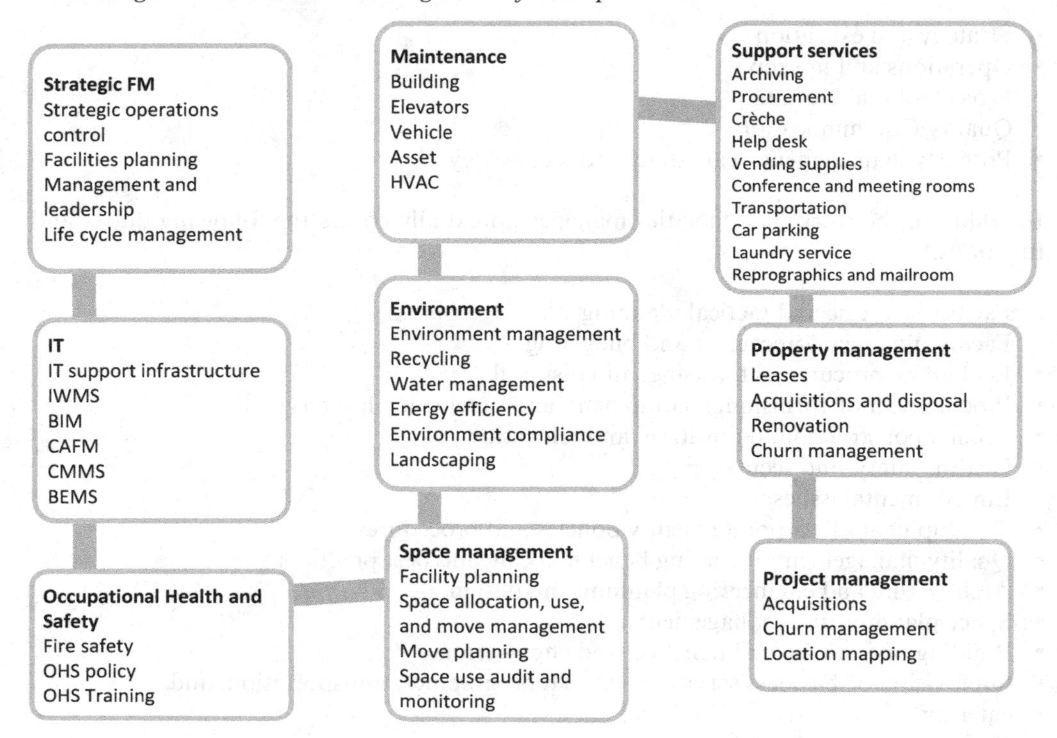

Figure 1.3 Scope of FM.

Source: Author

Table 1.1 Scope of FM in South Africa

Hard services	*Soft services*	*Business support services*
• Access control systems • Air conditioning systems • Audio-visual provision • Boilers • Building maintenance and repair • Building signage • Ceilings • Compactors • Construction • Data cabling • Diesel solution • Electrical service • Emergency generators • Energy • Energy optimisation • Escalators • Facilities information technology	• Anti-industrial espionage services and debugging • Furniture management • Gardening and landscaping • Gym management • Hygiene • Interior plants • Painting • Parking • Pest control • Relocation management • Risk management • Security • Space planning • Interior design • Sport and recreation • Sterile services • Storage	• Air quality • Asset management • Archiving and paper storage • Booking system • Budgeting • Car wash • Catering • Change management • Chauffeur services • Compliance audit • Concierge • Condition assessment • Conference facilities • Contract management • Courier services • Crèche provision • Design reviews • Distribution services

(*Continued*)

Table 1.1 (Continued)

Hard services	*Soft services*	*Business support services*
• Fire services • Handyman services • Health and safety audits • Infrastructure maintenance • Internal signage • Key and lock management • Lift maintenance • Lighting provision • Lubrication maintenance • Mechanical services • Meter reading • Metering and monitoring • Partitioning • Planned maintenance • Plumbing • Pool and fountain management • Pressurised gas • Refrigeration • Re-lamping • Security systems • Structural integrity inspections • Uninterrupted power supply • Utility management • Water treatment	• Tenant installation and management • Warehousing and logistics • Waste management • Water cooler and fresh drinking water supply	• Document management • Environmental management • ESG services • Finance • Florist and landscaping • Green building solutions • Healthcare and pharmacy • Help desk • Human resource management • Inventory control • Laundry • Life cycle management • Mailroom and postal services • Messenger, transport, and taxis • Non-core procurement • Occupational Health and Safety Act • Office and secretarial administration • Performance management and benchmarking • Printing, photocopying, and faxing • Professional consulting services • Project management • Quality management system • Real estate and property management • Space management • Stationery, consumables, and office Supplies • Switchboard and office communications • Technical library • Telecommunication • Training • Travel • Value engineering • Vending supplies • Video conferencing • Workplace planning • Workplace strategy

Source: Author

planning (SFP), facilities operation and management (FOM), facilities support services (FSS), and strategic property management (SPM) (Koleoso et al., 2018) (Figure 1.4). The tasks included under these headings include:

Cleaning, waste management, and sanitation (FSS), Management of maintenance operations (FOM) Facility planning (SFP),Security management (FSS), Fire and accident

Figure 1.4 FM tasks in Nigeria.

Source: Koleoso, H.A., Omirin, M.M., & Adejumo, F. (2018). Comparison of strategic content of facilities managers functions with other building support practitioners in Lagos, Nigeria. *Property Management*, *36*(2), 137–155.

prevention and response programmes (FSS), Safety of life and building users' personal security system (FOM),Operations of utilities and facilities (FOM), Management of parking facilities (FSS), Management of leases of existing building (SPM), Ground and environment maintenance (FOM), Building and facility condition assessment (FOM), Environmental sustainability tasks (SPM), Energy/power planning and management (SFP), Custodial services (FOM), Major maintenance and renovation (FOM), Land use management (SPM), Building risk management (SPM), Status report to client (SPM), Strategic operation cost control (SFP), Government relations (SPM), Building safety training and programmes (FSS), Space utilisation planning (SFP), Management of facility planning data (SFP), Strategic building design and construction control (SFP), Office material and general supplies control (FOM), Management of building health factors and first aid provision (FSS), Space allocation to individual staff members (SFP), Management of telecommunications (FSS), Regulatory financial data management (SPM), Interior designs (SFP), Procurement and purchasing of office supplies and equipment (FSS), Mail services (FSS), Real estate acquisition and disposal (SPM), Network and ICT coordination (SFP), Vehicle maintenance (FOM), Reception, internal/external porterage (FSS), Transportation and management of the company's fleet (FSS), Reprographics and stationery management (FSS), Catering and vending (FSS), Travel services (FSS).

1.7 Hard Services and Soft Services

In Africa, as elsewhere, FM is classified broadly into hard and soft services. Hard services are also described as hard support services. Organisations need well-organised buildings from which they operate and support systems to ensure their core activities can be performed without distraction or to prevent critical functions from suffering. Hard services ensure buildings are as required and that the business is supported with the correct support activities.

An example of a hard service is building support, whereby businesses are supported by ensuring that buildings are run properly. Here, responsibilities include:

a. Space planning
b. Installation of essential operational equipment
c. Maintenance
d. Energy management
e. Waste management
f. Customer services help desk
g. Contract management of supply, consultancy, or specialist involvement
h. Lease management
i. Operating stock control systems
j. Undertaking essential building work
k. Business continuity planning

Soft services are also known as soft support services. They allow working practices that are efficient. The facilities manager must undertake activities that enhance the building users' comfort levels. Soft services are responsibilities that could be more critical to how businesses function but also assist building users. The difference between soft and hard services is that the organisation's function will be fine without soft services. Soft services enhance work productivity and can include:

- Conference room organisation
- Mailroom services
- Reprographics service
- Storage and archiving
- Information technology and communications
- Operating a staff uniform provision service
- Cleaning
- Catering
- Storage lockers
- Changing facilities
- Providing multi-faith rooms
- Crèche facilities
- Gym facilities
- Security
- Cycling facilities
- Recycling
- Occupational health and safety
- Laundry service
- Firefighting equipment
- Car park control
- Reception service
- Visual site information service
- Visitor hygiene service
- Help desk
- Landscaping and grounds maintenance

There are differences and exceptions to the classification of soft and hard services, and services often categorised as soft services may be classified as hard services if they are crucial to the company. It's essential to have a solid grasp of the organisation's industry and the resources it requires to run properly.

Exclusions include the following examples:

- An ICT company will categorise technology as hard service work, although many organisations classify it as a soft service.
- Security is a difficult service for a company in a high profiled government institution since it is necessary for that institution.
- Cleaning would be challenging for a hospital because it is important to the organisation's operation and helps ensure patient well-being.

1.8 Levels of FM in the Organisation

For FM to function properly in organisations where it is used, it should function at all three levels. However, this is usually not the case, especially regarding the strategic level. Facilities departments are there to offer day-to-day services because facilities management is sometimes seen as a simple operational responsibility in organisations rather than a means to evaluate how facilities management could provide long-term help the main business.[8] The scope of FM should include all three levels of the FM decision pyramid:

- The strategic level is where FM functions' long-term goals and direction are discussed. This entails establishing goals in response to the FM functions' intended use and conducting long-term planning while considering the outside world's needs. Results and profitability are the responsibility of the strategic level. Examples of these tasks include planning, modelling, and simulation.
- The tactical (managerial) level is where one is responsible for ensuring that every aspect of the FM organisation operates well. This involves determining goals that deal with these demands and recognising needs. For example, managing, analysing, budgeting, and programming are all examples of tactical work, frequently done annually. Establishing standards, creating timetables, establishing procedures and processes, and procuring resources are all included in this area.
- The daily decisions made when running facilities are what one is concerned with at the operational level.

To further explain the role of strategic FM, we should start by explaining what strategic FM is. Strategic FM refers to FM that goes beyond operational concerns to consider strategic factors for future facilities and services. Strategic integration of FM inside the organisation is necessary for its effectiveness, and this may be achieved by highlighting its potential significance to the broader company processes.[9] Although professional and non-professional services are included in FM, a facilities manager's primary area of expertise is strategy-level FM, while also supervising operational issues. One should note that as FM clients become more informed and sophisticated, they place higher expectations on facilities managers and move the focus of FM to strategy.[10]

An organisation needs to have a vision for strategic FM to be effective. This depends on the facilities manager's knowledge of the organisation's objectives, economic considerations, productivity, workplace, and the ever-changing external market. Strategic facilities management involves anticipating change for employees to utilise space and facilities and being knowledgeable of technology, communications, laws, people's behaviour, risk, and demanding environmental and workplace standards.

A facilities manager who wants to perform a strategic FM role would be involved in the following:[11]

- Coming up with and communicating a facilities policy
- Planning and designing to make service quality better all the time
- Determining what the business and users need
- Negotiating service level agreements
- Making good purchasing and contract strategies
- Forming service partnerships
- Systematically evaluating service quality, value, and risk

Put another way, the manager would be involved in four strategic roles:[12]

- Sourcing support services (strategic outsourcing)
- Space management – the process of finding places to work
- Getting funds from building services (investment and project financing)
- Checking the quality and performance of facilities and how they are set up in institutions

Strategic planning may be used to determine strategic suitability between the organisation and FM.[13] The term "strategic facility plan" was defined by IFMA in its 2002 Project Management Benchmarks Survey (2009) as:

> A two-to-five-year facilities plan encompassing an entire portfolio of owned and leased space that sets strategic facility goals based on the organisation's strategic (business) objectives. In turn, the strategic facilities' goals determine short-term tactical plans, including prioritisation of and funding for annual facility-related projects.

A strategic plan serves as a framework for understanding and incorporating the organisation's drivers, organisational difficulties, industry benchmarks, and alternative work environments into a plan or road map. Highlighting the possible influence the actions may have on company performance, and vice versa, aids organisations in making well-informed decisions concerning their facilities. It enables managers to save real estate costs while using the process to enable change within the organisation.[14]

According to Klein (2003), the tools used during the process of strategic planning can be grouped into three areas:

1. Organisational development tools
    - Visioning sessions
    - Interviews and questionnaires
    - Focus groups
    - Macro-level programming (square feet per person)
    - Change management
    - Organisational analysis
    - Workplace mock-ups

2 Analytic tools
- Business mapping
- Four square analysis
- Scenario modelling
- Financial analysis
- Utilisation analysis (pre- and post-occupancy)

3 Industry norms
- Benchmarking
- Business profiles
- Industry standards
- Performance measures review

Many African organisations still view FM as low-value and unimportant to overall organisational value. The strategic FM job is one that many organisations have yet to embrace. The degree to which FM contributes to a firm primarily depends on FM's place in the organisational structure.[15] FM needs more recognition inside the company, especially by upper management. There are reasons why FM is not strategically integrated into organisations:

- Lack of a generally agreed definition of what facilities management means
- Shortage of qualified facilities managers to perform the FM role
- Limited national and regional policies that encourage the integration of FM
- Lack of emphasis placed on facilities management from other professions and the public outside the field of facilities management

1.9 The Role of FM in an Organisation

The three levels of FM – strategic, tactical, and operational – can be linked to the translator, processor, and demonstrator roles identified by CFM. This connection was used by Kaya et al. (2005) to create a world-class framework for FM. FM has three different responsibilities: facilities procedures entail converting business initiatives into FM projects, carrying out the necessary activities, and showing the outcomes after identifying the business needs. The result of these three responsibilities, impacted by stakeholders, shows the results of those needs. The processor function empowers users, the demonstrator role ensures the possible profit from resources, and the translator role supports FM strategically.

1.10 The Evolution of Facilities Management in Africa

Facilities management in Africa can be traced to the 1990s, when the need arose to rearrange space in companies and when certain professional bodies were established, such as IFMA and SAFMA.

1.10.1 FM Development in Africa

Some countries experienced the introduction of FM due to their participation in professional bodies such as IFMA. In some other countries, the activities of FM in neighbouring countries contributed to its development. The discovery of oil, for example, promoted FM in Ghana. The development of FM in South Africa contributed to the development of FM in neighbouring countries like Namibia and Botswana. In Egypt, the outsourcing of services has driven the introduction of FM; it emerged from property management or maintenance and now includes the integrated management of workplaces, industries, and communities.

The facilities management market in Africa is projected to expand at an annual growth rate of 11.2% for 2023–2028. Travel arrangements, waste removal, security, plumbing and drainage, air conditioning maintenance, handyman services, interior design, and maintenance, among other things, are all included in facilities management. The increase in infrastructure-building activity is the most important driver causing high market share in facilities management. Developing new infrastructure increases demand for FM while improving the market's overall growth. The market is expanding mainly due to rising demand for contracted facilities management services, infrastructure improvements, and private sector investments. However, challenges with operating expenses and a lack of infrastructure can delay market progress.

The African FM market is divided into End User (Commercial, Infrastructural, Institutional, and Industrial), Country, and Facility Type (Single Facility Management, Bundled Facility Management, and Integrated Facility Management). The well-served industries in Africa by FM include housing, commerce, industry, and public infrastructure.[16]

In Africa, professional organisations are the main drivers of FM. The presence of professional organisations shows how important they are becoming. SAFMA and HEFMA are organisations in South Africa. IFMA Nigeria and the Association of Facility Management of Nigeria are organisations in Nigeria. We also have IFMA Ghana and Botswana Facilities Management Association. The Egyptian Facility Management Association is present in Egypt. The African Facilities Management Network (AFM), founded in 2021, has 899 members.

1.10.2 Facilities Management in South Africa

"Facilities Management is an enabler of sustainable enterprise performance through the full life management of productive workplaces and effective business support services."[17] FM is a varied business that provides several work possibilities in South Africa. The GDP share of this sector is around 1.4%. It has evolved into a service business from a function.

SAFMA, or the South African Facilities Management Association, was established in 1998. In South Africa, the FM sector is still relatively new compared to other well-established professions. Before SAFMA was established, a facilities manager was considered a specialist to consult when there was an issue with the maintenance of buildings. Since then, due to SAFMA's attempts to organise the sector, industry has come to appreciate the worth of FM as a profession.[18] Due to these changes, clients' organisations have been forced to hire facilities managers or FM firms to manage their buildings.[19]

The 2008 global economic downturn forced organisations in the public and private sectors to take more extreme actions to manage their operations for sustainability. These cost-cutting strategies included outsourcing some of their FM-related tasks in order to refocus on core business operations (Frost & Sullivan, 2012). A significant number of game-changers may soon influence the FM industry in South Africa.[20] These include: digitisation and technology, including the utilisation of the most recent technological advancements, in order to deliver FM services more effectively and add value to all FM services; cost-saving measures; knowledge leadership, whereby end users look for FM teams that are more intelligent, empowered, skilled, and trained; employee experience, which involves creating an agile work environment and automating routine processes so that FM managers can focus on more intricate responsibilities; sustainability, whereby sustainable systems and procedures must balance FM optimisation; and quality data, which will help FM teams analyse and understand real-time data to make more intelligent judgments.

FM is governed by both national and international legislation. The SANS standards are regional, while ISO is global. The scope includes environmental, general FM regulations, emergency management, health and safety, and construction requirements. FM legislations are applied to many facilities to improve their operation and performance. SANS 1752, the first standard for facilities management systems in South Africa and a counterpart to ISO 41001, was recently produced. It supported ISO 9000 quality management concepts and ISO 55000 asset management principles. This standard is used in several situations (Molloy, 2019).

The FM sector is large and diversified, as the Servest report from 2019 demonstrated. Various primary, small, and medium-sized FM service providers offer support services among the stakeholders in FM. Another industry analysis conducted by SAFMA (2017) revealed the presence of outsourced FM service providers and in-house FM managers. It also includes property firms that offer FM services in the commercial, economic, and residential sectors. In particular, there needs to be more trained workers in the skilled trades, engineering, management, and executive positions, as well as in finance, administration, technicians, educators, and IT. Cost, service offers, reputation, long-term involvement, structural problems, and obstacles are the main determinants of industry performance. There are not many black FM sector stakeholders or business owners. According to Servest's 2019 research, the market should be worth around R51.2 billion. A significant part of the sector outsources its FM operations in one manner or another. Bidvest, Tsebo, Broll, and Servest are among the well-known and reputable integrated facilities management firms. According to the Servest report (2019), stakeholders and agents should promote security outsourcing, integrated FM, technical services, and parking. Soon, services are expected to rise by roughly 37%. The FM sector's clientele includes organisations in the banking, education, healthcare, hospitality, and tourism sectors, those in manufacturing, mining, and professional services, as well as those in the real estate, retail and distribution, telecom, and industrial and manufacturing sectors. These clients believed that to keep their business, FM service providers and contractors should embrace cost-saving efforts and provide integrated FM services.

Regarding their technical proficiency, involvement in B-BBEE, time management, and personnel quality, the clients were happy with the FM service providers or contractors. Users of FM service providers chose internal control, sound financial management, and better customer service as the top three most crucial qualities. A few jobs that enable FM managers to take on more strategic responsibilities and specialise in carrying out complicated duties will involve more automation.

According to Frost & Sullivan (2012), a strengths, weaknesses, opportunities, and threats (SWOT) analysis of the factors that affect the outsourcing marketplace for FM in South Africa include:

1 *Strengths and opportunities*
  - Cost advantages of outsourcing the FM service
  The necessity for organisations to focus their attention on the core business
  Infrastructure development and improvement within the private and public sectors
  Increased speciality in sustainability issues and "green" buildings

2 *Weaknesses and threats*

- Strong resistance to outsourcing FM services within the industrial sector
  The perception that the administration of FM services is achieved more cost-effectively in-house
  Delayed approval of state contracts

SWOT analysis and a market overview indicate that South Africa gradually appreciates the role and value of FM services. This is shown by Frost and Sullivan's (2012) finding that the industry has yet to receive official recognition from governmental agencies, which might be detrimental to business. The minimal market share held by the general public sector affects the limited credit of the FM sector in South Africa. Collaboration between the many stakeholders, including the facilities managers and service recipients, is necessary to organise the discipline. Additionally, there has been a drive toward insourcing, which many organisations need help with.[21]

The SAFMA analysis demonstrated that the challenges to the FM organisation include using other experts to handle FM and the increase in cleaning and security services, making it harder to get competent and reasonably priced services. Additionally, certain professionals need to upgrade their skills, and one must acknowledge that government legislation makes it more challenging to find competent workers.

1.10.3 Facilities Management in Nigeria

Chevron and Mobil launched facilities management in Nigeria in the 1980s as a component of their respective multinational corporations' relocation initiatives. The effects of globalisation, which forced industrialised nations to adopt the new profession of facilities management, are gradually making their way to the country. These adjustments include modifications to building design, urban planning, and construction materials, as well as improvements in transportation and communication technologies that have effectively made the globe into a small town. Economic policies supporting the unrestricted movement of commodities across borders have also increased investment in new markets like Nigeria.

Both the public and private sectors can use FM. Some organisations in Nigeria traditionally handle their management. They also frequently delegate the management of their capital assets to an administrative or finance officer who advises on real estate decisions and overseas operations and maintenance tasks. Building technicians and engineers, who subsequently became facilities managers, were in charge of running the buildings.[22] The inaugural Nigerian Institution of Estate Surveyors and Valuers conference, which addressed facilities management, took place in Kano in 1998. Major foreign and local companies entered the Nigerian FM industry between 2006 and 2016.[23] Today's government organisations, businesses, and nonprofits have realised how inadequate traditional organisational structures are for managing these tasks. Facilities management was developed to combat the disorganised administration of facilities. In Nigeria, there are several uses for facilities management. The National Theatre, NAL Towers, the Investment and Banking Trust Corporation building, Mobil, Chevron, and the sports complex at Adamasingba in Ibadan all use FM. Other real-world examples show how facilities management is used in the country's private and public sectors.

Janitorial services could be Nigeria's oldest and largest FM segment in terms of market size. In Nigeria, the janitorial industry has existed for approximately 50 years. In Nigeria, the first external cleaning company was founded in 1958. External security providers are another significant FM component. These guards are often uniformed and unarmed; Nigerian laws forbid private groups from having weapons. Due to the low barriers to entry, the cheap start-up capital needed, and the high unemployment rate, anybody may access it, lowering the service level. After that, there are engineering services for civil, mechanical, and electrical projects. Many active companies are available, and more technical services are provided depending on the business size. Most significant players first ran their operations out of their premises. However, they ultimately learned how much money could be made from doing so. This finally encouraged several companies to launch companies that offered maintenance services. Groundskeeping and landscaping services, administration services, and catering services are additional FM services.[24]

According to the IWFM study from 2019, the primary service types offered in Nigeria include insourced FM, out-tasking, integrated FM, and complete FM. While most services are supplied domestically in retail facilities, they are mainly outsourced in the office sector. Broll, Cushman and Wakefield, JLL, Knight Frank, Provast, Alpha Mead, Maxi Gold, Sodexo, proFM, Tsebo Rapid, and Filmo Realty are some prominent companies in the market.

Some of the problems that FM faces in Nigeria are:

- There is a crisis of identity among related professionals in the built industry, such as real estate agents, builders, and service companies
- Clients and contractors don't have enough industry knowledge and experience
- A price war against standards that can't be compared (low prices lower the quality of inputs)
- A lack of base standards and data
- Corruption in the contract handling system
- Trouble getting good subcontractors and supplies at the right price

Most organisations hire direct labour to provide and sometimes do their own facilities management services. However, they can only handle the modern challenges of maintaining high-quality buildings and public infrastructure if the job is getting more complex, systems are getting more complicated, and more professionals are needed for better productivity and living standards. Nigeria is also becoming more visible in the world's political and economic structures, so there is a need for transparency in organisational spending.[25]

In addition, threats to the market include un-unified approaches to standardisation, as many organisations work independently to achieve best practices and standards. Another threat is that of a poor maintenance culture, which influences the life span of assets. There is also poor enforcement of standards and regulations. Corporate organisations also have high standards and need help to obtain the right quality of service they demand from the market. Another challenge is low-cost visibility, as many facilities managers need to learn how to present the correct total FM cost breakdown to the management in their organisations, leading to wrong decision-making in facilities matters. Power availability is threatened as power cuts force many organisations to rely on backup generators. According to Wallbanks (2018), this has led to a significant increase in the cost of running facilities.

The IWFM report also identified IoT, drones, and BIM as the future of FM. The IoT is applicable in that it can be used to monitor the usage of energy and equipment in buildings. Drones can be used for aerial inspection. BIM is rarely used in Nigeria. Current trends in the practice of FM include visitor and community management to monitor and track visitors in buildings using software such as E-Estate and Gatepass. Also, space, as a service, is trending in that spaces are provided which encourage co-working and have constant power, internet, and workstations.

1.10.4 Facilities Management in Egypt

According to El-Motasem (2015) in Egypt, FM in Egypt is mainly seen as housekeeping and building maintenance. Egypt had roughly 20 facilities management companies in 2014; the first one opened its doors in 1999. Most FM companies were in the north (Lower Egypt), particularly in Cairo and Giza. The Nile Towers Mega Project outsourced its non-core services to an FM company in 2004 to save money and time, ultimately opening up the market for FM services in Egypt. Currently, FM organisations handle the non-core services for ordinary and big projects. In Egypt, FM firms exclusively offered building maintenance services between 1999 and 2004. Before 1999, Egyptian FM firms provided cleaning and security services.

The Nile City Towers Project was Egypt's first large-scale FM application in 2004. In 2014, FM was used in around 50% of large projects, 30% of typical projects, and 2% of minor projects. FM can be implemented for all mega and ordinary projects and 70% of minor projects. The vision of FM organisations in Egypt can be summed up like this: the organisation's vision is always chosen based on the client's wants.

- Manage and provide non-core services by simultaneously lowering costs and risks and raising quality and freedom
- Take care of buildings
- Bring valuable ideas to the area of FM in Egypt to make it better
- Treat people as the most valuable asset and promote a culture of ongoing personal growth through job advancement, learning new things, and improving your skills
- Encourage employees to keep improving their work for customers safely and professionally and interact with customers to find out how to best support their primary organisational goals

In the beginning, only maintenance and housekeeping companies offered their services. Later, in-house services appeared, and a combination of maintenance and housekeeping organisations was formed. This combination later developed into "Non-Comprehensive" and one or two "Comprehensive" FM organisations. The Egyptian FM firms offer Hard, Soft, and General Services.

No legal guidelines exist for employing FM during the design or pre-schematic phases. An integration, or a "collection," of design codes, fire and safety codes, health, safety, and environmental (HSE) regulations, ISO codes, maintenance codes, risk management codes, housekeeping codes, and catering rules might be referred to as the FM Comprehensive Code. Only a few maintenance regulations and ISO standards serve as FM's guidelines. An FM company may get ISO certified by applying international standards like the ISO 14001 Environment Management standard or the OHSAS 18001 standard. Additionally, integrated facilities management solutions are applied for various projects using Total Infrastructure and Facilities Management. Timberline and Archibus are the two most widely used FM applications in Egypt.

1.10.5 Facilities Management in Kenya

The Kenya Facilities Management Report (Mark Ntel Advisors, 2019) claimed that the expansion of the hotel industry in Kenya had been driven by tourism. An estimated 1.4 million foreign visitors came to Kenya in 2017. The government is also speeding up expenditures on transportation infrastructure, which may help provide better amenities to visitors. These efforts have influenced the growth of foreign hotels nationwide and raised the need for facilities management services. Additionally, many additional service providers from Europe and Africa hope to take advantage of the expansion possibilities in the business, healthcare, and educational sectors. Integrated, packaged, and single service providers supply services.

In 2019, single and bundled service providers dominated the market for facilities management. However, integrated facilities management services appeal to domestic and international end users. Large firms favour integrated management because of the simplicity of contract negotiations, cheaper costs, and best-in-class services. As a result, more people are anticipated to need integrated facilities management services.

The primary participants are Tsebo, G4S Kenya, Cushman and Wakefield, Excellerate Services, and others. Kenya's industrial automation market is divided into the following categories:

1 Type (hard, soft, risk, administrative)
2 Operating model (in-house, outsourced)
3 Service delivery (integrated, bundled, single services)
4 End users (commercial, residential, government, retail, education, healthcare, hospitality, others)
5 Enterprise size (large, mid-size, small)

The lack of sufficient funding to carry out planned activities, a lack of project management expertise for leasing and tenancy issues, improper use of the asset register, corruption, and a lack of FM staff motivation are some problems the facilities management industry faces.[26] Other issues include a poor work atmosphere, a lack of property management software, weak supervisory skills, and delayed client payments.

The Building Maintenance Policy, the Occupational Health and Safety Act, and the Building Code of Kenya, all published in 2009, are FM-related regulations. The Kenya Environmental and Sanitation Policy is governed by the Environmental Management Coordination Act of 1999, which also offers recommendations for sustainable environmental development. Contracts, agreements, acquisitions, expansions, and the creation and introduction of new products are recent market advancements and competitive strategies.

1.10.6 Facilities Management in Ghana

The FM sector is still young but crucial to changing the nation's maintenance culture. The country's infrastructure gap, which can need up to US$1.5 billion annually over ten years to close, has increased the importance of newly developing indigenous facilities management companies there. Business growth has aided in the development of the FM sector.

According to Alpha Mead (2017), Ghana's real estate market has expanded since the discovery of oil there in 2007. The Ghana Investment Promotion Centre registered building and construction projects valued at US$232 billion in 2016. An increasing profile of luxury residential assets and a projected 170,000 square metres of premium office and retail space are also helping the industry flourish. Additionally, the economy's growing real estate activity generates possibilities in facilities management. To assure a return on their investment, investors are becoming more interested in protecting the value of their assets, increasing occupancy, and lowering operating expenses due to developments in the Ghanaian real estate market.

The introduction of local FM companies has increased professionalism in the nation's FM service delivery. It is an opportunity for professionals working in the built environment or industry to develop and apply capacities. Like many other emerging sectors, local facilities management firms need help with several difficulties. This industry has few expert specialists because it's new in the country. Additionally, regulation differences mean that FM industry participants need more professionalism. For the administration of modern facilities, there is a need for some funding and competencies. The physical infrastructure development and the degree of public knowledge of these services affect the industry's growth. Some companies are undecided about using the services of capable facilities management firms. It is a sector with strong growth prospects, with a concentration in big cities, although the market is relatively small.

UT Properties, Handyman Facility Management Services, Facility Pro, City Facilities Management, SGS, and Bridgewave Facilities Management are some of the FM industry participants in Ghana.

National Building Regulations from 1996, the Labour Act from 2003, the Working Environment (Air Pollution, Noise and Vibration) Convention from 1977, the Hygiene (Commerce and Offices) Convention from 1964, the Persons with Disabilities Act from 2006, the Environmental Protection Agency Protection Act from 1994, and the Environmental Assessment Regulation from 1999 are some of the related FM regulations. The International Facility Management Association (IFMA) also maintains a local chapter in Ghana.

1.10.7 Facilities Management in Namibia

The origins of facilities management in Namibia still need to be discovered. However, facilities management generally began with small jobs like housekeeping, janitorial work, gardening that required organising and managing various aspects of a business or community. In 2001 the Ministry of Health and Social Services in Namibia (MOHSS) commissioned the Council for Scientific and Industrial Research (CSIR) in South Africa to investigate the problems of the ministry regarding maintenance and found insufficient planning and funding of maintenance initiatives and a lack of current, accurate, and reliable information addressing the demands of buildings' fabric and component care. Technical, financial, and communication considerations should be considered when determining the priority of maintenance work.[27] Although Namibia has made significant progress since the research's findings and is still changing, the facilities management industry there has expanded. It is getting more complex by the day. This includes the health sector, the real estate sector, and service providers, in addition to those mentioned above. A single point of contact for facilities management and computer-assisted facilities management solutions is now available through Namibia's FM, which began with routine physical management techniques.

Expanding sectors of the economy and globalisation impact facilities management in Namibia. Certainly, developed countries significantly influence Namibia. Therefore, the introduction of facilities management in Namibia was influenced by the global development of facilities management. One industry that facilities management is engaged in is real estate. For instance, expanding retail centres and rental buildings pushes investors and property owners to hire property managers and facilitators. A sizable unemployed population pressures businesses to hire and train more of their labour force in facilities management and employ them in various sectors. Unemployed people also take action and launch new companies like cleaning services, security firms, and gardening services, to name a few. Efficiency and productivity are also significant FM factors since professionals and property owners sometimes need more time to complete simple tasks.

The Department of Works in Namibia is the custodian of institutional buildings. At the same time, Namibia Facility Management, Equity Facility Management, and Broll Namibia are the country's most significant organisations in the facilities management market.

Some of the market's advantages are that the FM industry is now growing due to the influx of new competitors and the sector's rapid expansion. This is good since it shows that there is healthy competition in the market and that it is typically performing well. To increase cost and service efficiency, there will be some industry partnerships over the upcoming years, and the smaller service providers will join the larger ones. Because facilities managers have a sustainable perspective, their companies may operate more profitably and efficiently. Facilities management is essential in ensuring compliance with health and safety laws and reducing the risk of workplace hazards, for example due to COVID-19. Infrastructure delivery and planning are necessary to boost productivity and, among other risks, control those affecting a company's reputation, workers, suppliers, and facilities. Facilities management has proven to be a great asset to firms and has reorganised and assisted commercial operations.

Several new possibilities include adopting application-based facilities management, which has shown great potential as cloud-based operations become more widespread. These programmes offer adaptable solutions for companies whose stakeholders frequently travel. It makes it more straightforward for them to administer and interact on a cloud-based facilities management programme.

Shifting climatic conditions inevitably affect every global market, and facilities management is no exception. The globalisation-driven evolution of organisational cultures threatens the facilities management sector. Since most organisations are trying to hire fewer people and launching work-from-home initiatives, work culture and company operations are becoming more varied. Technological innovation has produced a mobile work culture. Namibia still needs to provide formal degree programmes in facilities management.

World events also have an impact on and pose a threat to the facilities management industry. For instance, the COVID-19 state of emergency caused most, if not all, businesses to shut down, halt all operations, or file for bankruptcy. This impacted facilities management because, with those businesses, facilities managers have jobs.

1.10.8 Facilities Management in Zambia

FM is still in its early stages in Zambia. Only a few organisations use it, and no local professional organisation or association oversees the practice. The Zambian facilities management market is anticipated to expand soon. Due to Zambia's politically stable climate and significant government expenditures on infrastructure projects, the market is growing alongside rising commercial, industrial, and residential building activity. It is, in turn, significantly boosting the market and increasing demand for facilities management services.

Additionally, given that the construction sector is one of the largest in the nation, it is highly likely to have a favourable influence on the demand for facilities management services in the coming years. In addition, due to the country's expanding visitor entry, the government is aggressively investing in the hospitality industry by building more hotels, retail centres, airports, and train stations, and embracing international businesses and investments. As a result, it creates sizable potential prospects for the industry's top companies to provide facilities management services and support market expansion.

There are four main market segments: Hard, Soft, Risk, and Administrative. The facilities management market in Zambia is expected to expand significantly, with a rising demand for soft facility management services. It primarily reflects the growing requirement to foster a well-run workplace by directly increasing a facility's production through professional knowledge.

Furthermore, it is anticipated that demand for soft services will increase due to the rising trend toward automation, minimum human help, and increased accuracy and time efficiency. Additionally, significant investments in the creation of corporate parks around Zambia by countries like China are boosting the need for soft facilities management services and, consequently, the expansion of the industry as a whole.

Knight Frank Zambia, Broll Zambia, Cushman and Wakefield, Excellerate, Tsebo Solutions Group, Central Estates, Sherwood Greene Zambia, and Pam Golding Zambia are examples of FM companies in the country.

One of the risks and weaknesses is a need for more knowledge about the market. It needs more support. Because only some individuals can afford consulting services, the economy's weakness has also contributed to companies' poor performance. The sector is prepared for expansion due to extensive development, and urbanisation might be an opportunity.

1.10.9 Facilities Management in Other Parts of the World

In developed countries like the US, FM services were provided for the first time in the 1950s and 1960s. In the 1970s, they got even better. Facilities management began in Europe in the 1980s. The UK, Germany, and France were the first places in Europe where facilities management began to grow. Currently, the Netherlands, the UK, the US, and the Nordic countries are considered the best FM countries. In Australia, the Facility Management Association of Australia created an accreditation system in 2000 that shaped how facilities management was done there. This method for granting accreditation sets three skill levels for people working as facilities managers in an organisation.

Asia's facilities management is less developed than in the UK, Europe, and the US. However, it is further developed than in Africa. Countries like Japan and Singapore are more advanced in facilities management than other Asian countries. Malaysia and Thailand, on the other hand, are developing countries in Asia where facilities management is advancing. Regarding facilities management, Hong Kong has become one of the best places in Asia. Hong Kong's facilities management began in 1994 with the formation of the International Facilities Management Association. The FM industry in Singapore uses facilities management services that include preventative maintenance. Korea and Malaysia need to improve their populations' awareness of facilities management.

In the US, the UK, and Europe, facilities management is generally a broader term than maintenance, and people are more aware of how vital the physical environment is for growing organisations than in developing countries. The IS0 9000 Quality Standards and EN 15221-3: 2008 Facility Management – Part 3 result from research in facilities management and set the standard for quality control in Europe. They advise on how to achieve quality in facilities management. There should be an agreement on this standard regarding facilities management services. This will help organisations be more productive and make people's lives better. The level of research in developing countries is not as significant as research in developed countries. Also, standards and quality are not as crucial in facilities management in developing countries as in developed countries.

In developing countries in Africa, maintenance is primarily reactive. However, some facilities management services in the US and Europe have already switched to integrative facilities management services, which use proactive maintenance instead of reactive maintenance. As part of integrative facilities management services, proactive maintenance brings together all the data about a building from its planning and construction stages to its operations phase. This makes it easier to control and maintain.

Most of the time, facilities management is done by outsourcing companies in developing Asian countries like Malaysia. In Africa, most plans have been outsourced, but now, local FM companies offer services in this area. Property companies in Malaysia take care of a lot of the buildings there.

Some problems arise when preventative repair is used, which facilities managers in the US and Europe also have to deal with. Preventive maintenance is the most common method used in developing countries. The first problem is that changing from reactive to proactive maintenance requires better technology and skills. The next problem is that building data is not organised from the planning and construction stages to the use and operation stages. This makes the proactive maintenance procedure hard. This might happen when Building Information Modelling (BIM) software is used. BIM is not used as much in developing countries as in developed countries. Another thing that needs to be addressed is the growing difference in quality between facilities management services in the US and the EU and those of developing countries, as this gap is more significant.[28]

1.11 Conclusion

This chapter has introduced the reader to the facilities management concept and the nature and scope of FM, including the differences between hard and soft services and the differences between various building support services. This chapter covered the development of FM within some African countries. It showed that FM still needs clarification with regard to other aspects of real estate management, such as building maintenance, property management, and asset management. The facilities manager's top priority is increasing the user's primary business effectiveness and productivity. To do this, the facilities manager must consider the entire life cycle of an asset. The property manager aims to increase the building's net operating income and value. FM definitions from Africa are similar to those from other parts of the world. The reason for this could be the influence of globalisation on FM practice. FM covers strategic, maintenance, environment, support services, property management, project management, space management, IT, and occupational health and safety. The strategic FM job is one that many organisations have yet to embrace in Africa as many organisations still view FM as low-value and unimportant to overall organisational value.

Some countries experienced the introduction of FM due to their participation in professional bodies such as IFMA. In some other countries, the activities of FM in neighbouring countries contributed to its development. In others still, it was through the discovery of oil. The FM market is expanding mainly due to rising demand for contracted facilities management services, infrastructure improvements, and private sector investments. However, there are challenges with operating expenses, and a lack of infrastructure can delay market progress.

The facilities management market in Africa is divided into end user (commercial, infrastructural, institutional, and industrial), country, and facility (single facility management, bundled facility management, and integrated facility management) type. The well-served industries in Africa by FM include housing, commerce, industry, and public infrastructure. The FM market in Africa is not as developed as in other areas since the practice focuses more on reactive maintenance, with less focus on research, technology, standards, and quality as compared to the practice in developed countries.

1.12 Guidelines

- Have a clear definition for facilities management
- Be clear about the scope of FM in the organisation
- Have enough funds for managing facilities
- Improve the awareness of the population and organisation about FM
- FM should get the right training from accredited institutions that provide training
- Involve the facilities manager at the building design stage to enhance the value added by FM
- FM should develop strategies to train and get more locals to participate in the sector
- FM must consider the life cycle of the assets in the management of facilities
- FM should understand the needs of users
- FM should perform a stakeholder analysis
- FM should quantify benefits from the process and demonstrate to organisations
- FM must embrace quality and measure the performance of services performed both by internal and external staff.
- The facilities manager should embrace strategic management of facilities
- FM should embrace standards and compliance with codes and regulations
- Encourage the use of green buildings to save cost
- Outsource services to encourage cost savings and transfer of risk
- Encourage benchmarking to promote transparency, quality and cost savings in the delivery of services
- The organisation and the FM should invest in the right technology for facilities
- FM should understand the local conditions in which they operate
- FM should embrace a practice that is resilient to uncertainties

1.13 Questions

1. What is the definition of facilities management?
2. What is the difference between facilities management and other real estate management disciplines?
3. What is the difference between hard and soft services?
4. Provide examples of hard and soft services
5. What is the term “core business”?
6. Describe the core business of your organisation
7. What is the scope of FM in the context of your business or country?
8. How has FM developed in selected countries in Africa?
9. What are the opportunities for the development of FM in different countries in Africa?
10. What are the problems or challenges of practicing FM in different African countries?
11. What legislations are in place that facilities managers will find useful in practicing FM in their countries?

Notes

1 Mewomo, M.C., Ndlovu, P.M., & Iyiola, C.O. (2022). Factors affecting effective facilities management practices in South Africa: A case study of Kwazulu Natal Province. *Facilities*, *40*(15/16), 107–124.

2 Jensen, P.A., Sarasoja, A.L., van der Voordt, T., & Coenen, C. (2013). How can facilities management add value to organisations as well as to society? In *Proceedings of the 19th CIB world building congress*. CIB.

3 Brackertz, N. (2004). A framework for the strategic management of facilities, balancing physical and financial considerations with service, customer utilisation and environmental requirements. In Pacific Rim Real Estate Society (PRRES) annual conference: International property investment. PRRES.

4 IFMA. (2003). *Scope and definitions of facilities management*. IFMA.

5 Chitopanich, S. (2004). Positioning facilities management. *Facilities*, *22*(13/14), 364–372.

6 Price, S., Pitt, M., & Tucker, M. (2011). Implications of a sustainability policy for facilities management organizations. *Facilities*, *29*(9/10), 391–410.

7 Junghans, A., & Olsson, N.O. (2014). Discussion of facilities management as an academic discipline, *Facilities*, *32*(1/2), 67–79.

8 Barrett, P. (EEd.) (1995). *Facilities Management – Towards Best Practice*. Blackwell Science.

9 Hinks, J. & Hanson, H. (1998). Facilities management's profound strategic potential. *Facilities Management World*. June, 30–32.

10 Tay, L., & Ooi, J.T.L. (2001). Facilities management: A "Jack of all trades"? *Facilities*, *19*(10), 357–362.

11 Alexander, K. (2003). A strategy for facilities management, *Facilities*, *21*(11/12), 269-274.

12 Yiu, C. (2008). A conceptual link among facilities management, strategic management and project management. *Facilities*, *26*(13), 501–511.

13 Jensen, P.A. (2010). The facilities management value map: A conceptual framework. *Facilities*, *28*(3/4), 175–188.

14 Klein, R.A. (2003). Strategic facilities planning: Keeping an eye on the long view. *Journal of Facilities Management*, *2*(4), 338–350.

15 Noor, M., & Pitt, M. (2009). A critical review on innovation in facilities management service delivery. *Facilities*, 27(5/6), 211–228.

16 Mordor Intelligence. (2023). *Africa facilities management report*. Mordor Intelligence.

17 SAFMA. (2017). *South African facilities management market and quantification and analysis study*. Retrieved 11 April, 2020 from https://www.abrafac.org.br/wp-content/uploads/2020/03/CE-ABRAFAC-2018-Kim-Veltman-SAFMA_FM-Industry-Assessment-Report_compressed.pdf.

18 Nkala, S.M. (2015). *Defining early facilities management involvement using the concepts of performance management*. Masters' thesis, University of Witwatersrand, South Africa.

19 Frost & Sullivan. (2012). *An assessment of the South African facilities management industry*. SAFMA.

20 Servest. (2019). 2018–2019 Facilities management market analysis survey: South Africa. SAFMA.

21 Wallbanks, S. (2018). *Changes in facilities management - Trends and progression*. Servest.

22 Sani, A.M. (1998). Emergent trends in facility management. In *28th NIESV annual conference proceedings*. NIESV.

23 IWFM. (2019). *IWFM report*. IWFM.

24 Alaofin, V. (2003). Overcoming the challenges facing FM operators in Nigeria to profit from hidden opportunities. *Facilities Management World*, *4*(1), 42–48.

25 Akintunde, F. (2019). Facility management as a defensive strategy to maximize value of real estate asset during the economic recession. University of Lagos Guest Lecture Series.

26 Kavinya, M.R. (2019). *The impact of organization strategy on facilities management: A case study of Postal Corporation of Kenya*. Masters' thesis, University of Nairobi, Kenya.

27 Prestoruis, M., & Hauptfleisch, C. (2010). A comparative overview of facilities management Namibia and South Africa regarding building maintenance. ASOCSA2010-26. Accessed December 2023 at Microsoft Word – CD contents page.doc (irbnet.de).

28 Sari, A.A. (2018). Understanding facilities management practices to improve building performance: The opportunity and challenge of the facilities management industry over the world. In *MATEC web of conferences* (Vol. 204). EDP Sciences.

Bibliography

Adewunmi, Y.A. (2014), *Benchmarking practice in facilities management in selected cities in Nigeria.* PhD thesis, University of Lagos, Nigeria.

Akintunde, F. (2019), Facility management as a defensive strategy to maximise value of real estate asset during the economic recession. University of Lagos Guest Lecture Series.

Alaofin, V. (2003). Overcoming the challenges facing FM operators in Nigeria to profit from hidden opportunities. *Facilities Management World*, *4*(1), 42–48.

Alexander, K. (2003). A strategy for facilities management. *Facilities*, *21*(11/12), 269–274.

Alpha Mead. (2017). *Facilities management opportunities in Ghana*. Accessed July 2022. Retrieved from How to take advantage of the facilities management opportunities in Ghana – Best facilities management training in Nigeria, Africa (alphameadtraining.com).

Amos, D., Musa, Z.N., & Au-Yong, C.P. (2019). A review of facilities management performance measurement. *Property Management*, *37*(4), 490–511.

Barrett, P. (Ed.) (1995). *Facilities management – Towards best practice*. Blackwell Science.

Brackertz, N. (2004). A framework for the strategic management of facilities, balancing physical and financial considerations with service, customer, utilisation and environmental requirements. In *Pacific Rim Real Estate Society (PRRES) annual conference: International property investmen*, PRRES.

Chitopanich, S. (2004). Positioning facilities management. *Facilities*, *22*(13/14), 364–372.

Dettwiler, P. (2008). Modelling the relationship between business cycles and office location: The growth firms. *Facilities*, *26*(3/4), 157–172.

El-Motasem, S.M. (2015). *Integrating value engineering and facility management as an approach to face risks*. Masters' thesis, Ain-Shams University, Cairo, Egypt. Retrieved from http://www.cpasegypt.com/pdf/Samar_ElMotasem/Samar_ElMotasem_Thesis.pdf

Frost & Sullivan. (2012). *An assessment of the South African facilities management industry*. SAFMA.

Hinks, J. & Hanson, H. (1998). Facilities management's profound strategic potential. *Facilities Management World*. June, 30–32.

IFMA. (2003). *Scope and definitions of facilities management*. IFMA.

IFMA. (2009). *Strategic facility planning white paper*. IFMA.

IWFM. (2019). *IWFM report*. IWFM.

Jensen, P.A. (2010). The facilities management value map: A conceptual framework. *Facilities*, *28*(3/4), 175–188.

Jensen, P.A., Sarasoja, A.L., van der Voordt, T., & Coenen, C. (2013). How can facilities management add value to organisations as well as to society? In *Proceedings of the 19th CIB world building congress*. CIB.

Junghans, A., & Olsson, N.O. (2014). Discussion of facilities management as an academic discipline. *Facilities*, *32*(1/2), 67–79.

Kavinya, M.R. (2019). *The impact of organization strategy on facilities management: A case study of Postal Corporation of Kenya*. Masters' thesis, University of Nairobi, Kenya.

Kaya, S., Heywood, C.A., Arge, K., Brawn, G., & Alexander, K. (2005). Raising facilities management's profile in organisations: Developing a world-class framework. *Journal of Facilities Management*, *3*(1), 65–82.

Klein, R.A. (2003). Strategic facilities planning: Keeping an eye on the long view. *Journal of Facilities Management*, *2*(4), 338–350.

Koleoso, H.A., Omirin, M.M., & Adejumo, F. (2018). Comparison of strategic content of facilities managers functions with other building support practitioners in Lagos, Nigeria. *Property Management*, *36*(2), 137–155.

Lapides & Frank (1991). Lapides and Frank, An Overview of Corporate Real Estate Management, 4.

Leväinen, K. I. (1997). Building sites as a city facility. *Facilities Management –European Practice*, 44–47.

Mark Ntel Advisors. (2019). *Kenya facility management report*. Mark Ntel Advisors.

Maxi Gold Consulting. (2019). *The differences between asset, facilities, and property management*. Maxi Gold Consulting.

Mewomo, M.C., Ndlovu, P.M., & Iyiola, C.O. (2022). Factors affecting effective facilities management practices in South Africa: A case study of Kwazulu Natal Province. *Facilities*, *40*(15/16), 107–124.

Molloy. S. (2019). Safety presentation. University of Witwatersrand Lecture Notes.

Mordor Intelligence. (2023). *Africa facilities management report*. Mordor Intelligence.

Nkala, S.M. (2015). *Defining early facilities management involvement using the concepts of performance management*. Masters' thesis, University of Witwatersrand, South Africa.

Noor, M., & Pitt, M. (2009). A critical review on innovation in facilities management service delivery. *Facilities*, *27*(5/6), 211–228.

Prestoruis, M., & Hauptfleisch, C. (2010). *A comparative overview of facilities management Namibia and South Africa regarding building maintenance*. ASOCSA2010-26. Accessed December 2023 at microsoft word – cd contents page.doc (irbnet.de).

Price, S., Pitt, M., & Tucker, M. (2011). Implications of a sustainability policy for facilities management organisations. *Facilities*, *29*(9/10), 391–410.

SAFMA. (2017). *South African facilities management market and quantification and analysis study*. Retrieved 11 April, 2020 from https://www.abrafac.org.br/wp-content/uploads/2020/03/CE-ABRAFAC-2018-Kim-Veltman-SAFMA_FM-Industry-Assessment-Report_compressed.pdf

Sani, A.M. (1998). Emergent trends in facility management. In *28th NIESV annual conference proceedings*. NIESV.

Sari, A.A. (2018). Understanding facilities management practices to improve building performance: The opportunity and challenge of the facilities management industry over the world. In *MATEC web of conferences* (Vol. 204). EDP Sciences.

Servest. (2019). *2018–2019 Facilities management market analysis survey: South Africa*. SAFMA.

Tay, L., & Ooi, J.T.L. (2001). Facilities management: A "Jack of all trades"? *Facilities*, *19*(10), 357–362.

Tuomela, A., & Puhto, J. (2001). *Service provision trends of facility management in Northern Europe*. Research paper of the Department of Construction Economics and Management of Helsinki University of Technology.

Wallbanks, S. 2018. *Changes in facilities management - Trends and progression*. Servest.

Waly, A. F., & Helal, D.M. (2010). The impact of facility management on office buildings performance in Egypt. In *Second international conference on construction in developing countries. Advancing and integrating construction education, research & practice*.

Yiu, C. (2008). A conceptual link among facilities management, strategic management and project management. *Facilities*, *26*(13), 501–511.

Part I

Strategic Issues in the Management of the Workplace

2 Work Productivity

2.0 Introduction

One of the main focuses of facilities management is work productivity. In strategic FM, the facilities manager seeks to align work to meet the organisation's strategic goals. This can be done by considering how the current and future requirements for doing work in terms of the nature of work help the employee to be productive. For example, whether there is a need to focus on remote working, as happened during the COVID-19 pandemic, or combine remote and physical work, or opt for work shift days. The arrangement of space to match the demands of users is in the job description of facilities management. Once the nature of work has been decided upon, there is a need to plan how work will be done and how space will be arranged. Work productivity is also the goal of sustainable facilities management. Many of the things sustainable FM considers promote work productivity, such as indoor environmental quality and energy efficiency. If the indoor environmental quality is poor, there will be absenteeism from ill health, which will negatively influence productivity. If energy concerns are a problem, there will be interruptions in work, and productivity will be affected. If there is a limited water supply in the workplace, employees cannot report to work. As people have returned to work after the pandemic, work productivity issues have been more pronounced in Africa due to the greater reliance on technology; there is a shifting emphasis to support women at work, concerns to create more jobs to solve unemployment problems though some sectors have skills shortage, and an increasing focus on the health and wellbeing of employees. Having appropriate guidelines to manage workplaces can improve productivity. This chapter focuses on work productivity.

2.1 Concept of Workplace Productivity

Productivity is one of the most crucial indicators of economic growth. It is essential to find important measures of the workforce's health (or lack of health) to describe social and business perspectives. Productivity is hard to define. The standard way to think about production is from an economic point of view, which looks at the relationship between inputs (like the number of hours worked) and outputs (like the number of units made). Work productivity means how well someone can do their job by making things or providing services that are needed for that job. The skill can be observed by looking at the number of calls for customer service or the number of units completed each workday. The workers' views about their ability to create things may also show this skill (Harris, 2019). Productivity is getting the most out of output with minimum effort or expenditure while realising the

DOI: 10.1201/9781032656663-3

desired outcomes.[1] Efficiency, or the number of output units produced with the usual or fewer input hours, is another word for productivity.[2]

Productivity is closely related to both effectiveness and efficiency. When the proper procedures are completed throughout a work process, all activities contribute to achieving the specified goals, and the output is near to what was anticipated. When something is done well, the least resources are used to get the desired result. Regarding productivity, efficiency is linked to the input (the fewest resources possible). In contrast, effectiveness is mainly related to the output (the most critical outcome). Expenditure on buildings and facilities is usually 10% less than expenditure on the labour force, which can make up 80% of a budget. This suggests that enhancing facilities to increase knowledge workers' output may be economical.[3]

In addition, it is essential to know the difference between the productivity of a single worker, the productivity of teams, departments, and the organisation as a whole, and the productivity of the whole industry.[4]

Support services, often office jobs or in the service business, like FM, prefer to make support services more productive. What makes them different is how output in the office is measured, which shows how much a support service (as an input) assists with a primary task. "Productivity in the workplace" refers to how much work every individual gets done. FM and customer productivity can be monitored directly and indirectly by measuring performance.

Research from various fields (environmental psychology, corporate and public real estate management, facilities management, and business administration) summarises many ways to measure the productivity of knowledge workers (De Been et al., 2016).

Examples of productivity measures:

1. Actual output versus actual input. For example:
   - i the number of translated words per team or per employee per unit of time (translation agency), the number of phone calls per day (call centre), or the number of manufactured products per full-time equivalent in the industry sector; and
   - ii the impact of facilities on the outcomes of cognitive performance tests (such as working memory, processing speed, and concentration).
2. Actual input; for example, monitoring computer activity (keystrokes, mouse clicks) used to produce the output.
3. Amount of time spent or saved; for example, the amount of time gained by implementing a new computer system that works faster or the amount of time lost by logging on frequently to a time-consuming computer system.
4. Absenteeism due to illness or other reasons and thus being non-productive, or the opposite: presence. Related measures include the reported frequency of health issues.
5. Satisfaction is based on the assumption that a satisfied worker is productive. Related measures include job satisfaction, job engagement, satisfaction with facilities, or the intention to stay or quit.
6. Perceived productivity support, i.e. the perceived support of productivity by the current work environment, the estimated percentage of time being productive, the perceived productivity gain when all facilities are excellent, or the perceived increase or loss of productivity after a change.
7. Indirect indicators include the extent to which people can concentrate adequately, the frequency of being distracted, the ease with which employees can solve a problem, or the lack of knowledge through insufficient collaboration with colleagues.

2.2 Work in Africa

According to the Statista, about 466 million people worked in Africa in 2021. Compared to the previous year, the amount went up, following a general growth trend seen during the period under study, with only about 366 million jobs available in 2010. In 2022 Nigeria and Ethiopia had the most jobs in Africa, and they are also two of the most populace countries in Africa. There were about 60 million working people in India and 56 million working in China.

How the so-called "Fourth Industrial Revolution" will affect labour markets and employment also needs to be clarified: will it result in the loss of jobs for people or the emergence of numerous new goods, services, and employment opportunities that we can only now imagine? These may also present new employment opportunities in the FM sector, which technology may develop. Sub-Saharan Africa's meeting with the rest of the world might be helped by technology if it makes people's work easier. However, Sub-Saharan Africa may separate from the rest of the world if technology takes jobs from people. In Sub-Saharan Africa, employment is less susceptible to automation than in advanced economies. The internet cost is high compared to other parts of the world. There has been a conversation within the FM sector about using automation to enhance FM functions like cleaning and other monotonous duties. Countries must continuously create new employment since working populations are expanding quickly. The working population tends to be more of the younger age group. The conclusion for the FM sector is that they should think about creating actual jobs rather than employing technology to replace employees. Governments and business owners in Sub-Saharan Africa are already making use of the Fourth Industrial Revolution's creations. First and foremost, East Africa leads in developing mobile money, giving millions of previously excluded people access to financial services. But, technology is used not only in the financial sector. Technologies like CAFM, BIM, IoT, CMMS, and others used internationally in the FM sector are being used in Africa. Cost constraints, connectivity limitations, and a need for more data access are some difficulties encountered while employing such technologies in Africa.

The growth in the working population in Africa could improve investment and drive economic growth. The need for agglomeration economies in urban areas is one reason income per capita growth needs to catch up to urbanisation. On average, Sub-Saharan African cities are economically dense but congested due to substandard residential housing close to city centres, highly fragmented due to unreliable transportation, which prevents the efficient clearing of labour and product markets, and expensive regarding food and transport compared to income levels, which puts pressure on formal employment income. Most facilities managers in Africa are employed and based in urban areas. Sub-Saharan African cities have the highest urbanisation rates and the non-tradable, informal services that explain the highest degrees of employment there.

Technology has been used to assist in reducing some of the effects of climate change. More effective use of the limited water resources was made possible by intelligent water management technologies. Technology offers chances to accelerate growth and reduce negative consequences, particularly those caused by climate change, with the necessary infrastructure and skill sets.

This suggests that facilities managers need to be ready to handle changes brought on by climate change, including using water-saving techniques, improving the water quality provided in managed facilities, and ensuring that indoor air quality and thermal comfort levels function at their best throughout the year. It's also crucial to consider the health and safety of building occupants. Buildings should be maintained to reduce the spread of diseases like malaria and other vector-borne illnesses.

Sub-Saharan Africa still has the lowest internal and external integration levels regarding trade. Small-scale firms in the services sector that cater to a worldwide market or multinational corporations incorporated into global value chains make up most of the workforce. For instance, design companies could quickly establish connections with international brands. The most predominant professions are professional services, IT entrepreneurs, and low-skilled services. Despite the fast economic growth of many countries, people frequently change jobs due to shifting global trends. These multinational companies also employ facilities managers, and to conduct their jobs effectively, they must adapt to changing international trends.

Information technologies help to reduce the inefficiencies of an informal economy. However, given the high unemployment rate and low job security, employees remain exposed to economic downturns. People must rely on themselves to weather the storms without a robust social security safety net. Because of this, poverty is still common, mostly in rural regions. Sub-Saharan Africa has taken advantage of the opportunity created by technical advancement and global integration, resulting in a well-to-do middle class. However, employment instability is typical in the gig economy. This has allowed facility managers to use gig platforms to hire people from the informal sector, preventing systemic inefficiencies.

A flexible education is a secret to allowing individuals to work with and profit from new technologies without being displaced. Flexible education gives people in the FM sector a chance to acquire skills at every level, enabling them to compete with global trends.

2.3 Workplaces in Africa as a Result of COVID-19

2.3.1 Growth of Remote Work

Most workers need to conduct more remote work. Whether remote work was done frequently depended on the individual's industry. Employers in higher-paying service industries use more remote employees than those in lower-paying sectors such as manufacturing, retail, hotels, and restaurants. The health risk to a particular worker influenced remote employment in another way. The requirement for a physical presence at work was one of the reasons why an organisation allowed remote work. While this was more common with large organisations, most remote employers paid for some or all of their employees' related expenses. Most organisations still maintain an in-person dimension to their operations and aim for workplaces to remain in-person or hybrid despite remote work being a unique part of the pandemic. This implies that in workplaces, there is still a need to have a physical workspace, and workplaces need to be prepared for future pandemics.

2.3.2 Remote Work and Recruitment

Organisations had more freedom to choose employees who would work best for their company due to the COVID-19 pandemic, which has significantly changed how they think about their future workforce. Some companies in Africa changed their recruiting standards to take new worker categories into account, such as entirely remote employees who don't live close to their places of employment. Both information and communication technology and administrative and support services showed a lot of this tendency. African companies have both possibilities and problems due to the failing link between the place of residence and employment. They will increasingly face competition from organisations worldwide for employees with crucial skills. This might increase salaries, which would be difficult for SMEs in particular. As a result of the development, people now have access to remote employment options locally and overseas without having to relocate. This development will also affect trade unions, with unions specifically seeing new opportunities and obstacles when attempting to organise and engage a geographically spread workforce. Since the FM industry is an essential services industry, hiring managers locally is still needed since many of their duties cannot be performed remotely. Managers will, however, have to manage a more diverse workplace if there are recruitments from overseas and also learn to use ICT to meet users' needs.

2.3.3 *Measures for Health and Safety at Workplaces*

During the pandemic, there were several physical changes made to businesses. These included setting up sanitation stations, constructing safety barriers, checking worker temperatures, changing workstations, and giving consumers or employees access to personal protective equipment (PPE). Despite the difficulties, several companies indicated that these workplace changes favoured worker morale, customer satisfaction, and production efficiency (Weijs-Perree et al., 2019). The use of such measures in the workplace brought about the need for facilities managers to be versatile in health and safety compliance and training, and also the need for change management skills to quickly respond to changes so that the business continued to function with minimal disruptions.

2.3.4 *Priority Skills*

Four main skill sets were identified as essential to the future success of businesses due to the pandemic. Digital and communications skills were the foremost. How organisations train, share information, and interact have also changed. Many organisations have widely implemented digital training programmes. The cost of reskilling and upskilling employees is a concern for organisations (ILO, 2022). The facilities managers also need these skills and training in these areas.

2.3.5 *Improved Productivity*

Gains in productivity were mainly established in the service sector. Productivity increases have been helped by efforts to determine digital processes and change similar ones. Numerous companies also redesigned their workstations, occasionally resulting in more efficient workspace organisation. Employee commitment promotes productivity (ILO, 2022). With improved productivity in the workforce, FM service delivery needs to be more effective and efficient to meet these changes.

2.3.6 *Increased Workplace Gender Inequity*

The pandemic made it difficult for women to work from home. Women are more likely to have temporary and part-time jobs that are more risky. Additionally, women are more likely to work in struggling industries like leisure, hospitality, and retail. The pandemic has also contributed to increased employment-related hazards that have affected women, such as domestic violence against women working from home. The pandemic worsened difficult circumstances for African women, including their inclination to work in low-skilled jobs, employment in the informal sector, and exposure to less favourable working conditions than males (ILO, 2022). In the future, the facilities manager should consider how to manage workplace infrastructure for inclusion; for example, providing support for women who work in low-skilled jobs and women who work from home.

2.3.7 Improvement in Social Dialogue

The pandemic improved public dialogue about health and safety problems. Due to the various concerns organisations need to consider, such as workplace health and safety and the regulation of remote work, labour laws are needed to help keep up with workplace changes.[5]

2.4 The Nature of Work in Africa

2.4.1 Remote Working

Kirk and Belovics (2006) believe that e-workers are non-physical employees who do their work away from their place of employment and mostly communicate with their co-workers online. The way people work has changed and is still changing because of online work. Employees no longer have to be in the office to do their jobs; they can work from home or elsewhere within the company.

There are some benefits to working from home, but there are also some issues employers should always be aware of. For instance, some workers who work from home tend to overwork themselves by working longer hours than they would if they were only in the office during regular business hours. So, managers must care about their workers' health and how much work they do. Certain internal and external factors are related to the social processes that remote workers face.[6]

2.4.2 External Features

Social structures have a close relationship with external influences. These variables impact respondents' ability to work from home but do not influence the nature of their actual jobs. External issues include energy availability and the ability to connect to the internet. For instance, power outages in some countries are a problem that neither the company nor the employee can manage. Additionally, although employers may be able to influence this, the quality of internet coverage is eventually a matter that is in the hands of the government. The quality of internet coverage is outside the control of the workers.

2.4.3 Internal Features

Remote workers have some degree of control on internal features. To maintain a healthy lifestyle, remote employees must schedule time for their families and extracurricular activities besides their employment. Even the most diligent, driven, ambitious, passionate, and dedicated remote worker must balance work and personal obligations. If not, sadly, one of the two will suffer. For example, remote employees can and should establish rigid deadlines and try to stick to them to have time for other activities.

Most workers need to conduct more remote work. Whether remote work is done frequently depends on the individual's industry. Employers in higher-paying service industries use more remote employees than those in lower-paying sectors like manufacturing, retail, hotels, and restaurants. The health risk to a particular worker influences remote employment in another way. The requirement for a physical presence at the place of business may be one of the reasons why an organisation does not allow remote work. However, there is a need to maintain a work-life balance.

Furthermore, some employees' home offices can experience occasional disruptions. Evidence, for instance, suggests that household duties might prevent someone from concentrating on their work at specific hours when there are lockdown conditions. Social systems make the resource gap for many distant employees even more evident. Governments should address the issue of social structures and the provision of adequate infrastructure to establish an atmosphere that makes it easy for remote employees to work. For instance, a problem with an unpredictable electrical supply, such as load shedding or load reduction, hinders remote employees from accessing their employer's information systems, which stops them from carrying out their obligations.

Additionally, the infrastructure for network connections has to be addressed. Compared to employees living in locations with poor internet infrastructure, some workers have high-quality internet connections in their homes. As a result, governments must collaborate with network service providers to encourage growth in underdeveloped regions of a country (Matli, 2020).

2.5 Co-working Spaces

A new kind of office environment, called co-working spaces, is starting to emerge. It is a reaction to the knowledge-based economy of the present. Co-working is sharing physical workstations and resources with other users or co-users who aren't necessarily co-workers but have joined together to share a common area. Co-working aims to create a network of professionals who communicate and exchange expertise, including independent contractors, new business owners, freelancers, and company staff members from various sectors and industries.

Users collaborate while working individually in shared offices at co-working locations. There are no social barriers, and the workplace fosters innovation and creativity by offering support networks. Users work together and advance their skills by drawing on the support of their peers in this community-driven environment. Co-working, therefore, aims to establish an open and approachable network of motivated employees. Professionals have access to workplaces that offer them possibilities for contact, building social and professional networks, information sharing, collaboration, learning from the ideas and experiences of others, and more. The shared basic principles of community, openness, accessibility, sustainability, and trust are its foundation.

The jobs in the spaces are often ones that might be performed from home or in a conventional office. Still, the co-workers choose to complete such duties in a room with full office facilities since it fosters more social interactions and collaborations. It offers customers an alternate workplace to satisfy their social and professional needs.

There are six different types of co-working spaces, which are defined by the business model and level of user access: third places, which are public spaces that require paid services (like cafes); public offices, which are free co-working spaces (like libraries); co-working hotels, which are shared workspaces with short-term leases and minimal service packages; incubators, which are shared workspaces that emphasise entrepreneurship; collaboration hubs, which are public workspaces that emphasise worker collaboration; and shared studios, which are shared workspaces that an organisation or entrepreneur rents based on flexible-lease contracts.[7]

Characteristics of co-working spaces (see Figure 2.1) include internet access 24 hours a day, a co-working host, access to resources, tools, and a social network to work with, the

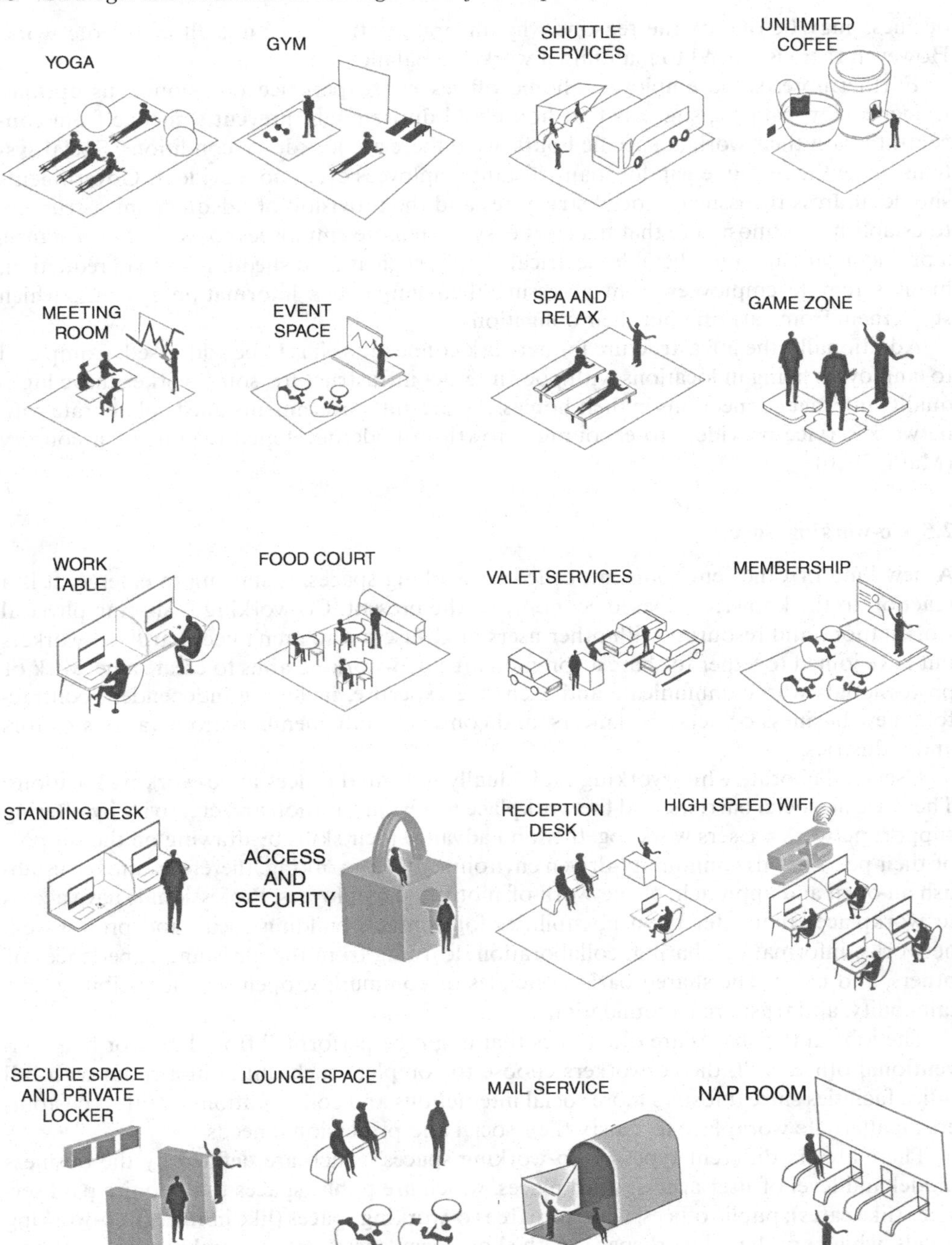

Figure 2.1 Co-working office design with the typical international characteristics of co-working spaces such as meeting rooms, an event space, work tables, a food court, reliable internet, etc.

Source: Author

space's aesthetics, collaborative spaces, concentration rooms, a variety of tenants, event spaces, flexible shared workspaces, flexible lease contracts, easy access to parking and public transportation, kitchens, meeting rooms, printers and copiers, a lounge area, networking events and workshops, a virtual platform, and an open layout (Weijs-Perree et al., 2019).

Example of a co-working space design: AfricaWorks, 33 Baker Street, Rosebank, Johannesburg, South Africa.

AfricaWorks is a co-working space organisation found in seven African countries. It received the *Financial Times* Award for Top 5 Growing Companies in Africa 2022 and 2023. The AfricaWorks business community comprises over 400 corporate members – from Fortune 500 companies to fast-growing start-ups – and over 2,000 individual members.

This space is designed for young, educated business people and digital nomads. Its design and furnishing are a combination of traditional and modern. It promotes an atmosphere of community and a place for corporate work. It blends an industrial look with African craftsmanship, a mix of cutting-edge iPad check-in or QR Code payment with ancient wooden stools. The common areas are a lobby, co-working cafe, community living room, and community kitchen, while the private areas are studios and suites. The facilities provided are lifts, parking, generators, load-shedding proof, a terrace, a cafe, events spaces, meeting rooms, wellness rooms, and a media and podcast room. The design of the AfricaWorks co-working space in Johannesburg can be seen in Figures 2.2 and 2.3.

2.6 Co-working Spaces in Africa

A 2022 Allwork report report showed 1,158 co-working spaces were operating in Africa (27,000 globally) (Michaelides, 2022). The expansion of the co-working sector here is on a parallel course with growth in the start-up sector. According to Co-working Africa, a widely acknowledged link exists between the two. Africa has several features that give it a competitive edge within the global co-working industry. It is a vast continent comprising

Figure 2.2 AfricaWorks design concept, as seen in a co-working space in Rosebank, Johannesburg, South Africa.

Source: AfricaWorks

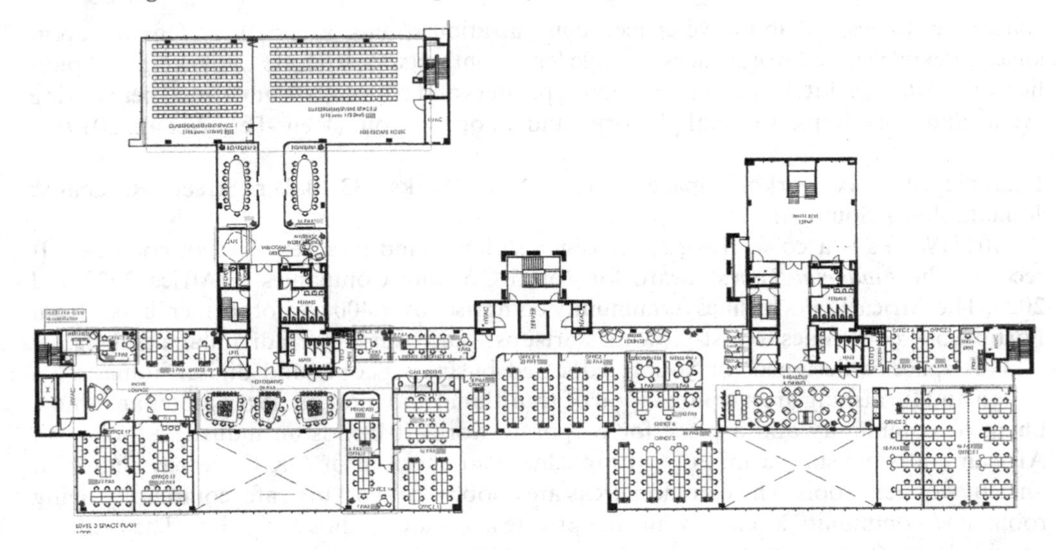

Figure 2.3 Floor plan showing private spaces, events spaces, and meeting rooms for AfricaWorks, Rosebank, Johannesburg, South Africa.
Source: AfricaWorks

54 countries with a highly diverse population that includes speakers of more than 2,000 languages. Africa also has a young populace: 70% are under 30. Youthfulness could be its greatest asset if young Africans receive the education, skills, and training required to ensure future economic growth. Many co-working spaces in Africa are based on a climate of cooperation and offer opportunities for their members to join a community (several offer co-living spaces). Values such as *collaboration*, *cooperation*, and *Ubuntu* (a Nguni term for *humanity*) feature on the websites of many African co-working spaces. Partnerships with start-up companies are also crucial to the success of many co-working spaces in Africa.

Twelve Common Elements of Co-working Spaces Across Africa

1. Collaboration with start-up companies, start-up hubs, and incubators
2. An active online presence (including social media, blogs, etc.)
3. Flexible membership packages
4. High-quality, modern office designs
5. Eco-credentials and eco-friendly amenities and practises
6. 24/7 access
7. Fast, reliable broadband connections
8. Access to marketing opportunities and business networking events
9. Flexible support for a variety of business needs (beyond start-ups)
10. Opportunities for social events (networking, informal meetings, talks, etc.)
11. A creativity and innovation hub for children
12. A podcast studio for digital companies

The problems that co-working spaces in Africa face include a need for more supporting facilities like internet access, proper power supply, inconsistent government regulations, and a market data shortage, further worsening the African market's peculiarities. Since co-workers are free to choose their workstations depending on preferences for services provided, co-working space investors must build a strong customer orientation, in addition to the difficulty of encouraging more employee mobility. Additionally, the space users change daily due to the need for more commitment from co-workers and membership flexibility. There are two effects of the rapid user turnover in space. First, the upkeep and administration of the co-working spaces' operations can be demanding. Due to a need for clarity on the degree of customer patronage, it may be challenging to estimate revenues. For investors and commercial property owners, the financial feasibility of the co-working business model becomes relevant. However, co-working spaces are lucrative when 80% to 85% occupancy rates can be reached.

Flexible office space is becoming more popular in Africa because it is flexible, affordable, and cost-effective, encourages people to be businesses, and allows people to share risks. This backs up the main reasons why there are more co-working places worldwide.

For example, according to Sankari et al. (2018), co-working spaces are advantageous for academic space users since they are appealing and well-liked by the community. Another research work conducted in Namibia by Lahti et al. (2022) utilising the Future Tech Lab at Namibia University of Science and Technology as a case study revealed that the co-working and co-learning environment has been effective even if revolutionary technology is not yet accessible. The room is different from everyday Namibian workspaces regarding furnishings and interior design. In contrast to the more conventional emphasis on autonomous, individual work on university campuses, the established environment emphasises teamwork.

Co-working spaces in Africa are used by young adults who are working on their own, employees of new businesses, mobile and smart business owners, business/service owners, specialised service professionals, field workers, IT and telecommunications operators, young high school artisans and mobile analysts, individual workers, and staff members of new businesses. This is similar to what happens in developed countries.

Users' expectations focus on available internet services, suitable furniture and fittings, convenience and support, peace, good communication among co-users, and consistent power supply in the working environment/environment of a co-working station. Users can get complementary benefits or economies of scale. Economic benefits can reduce operating costs and the responsibility of maintaining shared facilities while also modernising old customs through a channel of communalism and sharing. This is also comparable to what happens in developed countries.

Co-working spaces face various African management issues, such as operating and administrative costs, maintenance, acceptability and awareness, unpredictability, patronage and customer relationships, people management, theft and rule-breaking, electricity and support infrastructure, balancing competing needs, and property conversion. Consequently, issues arise from plans that connect the investment rate to possibilities for profitability. Additionally, there are issues with the infrastructure, a need for understanding among users, and high operating costs.[8]

During the COVID-19 pandemic, people had to follow specific rules and regulations, like staying away from sick people, washing their hands, and wearing face and mouth masks. The occupancy ratio decreased because of the pandemic and the need for personal health and safety. Users' feelings about returning to workstations after COVID-19 are mixed.

2.7 Differences Between Co-working Spaces and Traditional Office Spaces

The differences between co-working and traditional office spaces are shown in Table 2.1 and Figure 2.4.

Table 2.1 Differences Between Co-working and Traditional Office Spaces

Co-working Spaces	Traditional Office Spaces
It is a membership-based workspace.	It is not membership-based.
Co-working spaces are more cost-effective and provide greater flexibility without long-term commitments.	It could cost more, has long- and short-term commitments, and comes with a conventional lease.
Such spaces are designed to foster collaboration and productivity among creative minds.	They are not necessarily designed to foster collaboration and productivity.
Your fully equipped space can be occupied immediately. It has furnishings, internet and phone connections, various amenities, and on-demand meeting space arrangements.	It may not be furnished.
There is an inbuilt office management.	There will be a need to hire an office manager and reception staff to keep the office running smoothly.
Suitable for a hybrid working model of working in the office and working from home.	Designed to take care of working in the office.

Source: Author

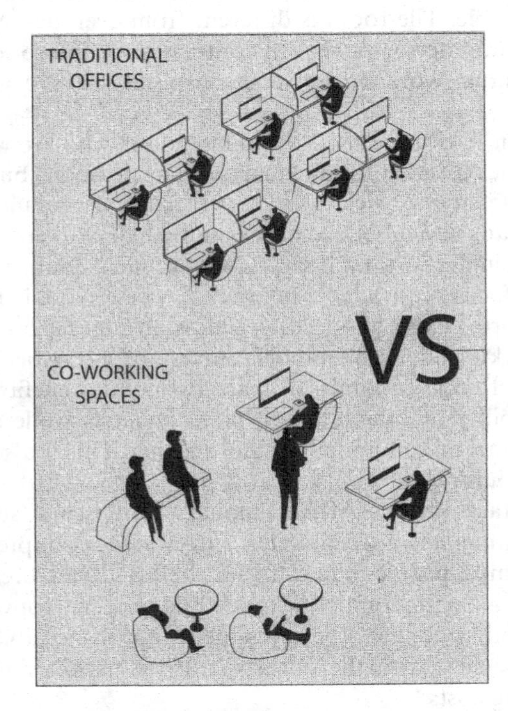

Figure 2.4 Representation of the differences between co-working spaces and traditional office spaces.

Source: Author

2.8 What to Consider in the Management of Co-working Spaces

2.8.1 Meet Users' Needs

Users know what they need in an office, so facilities managers should determine the type of users they want to attract. Also, the space should attract diverse tenants so they can socialise with others from diverse backgrounds.[9]

2.8.2 Ensure the Space Promotes Community

A lively community is one of the critical distinctions between a co-working space and a conventional workplace. Management has to make sure it takes care of a sense of inclusion. Implementing an application and interview procedure rather than an open-door approach can be helpful to ensure that each professional resident is a suitable match for the culture.

- You could also think about using the following strategies to encourage socialisation: Install a photo board (or create an online forum) where renters may upload images, stories, advertisements, and other content
- Distribute a regular email newsletter that highlights residents
- Offer chances for learning, including seminars or lunch-and-learns. Invite clients to after-hours activities like happy hours
- Interact with clients on social media

2.8.3 Spatial Arrangement

Tenants like different workspaces. They choose co-working spaces for the social component; therefore, workstation layouts should consider this. The arrangement of your co-working space should be planned to encourage spontaneous conversations throughout the day, such as in an open office. Additionally, it's crucial to provide private areas for tenant meetings and places with partial privacy, as tenants need. Several offices feature tiny rooms or private phone booths for making personal calls. It is also best to have conference rooms isolated from the main working space.

2.8.4 Use of Technology

Equipment that promotes collaboration should be provided, like HD-quality video conferencing, whiteboard stations, projectors, and multimedia equipment, which may not be considered a selling feature. These tools offer advantages over a home office's drawbacks but are also increasingly used in office communication. They are thus anticipated in a contemporary workplace. This applies if the millennial generation or the burgeoning Generation Z are the target renters.

2.8.5 Include Unique Features

Think of potential ways to make the facility stand out and provide value to membership. This can be its physical location or a free concierge service that takes care of lunch orders, restocks the snack bar, and refills the coffee.

2.8.6 Monitor Utilisation

The space should be monitored so that it meets the needs of tenants. The usage should be monitored to ensure the area is ideal for professionals. All resources should be available to tenants equally, and a mechanism for scheduling rooms should be in place to ensure this occurs. The infrastructure of the co-working space must also be closely checked since any downtime or equipment issues might hinder the tenant's capacity to work, which would cause complaints. Managers of co-working spaces should consider utilising space management software to visualise better space utilisation, occupancy, and power (mainly if they manage many co-working spaces). Software for space management will also assist management in predicting spatial requirements based on changing populations.

2.9 Best Co-working Spaces in Africa

These office spaces offer facilities comparable to other co-working facilities in other regions. These co-working spaces were selected based on their online reviews or the awards that they have received in the industry. Co-working spaces in Africa provide services to cater for infrastructure gaps, such as generator services and special security. This section discusses the characteristics of some of the best co-working spaces in Africa and one in the USA.

2.9.1 Makanak, Cairo, Egypt

Makanak is known as the best workspace in North Africa from the Northern Africa Start-up Awards and the 2019 Co-worker Member Choice Awards. The space complies with ISO 9001: 2015. It is an incubator space that provides co-working space for start-ups to meet their budgets. It also offers a range of facilities. An in-house facilities management team manages the space, and services provided include soft services like quality assurance, cleaning, housekeeping, catering, reception, car parking, maintenance and procurement, and management services such as brokerage, establishment registration services, and property management. This space assists SMEs in meeting their goals through the team, the location of the spaces, and the facilities it provides. The space encourages collaboration and innovation among professionals who use it. It further has mentorship, incubator, and accelerator programmes as networking events. The kitchen area provides tea and coffee services. The facilities provided in this space are found in Table 2.2.

2.9.2 BaseCamp Initiative, Accra, Ghana

BaseCamp Initiative is a co-working hotel and an incubator space for remote workers and lifestyle enthusiasts. It is comprised of facilities found in Table 2.2 and also has an art gallery. It promotes culture, creativity, and collaboration. It hosts networking events for entrepreneurs, such as mentorship programmes and incubator and accelerator programmes. It also uses community apps and has an Airbnb nearby. The kitchen space caters for snacks and alcohol.

2.9.3 PAWA254, Nairobi, Kenya

This is an incubator hub and co-working hotel with a focus on arts and activism based in Nairobi. It targets photographers, graphic artists, journalists, musicians, creators, bloggers, social media influencers, designers, playwrights, illustrators, filmmakers, animators, and poets. PAWA254 empowers young professionals through workshops, training, forums, and events to create a social impact and change in Africa. It fosters creativity and collaborations among those in the creative industry. In addition to having a lounge as a relaxation area, it has a meditation area. There is also an event space to rent to those outside the community. It has unique networking and incubation activities such as Facebook groups, pitching events, sporting activities, and community drinks. It also has a library. This space has transitioned to a community hub.

Table 2.2 Facilities in Co-working Spaces in Selected African Countries and in Workville, New York, USA

Co-working space characteristics	*Makanak, Egypt*	*BaseCamp Initiative, Ghana*	*PAWA 254, Kenya*	*AfricaWorks, Nigeria*	*Office phase, Kigali*	*Workville, New York, USA*
Co-working host	X	X	X	X	X	X
24-hour internet	X	X	X	X	X	X
Access to resources	X	X	X	X	X	X
Aesthetics	X	X	X	X	X	X
Collaborative spaces	X	X	X		X	X
Concentration rooms	X	X	X	X		X
Diversity of tenants	X	X	X	X	X	X
Event spaces	X	X	X	X	X	X
Flexible shared spaces	X	X	X		X	X
Flexible lease contract	X	X	X	X	X	X
Accessibility to transport	X	X	X	X	X	X
Reception area	X	X	X	X	X	X
Kitchen areas	X	X	X	X	X	X
Meeting rooms	X	X	X	X	X	X
Printer and photocopying	X	X	X	X	X	X
Lounge space	X	X	X	X	X	X
Networking events	X	X	X	X	X	X
Open layout	X	X	X	X	X	X
Virtual platform	X	X	X	X		X

Source: Author

Figure 2.5 Meeting room and reception area in Makanak co-working space, Egypt.

Source: www.makanakoffice.com

Figure 2.6 BaseCamp Initiative collaborative spaces.

Source: www.basecampinitiative.com

Figure 2.7 Networking event, PAWA254.

Source: www.pawa254.org

2.9.4 AfricaWorks, Victoria Island, Lagos, Nigeria

This AfricaWorks facility is a fully serviced co-working space consisting of private offices, shared offices, and customisable enterprise solutions that encourage innovation and collaboration among large corporations. It offers premium amenities and flexible leases to match users' requirements. Some facilities provided are available 24 hours for members, and there is uninterrupted electricity and armed security personnel. It has access to restaurants, boutiques and cafes. The facilities in the space are listed in Table 2.2. The space is managed in-house by technical and non-technical support staff to quickly adapt services, amenities, and even the layout to the company's needs.

2.9.5 Officephase, Kigali, Rwanda

This is a fully serviced co-working space that can be customised to meet the requirements of organisations that use it. It has a booking system for meeting room reservations and a library. It is also has access to the convention centre, public transport, government institutions, and shopping centres. It creates opportunities for innovation and collaboration.

2.9.6 Workville – One of the Best Co-working Spaces in New York

Workville is one of the best boutique co-working spaces in New York. It provides a collaborative space for events and conferences and is a productive space. It is an enterprise solution for large teams since it has customisable spaces. Some of the features of this co-working space include move-in ready offices, meeting rooms, individual workspaces, flexible lease terms, a 24/7 doorman, café lounges, and outdoor terraces. This space also complies with Centre for Disease Control (CDC) requirements for housekeeping and it is being managed in-house. It places a lot of emphasis on customer relations management. Facilities in the space are listed in Table 2.2.

Figure 2.8 Lounge area of AfricaWorks, Lagos, Nigeria.

Source: AfricaWorks

Figure 2.9 The community kitchen, Officephase, Kigali, Rwanda.

Source: Officephase

Figure 2.10 Outdoor terraces and cafe area at Workville, co-working space in New York.
Source: www.workvillenyc.com

2.10 Measurement of Workplace Productivity

2.10.1 Physical Conditions

2.10.1.1 Indoor Climate

An uncomfortable degree of heat or cold at work can make people less productive. The temperature, social and cultural norms (like dress codes), and organisational features (like managing expectations, explaining and imagining how to use the installation, and handling complaints correctly) can also affect how much value is put on the interior environment.

2.10.1.2 Personal Control

Many research studies have shown that a person's ability to control things like weather, air quality, light, and noise in their environment affects how productive they say they are.

Objective productivity data shows that having the most control over the temperature setting boosts productivity by 2.7% for thinking logically, 7% for typing, 3.4% for skilled office work, and 8.6% for routine office work. Leaman et al. (1999) found a significant link between self-assessed output and perceived control in seven of 11 buildings. The lack of environmental control was the primary concern of people who worked there (De Been et al., 2016).

2.10.1.3 Light (Daylight, Windows, Lighting)

According to research, individuals prefer daylight in their workspaces. Windows are usually preferred, and low occupant acceptability is seen for fully automated systems. Users of buildings have been shown to value individual control over lighting systems, and they do so if the systems are straightforward. People in workspaces with daylight reported varying amounts of chosen illumination and uncomfortable glare. According to research by Katabaro and Yan (2019) on office buildings in Tanzania, most tenants were less than happy with the lighting in their workspaces. Some respondents said that it significantly impacted their welfare and job effectiveness.

2.10.1.4 Greenery

Having plants in the office has been shown in many studies to boost employee confidence, mental health, and productivity. Light, temperature, relative humidity, air quality, noise, and static electricity are all changed by plants, which thus alter the quality of the surroundings. People's reactions can be mental, emotional, physical (like headaches and high blood pressure), or a combination (like being able to focus better). Plants may indirectly boost production by affecting people's health. Biophilic design in office buildings affects how much work gets done.[10]

2.10.1.5 Sound

Overheard discussions are some of the most annoying noises in the office.[11] Compared to working in a calm atmosphere, background music and conversational sounds tend to hurt performance. Regarding distraction and its effect on productivity, the layout is crucial.

2.10.1.6 Thermal Comfort

Adapting environmental control systems changes the temperature and humidity in office buildings to make people more comfortable. A study from 2010 by Koranteng and Mahdavi says that changing the structure and management of buildings could lower their cooling needs by 20–35% and their CO2 emissions by 27%. These changes could pay for themselves in 3–12 years. Also, the studies showed that 45% of people living in naturally aired buildings and 70% of people living in mixed-mode buildings were unhappy with the air quality during the dry season. At 85% of the naturally ventilated buildings, people said they were the most dissatisfied with their indoor surroundings. Windows and shades were given a lot of weight – between 55% and 80%, depending on the style of the building.

2.10.2 Space

2.10.2.1 Communication and Concentration

The office idea affects (perceived) productivity. The realisation of focused tasks and outstanding cooperation are essential components of organisational effectiveness, and making a space that encourages both is difficult.

2.10.2.2 Ergonomics

The physical layout of workstations and other office furniture is crucial for ergonomics, impacting productivity and health. Sit-stand workstation implementation will possibly lead to reduced levels of physical pain and may also improve performance in an office setting. Specific ergonomic concerns may be made in the design of the desk and chair for office workers who use the same desk for all their work-related tasks. Therefore, a unique ergonomic solution may be created for a specific office user.[12] However, group interactions and collaborations at multi-user workstations are the current trends in today's modern office setting. Various users may use these multi-user workstations during a working day, so an ergonomic design that best fits multiple users must be considered.[13]

2.10.2.3 Aesthetics

The architectural designs of the outside and inside of buildings affect how they look. Colour is one thing that can change how people act, feel, and behave. As a result, colour may also affect how productive people are. Research shows that red makes people better at jobs that require attention to detail, while blue makes people better at artistic jobs. Despite this, a study by Bakker et al. (2013) found that the colour of the conference room (red, blue, or neutral) had no significant effect on the success of meetings, the unity of groups, or people's health. The quality of the indoor surroundings has been linked to resident satisfaction by research.[14]

2.10.3 Psychosocial

2.10.3.1 Social Interaction

Productivity may be enabled by creating spatial arrangements that encourage unprepared interaction and cooperation. Since there is so much visible accessibility in open workspaces, it can be easier for workers to communicate effectively with one another and connect with co-workers more frequently. A lot of the interaction between co-workers happens naturally (e.g. in a corridor, canteen, or shared service area). Proximity, visibility, and flow increase the relations between knowledge workers, promoting creativity. Compared to contained space contexts, moments of conversation in open office settings appear to be more frequent and shorter in duration.[15] Provided these areas are located close to the workstations, including designated meeting places, the volume of contact and the perceived support of cooperation increases. People prefer using meeting venues that are accessible and provide some degree of privacy. There are signs that communication may suffer when working in activity-based office settings with numerous unassigned workspaces and remote working options, even when many communication channels are available. This is because it may be harder for people to find one another and form strong social connections.

2.10.3.2 Distraction

Compared to working in a cubicle office, working in an open environment frequently results in disturbance and distraction, which reduces support for perceived productivity. The openness might cause distractions, especially while performing tasks that call for focus, and when performing creative work, as much creative thought occurs alone. To prevent workers at close workstations from being distracted, open spaces, conference rooms, and communication places must have the right acoustic control and be appropriately positioned.

2.10.3.3 Task Complexity and Personal Characteristics

The activities that building occupants engage in and the task difficulty that goes along with them appear to be critical factors in the connection between the environment and productivity. While complicated task workers were shown to be happier and more productive in a private office, essential task workers seemed to do better in a non-private setting. Introverts appear to have even more difficulty focusing on difficult activities when interrupted than extroverts.

2.11 Work Environment

An organisation's environment consists of its systems, methods, practices, attitudes, and philosophies. The atmosphere of the company is created by management. Employees will have little motivation to assist those seeking to enhance the quality of the work because the measurement system rewards quantity rather than quality. As a result, concerns about the workplace environment affect employees' productivity.[16]

Opperman (2002) says that working environments comprise three main sub-environments: the technology environment, the human environment, and the social environment. The technical environment includes technical equipment, tools, machines, and other physical or technical components. The technological environment creates the tools employees need to do their jobs and tasks. The phrase "human environment" refers to co-workers, clients, team members, relationships at work, management, and leadership. The goal of this space was to encourage a casual touch at work so that people could share information and ideas more easily. This is what should be done to be as productive as possible.

From another perspective, the working environment can be measured using three dimensions:[17]

- The employee's physical working environment includes their general health and safety, such as where they work and what causes accidents and illnesses
- Several job factors affect an employee's psychological working environment. These include how people connect, their work, and the company. Researchers have found that job stress, unhappiness, and surroundings affect how much work gets done
- The well-being of employees is more clearly shown by work-related illnesses, accidents, and conditions, as well as their overall health

Many of these things about the physical environment have to do with how it affects the health and well-being of workers. But they also include how the space can support different working methods and be used for learning and sharing knowledge. They also help tell the difference between how the workplace is designed and managed. Having the right physical environment should help with job activities, the ability to communicate and concentrate, mood, and both formal and informal meetings. The following physical features affect work productivity.[18]

i Being able to control the immediate environment (e.g. lighting or temperature)
ii Having as much natural light as possible
iii Having few visual distractions (when focus is needed)
iv Having the proper lighting for the job
v Having a variety of workspaces that meet business needs
vi Being able to adapt design and infrastructure to change
vii Having good air quality inside
viii Having places for socialising, relaxing, and "psychological restoration"

A good psychosocial workplace helps occupants manage and encourage themselves to be as productive as possible. These places help with self-control, communication, and concentration, increasing productivity (Harris, 2019). The features of such workplaces are as follows:

- A variety of spaces that fit the different ways that people work, including sufficient space for concentration and reflection
- Spaces for planned and unplanned communication and collaboration
- Shared amenity areas to support working and collaborating on the spot
- Technology and other resources that allow flexible access to and sharing of ICT
- The right choice when choosing the right place and conditions to work
- Acoustic and visual controls to enable effective use of each workspace

Individuals generally agree that stress and illness at work increase costs. Still, the place where many people spend most of their waking hours gets little attention. It makes financial sense to plan the workplace to improve employees' well-being and happiness because even a slight increase in employee happiness can significantly impact companies' bottom lines, much bigger than any savings from lower property costs. Employees must be able to work in a healthy environment. The workplace needs good environmental rules, like ergonomic furniture, a clean environment, easy access to food and drink, and chances to be active. People's physical and mental health must be supported so that they can work at their best. However, substandard design and bad natural conditions often have the opposite effect.[19] The defining features of a healthy workplace are as follows:

- Support and improve the individual's well-being at work
- Secure and safe space
- Active design features that encourage movement
- Ergonomic workstations and furniture that accommodate different ways of working
- Comfortable level of light with access to natural light
- Connection to nature through raw materials, views, and green spaces
- Best possible indoor air quality and temperature range
- Neat and clean space
- Access to good nutrition and water

Table 2.3 Measures of the Working Environment

Physical Measures	
Measures	*Sub measures*
Air quality	CO_2 levels Relative humidity Ventilation
Air temperature	Thermometer readings
Cleanliness	Inspections Tests
Decor	Measurements of the space, such as length and width
Furniture	Dimensions of furniture
Lighting	Visual perception Test of candles, foot candles, lumens, and lux
Office layout	Measurement of office space Number of people occupying a space The equipment in the space The team in the space The flow of information in the space The comfort of the space Safety of the space
Thermal comfort	Lighting temperature Air temperature Air relative humidity Airspeed
Views	Visual disturbance
Psychosocial Measures	
Measures	*Sub measures*
Distraction	Interruptions, crowding, noise, privacy, the overall atmosphere
Psychosocial	Leadership, goal-setting cooperation, loyalty, control, perceived performance
Informal interaction	Frequency, duration, focus and intensity of interactions
Informal meeting areas	The capacity of the space Safety of the space Social interactions Space dimensions Comfort levels
Degree of openness	The ratio of the total square footage of the office to the full length of its interior walls and partitions

Table 2.3 (Continued)

Health Measures	
Measures	*Sub measures*
Physical health	Physical symptoms, workability, illnesses, diseases, limiting conditions, sick leave, time management, physical demands, interpersonal demands, disease, chronic conditions, smoking, drinking, mental well-being, attention/concentration, energy levels/sleep, sick days, health limitations on work, hours/days absent, self-assessed health, attitudes, disorders
Mental health	Mental health impacts on workability, job-related stress
Fatigue	Mental fatigue, subjective fatigue

Source: Adapted from Bushiri, C.P. (2014). *The impact of working environment on employees' performance, the case of Institute of Finance Management in Dar es Salaam*. PhD thesis, University of Tanzania, Tanzania; Bortoluzzi, B., Carey, D., McArthur, J.J., & Menassa, C. (2018). Measurements of workplace productivity in the office context: A systematic review and current industry insights. *Journal of Corporate Real Estate*, *20*(4), 281–301

2.12 Conclusion

This chapter found that the Fourth Industrial Revolution will affect labour markets and employment, resulting in the loss of jobs for people. It may also present new employment opportunities in the FM sector, which technology may develop. There has been a conversation within the FM sector about using automation to enhance FM functions like cleaning. The FM sector should focus on using technology to assist employees in their work rather than replace to them.

Technologies like CAFM, BIM, IoT, CMMS, and others utilised internationally in the FM sector are being used in Africa. Cost constraints, connectivity limitations, and a need for more data access are some difficulties encountered while employing such technologies in Africa. Most facilities managers in Africa are employed and based in urban areas. Sub-Saharan African cities have the highest urbanisation rates, and non-tradable, informal services account for the highest employment rate there. Facilities managers need to be ready to handle changes brought on by climate change, including using water-saving techniques, improving the water quality provided in managed facilities, and ensuring that indoor air quality and thermal comfort levels function at their best throughout the year. It's also crucial to consider the health and safety of building occupants. Since multinational companies also employ facilities managers, they must adapt to changing international trends to conduct their jobs effectively.

A flexible education is a secret to allowing individuals to work with and profit from new technologies without being displaced. Flexible education gives people in the FM sector a chance to acquire skills at every level, enabling them to compete with global trends.

The nature of work in Africa is remote, physical, and co-working spaces.

The chapter further found that Africa has several features that give it a competitive edge within the global co-working industry. Many co-working spaces in Africa are based on a climate of cooperation and offer opportunities for their members to join a community (several offer co-living spaces). Partnerships with start-up companies are also crucial to the success of many co-working spaces in Africa.

The problems that co-working spaces in Africa face include a need for more supporting facilities like internet access, proper power supply, inconsistent government regulations, and a market data shortage, further worsening the African market's peculiarities. Co-working spaces face various African management issues, such as operating and administrative costs, maintenance, acceptability and awareness, unpredictability, patronage and customer relationships, people management, theft and rule-breaking, electricity and support infrastructure, balancing competing needs, and property conversion. Consequently, issues arise from plans that connect the investment rate to possibilities for profitability. Additionally, there are issues with the infrastructure, a need for understanding among users, theft, and high operating costs. Flexibility, affordability, cost-effectiveness, entrepreneurship motives, and the possibility for risk sharing drive the growth of flexible office space in Africa. This supports the primary factors contributing to an increase in co-working spaces worldwide.

Independent young adults, corporate start-up employees, business/service operators, specialised service professionals, field workers, and information and communication technology operators are among those who use co-working spaces. This is also comparable to what happens in developed countries.

Users' expectations focus on available internet services, suitable furniture and fittings, convenience and support, peace, good communication among co-users, and consistent power supply in the working environment/environment of a co-working station. Users can get complementary benefits or economies of scale. Economic benefits can reduce operating costs and the responsibility of maintaining shared facilities while also modernising old customs through a channel of communalism and sharing. This is also comparable to what happens in developed countries. In addition, workplace productivity measurement is physical, space, ergonomics, aesthetics, psychosocial and work environment.

2.13 Guidelines

- Use technology and innovation to improve productivity
- Facilities management must be vast in property management
- FMs should ensure the organisation invests in fast-speed internet
- Buildings should consider climate change in operations
- Facilities should invest in water-saving practices
- Maintenance of buildings to consider reducing vector-borne diseases
- FMs should fashion supply chains to create jobs locally
- FMs should have entrepreneurship and business skills
- FMs should plan and encourage the right use of work nature that aligns with the organisation current needs
- FMs should have plans in place for backup power supply where there is a power cut
- FMs must be versatile in flexible space management
- FMs must be creative
- FMs should create work spaces that take well-being seriously
- FMs must have competencies in providing comfort and aesthetics
- FMs should understand furniture arrangements
- FMs should have competencies in service delivery options
- FMs should engage in cost-saving practices
- FMs must communicate with all stakeholders
- FMs must be able to interact with people from diverse backgrounds

2.14 Questions

1 What is work productivity?
2 Describe the nature of remote working in your country

3 What are co-working spaces?
4 What are the developments in the co-working space market in your country?
5 How can workplace productivity be measured?
6 What are the measures of the physical working environment?

Notes

1 Harris, R. (2019). Defining and measuring the productive office. *Journal of Corporate Real Estate, 21*(1), 55–71.
2 Escorpizo, R. (2008). Understanding work productivity and its application to work-related musculoskeletal disorders. *International Journal of Industrial Ergonomics, 38*(3/4), 291–297.
3 De Been, I., van der Voordt, T., & Haynes, B. (2016). Productivity. In P.A. Jensen & T. van der Voordt (Eds.), *Facilities management and corporate real estate management as value drivers* (pp. 174–192). Routledge.
4 Jensen, P.A., & van der Voordt, T. (2020). Productivity as a value parameter for FM and CREM. *Facilities, 39*(5/6), 305–320.
5 International Labour Office (ILO). (2022). The next normal: The changing workplace in Africa: Ten trends from the COVID-19 pandemic that are shaping workplaces in Africa. ILO.
6 Matli, W. (2020). The changing work landscape as a result of the Covid-19 pandemic: Insights from remote workers life situations in South Africa. *International Journal of Sociology and Social Policy, 40*(9/10), 1237–1256.
7 Weijs-Perrée, M., Van De Koevering, J., Appel-Meulenbroek, R., & Arentze, T. (2019). Analysing user preferences for co-working space characteristics. *Building Research & Information, 47*(5), 534–548.
8 Ayodele, T.O., Kajimo-Shakantu, K., Gbadegesin, J.T., Babatunde, T.O., & Ajayi, C.A. (2022). Exploring investment paradigm in urban office space management: Perspectives from co-working space investors in Nigeria. *Journal of Facilities Management, 20*(1), 19–31.
9 iOFFICE Eptura. (2016). Six things executives should know about co-working and facility management. iOFFICE Eptura.
10 Aduwo, E.B., Akinwole, O.O., & Okpanachi, P.O. (2021). Assessing workers' productivity through biophilic design as a measure of sustainability in selected office buildings in Lagos state, Nigeria. In *IOP conference series: Earth and environmental science* (Vol. 665, No. 1, 012047). IOP Publishing.
11 Sundstrom, E., Town, J.P., Rice, R.W., Osborn, D.P., & Brill, M. (1994). Office noise, satisfaction, and performance. *Environment and Behavior, 26*(2), 195–222.
12 Sauter, S.L., Schleifer, L.M., & Knutson, S.J. (1991). Work posture, workstation design, and musculoskeletal discomfort in a VDT data entry task. *Human Factors, 33*(2), 151–167.
13 Mahoney, J.M., Kurczewski, N.A., & Froede, E.W. (2015). Design method for multi-user workstations utilizing anthropometry and preference data. *Applied Ergonomics, 46*, 60–66.
14 Huang, L., Zhu, Y., Ouyang, Q., & Cao, B. (2012). A study on the effects of thermal, luminous, and acoustic environments on indoor environmental comfort in offices. *Building and Environment, 49*, 304–309.
15 Appel-Meulenbroek, H.A.J.A., de Vries, B., & Weggeman, M.C.D.P. (2014). Layout mechanisms that stimulate innovative behaviour of employees. In *13th EuroFM research symposium: EFMC conference* (pp. 5–17). EuroFM.
16 Bushiri, C.P. (2014). *The impact of working environment on employees' performance, the case of Institute of Finance Management in Dar es Salaam*. PhD thesis, University of Tanzania, Tanzania.
17 Foldspang, L., Mark, M., Hjorth, L.R., Langholz-Carstensen, C., Poulsen, O.M., Johansson, U., & Rants, L.L. (2014). *Working environment and productivity: A register-based analysis of Nordic enterprises*. Nordic Council of Ministers.
18 Harris, R. (2019). Defining and measuring the productive office. *Journal of Corporate Real Estate, 21*(1), 55–71.
19 Bortoluzzi, B., Carey, D., McArthur, J.J., & Menassa, C. (2018). Measurements of workplace productivity in the office context: A systematic review and current industry insights. *Journal of Corporate Real Estate, 20*(4), 281–301.

Bibliography

Aduwo, E.B., Akinwole, O.O., & Okpanachi, P.O. (2021). Assessing workers' productivity through biophilic design as a measure of sustainability in selected office buildings in Lagos state, Nigeria. In *IOP conference series: Earth and environmental science* (Vol. 665, No. 1, 012047). IOP Publishing.

Appel-Meulenbroek, H.A.J.A., de Vries, B., & Weggeman, M.C.D.P. (2014). Layout mechanisms that stimulate innovative behaviour of employees. In *13th EuroFM research symposium: EFMC conference* (pp. 5–17). EuroFM.

Ayodele, T.O., Kajimo-Shakantu, K., Gbadegesin, J.T., Babatunde, T.O., & Ajayi, C.A. (2022). Exploring investment paradigm in urban office space management: Perspectives from co-working space investors in Nigeria. *Journal of Facilities Management*, *20*(1), 19–31.

Bakker, A., Shimazu, A., Demerouti, E., Shimada, K., & Kawakami, N. (2013). Work engagement versus workaholism: A test of the spillover-crossover model. *Journal of Managerial Psychology*, *29*(1), 63–80.

Bortoluzzi, B., Carey, D., McArthur, J.J., & Menassa, C. (2018). Measurements of workplace productivity in the office context: A systematic review and current industry insights. *Journal of Corporate Real Estate*, *20*(4), 281–301.

Bushiri, C.P. (2014). *The impact of working environment on employees' performance: The case of Institute of Finance Management in Dar es Salaam*. PhD thesis, University of Tanzania, Tanzania.

De Been, I., van der Voordt, T., & Haynes, B. (2016). Productivity. In P.A. Jensen & T. van der Voordt (Eds.), *Facilities management and corporate real estate management as value drivers* (pp. 174–192). Routledge.

Escorpizo, R. (2008). Understanding work productivity and its application to work-related musculoskeletal disorders. *International Journal of Industrial Ergonomics*, *38*(3/4), 291–297.

Foldspang, L., Mark, M., Hjorth, L.R., Langholz-Carstensen, C., Poulsen, O.M., Johansson, U., & Rants, L.L. (2014). *Working environment and productivity: A register-based analysis of Nordic enterprises*. Nordic Council of Ministers.

Harris, R. (2019). Defining and measuring the productive office. *Journal of Corporate Real Estate*, *21*(1), 55–71.

Huang, L., Zhu, Y., Ouyang, Q., & Cao, B. (2012). A study on the effects of thermal, luminous, and acoustic environments on indoor environmental comfort in offices. *Building and Environment*, *49*, 304–309.

International Labour Office (ILO). (2022). The next normal: The changing workplace in Africa: Ten trends from the COVID-19 pandemic that are shaping workplaces in Africa. ILO.

iOFFICE Eptura. (2016). Six things executives should know about co-working and facility management. iOFFICE Eptura

Jensen, P.A., & van der Voordt, T. (2020). Productivity as a value parameter for FM and CREM. *Facilities*, *39*(5/6), 305–320.

Katabaro, J.M., & Yan, Y. (2019). Effects of lighting quality on working efficiency of workers in office building in Tanzania. *Journal of Environmental and Public Health*, 1, 3476490.

Kirk, J., & Belovics, R. (2006). Making e-working work. *Journal of Employment Counseling*, *43*(1), 39–46.

Koranteng, C., & Mahdavi, A. (2010). An inquiry into the thermal performance of five office buildings in Ghana. In *Clima 2010 - 10th REHVA world congress*. Verlag/Organisation/Universität mit wissenschaftlichem Lektorat.

Lahti, M., Nenonen, S.P., & Sutinen, E. (2022). Co-working, co-learning and culture–co-creation of future tech lab in Namibia. *Journal of Corporate Real Estate*, *24*(1), 40–58.

Leaman, A., Cassels, S., & Bordass, B. (1999). The new workplace: friend or foe? *Environments by Design*, *3*(1).

Mahoney, J.M., Kurczewski, N.A., & Froede, E.W. (2015). Design method for multi-user workstations utilizing anthropometry and preference data. *Applied Ergonomics*, *46*, 60–66.

Matli, W. (2020). The changing work landscape as a result of the Covid-19 pandemic: Insights from remote workers life situations in South Africa. *International Journal of Sociology and Social Policy*, *40*(9/10), 1237–1256.

Michaelides, S. (2022). The Growing Coworking Industry is Revitalising Communities in Africa, Accessed January, 2023 https://allwork.space/2022/10/the-growing-coworking-industry-is-revitalizing-communities-in-africa/

Opperman C.S. (2002). Tropical business issues. PricewaterhouseCoopers.
Sankari, I., Peltokorpi, A., & Nenonen, S. (2018). A call for co-working–users' expectations regarding learning spaces in higher education. *Journal of Corporate Real Estate*, *20*(2), 117–137.
Sauter, S. L., Schleifer, L.M., & Knutson, S.J. (1991). Work posture, workstation design, and musculoskeletal discomfort in a VDT data entry task. *Human Factors*, *33*(2), 151–167.
Sundstrom, E., Town, J.P., Rice, R.W., Osborn, D.P., & Brill, M. (1994). Office noise, satisfaction, and performance. *Environment and Behavior*, *26*(2), 195–222.
Weijs-Perrée, M., Van De Koevering, J., Appel-Meulenbroek, R., & Arentze, T. (2019). Analysing user preferences for co-working space characteristics. *Building Research & Information*, *47*(5), 534–548.

3 Strategy in the Management of Workplace Facilities

3.0 Introduction

Strategic management in facilities management is concerned with the long-term business plan and goals. The facilities manager performs market research, benchmarking, and customer management tasks at the strategic level. The manager obtains information about the competition and innovates, performs cost-benefit analysis, budgets funds for FM, decides the management structure, and supervises staff and contractors. A corporate strategy informs the FM strategy, in which space and service requirements match operational requirements. This translates to space and service planning, sourcing and delivery. The subsequent review procedure enhances the effectiveness and efficiency of the space and services. Research has shown that FM's strategic integration level is low in Africa, as FM is a fairly new discipline. The maintenance culture in many organisations is lower than that found in developed countries. As more facilities managers assume strategic positions, the position of strategic FM improves. Some benefits are aligning FM processes and practices to organisation goals for cost savings and efficient use of resources to deliver competitive and sustainable services. The focus of the chapter is on strategy.

3.1 Meaning of Strategy

According to David (2011), strategy is "the direction and scope of an organisation over the long term, which achieves advantage in a changing environment through its structure to achieve resources and competencies to meet stakeholder expectations."

For the organisation in charge of facilities management, the strategic responsibilities include:

- Developing and communicating a facilities policy
- Planning and designing to improve service quality all the time
- Determining the needs of the business and users
- Negotiating an understanding of service levels
- Making effective purchasing and contracting plans
- Forming service partnerships
- Service appraisal in a way that takes quality, value, and risk into account

According to Barrett (2000), an FM strategy is developed through:

- Communicating consistently with the core business to find out about and understand any changes that may happen to the business due to external factors, like competition strategies

DOI: 10.1201/9781032656663-4

- Examining potential longer-term developments in understanding the specialised subject of facilities management
- Establishing a strategy that is a policy framework that surrounds decision-making inside the facilities management organisation while fully considering external factors that influence the core business and the industry as a whole

To understand what strategy is, it is essential to consider the following defined concepts:

- A *mission* is "a general expression of the organisation's overall purpose, which, ideally, is in line with the values and expectations of significant stakeholders and concerned with the scope and boundaries of the organisation"
- A *vision* "is the organisation's desired future state. It is an aspiration around which a strategist, perhaps a chief executive, might seek to focus the attention and energies of members of the organisation"
- A *goal* is "generally aligned with the mission and may be qualitative. On the other hand, an *objective* is likely to be quantified"
- The *strategy concept* is "the organisation's long-term direction. It is typically expressed in broad statements about the direction the organisation should be taking and the action required to achieve objectives"[1]

3.2 Mission and Vision Statements

Many organisations today develop a *vision statement* that answers the question, "What do we want to become?" A vision statement is frequently believed to be the first stage in strategic planning, with a mission statement coming in second. Many vision statements are just one line long. For example, the FM company Broll's vision statement is, "Our vision is to take care of your vision."

For David (2011), *mission statements* are "enduring statements of purpose that distinguish one business from other similar firms. A mission statement identifies the scope of a company's operations in product and market terms." It deals with the question: "What is our business?" A clear mission statement tells what an organisation values and its top responsibilities. When strategy developers write a mission statement, they must consider the type and scope of their current operations and the potential appeal of new markets and activities.

3.3 Mission Statement Components

Mission statements have different lengths, contents, formats, and levels of detail. There should be nine parts to a good mission statement, and a maximum of 250 words. Most of the time, the mission statement is the most well-known part of strategic management. For that reason, it needs to have the following nine elements:

1 *Customers* – find out who the organisation's customers are
2 *Products or services* – find out what the organisation's primary products or services are
3 *Markets* – determine where the company competes geographically
4 *Technology* – find out if the company is up to date on technology
5 *Concern for survival, growth, and profitability* – find out if the organisation wants to grow and stay financially stable
6 *Philosophy* – find out the organisation's core views, values, goals, and ethical concerns

7 *Self-concept* – find out if the organisation has a unique skill or a big edge over its competitors
8 *Concern for public image* – determine if the organisation is concerned about neighbourhood, environmental, and social issues
9 *Concern for employees* – find out if employees are an advantage to the organisation

The benefits of strategic FM, according to IS0 41014, are:

- A better understanding of the organisation's goals, needs, and limitations, as well as the best way to handle FM and facility services
- A lower likelihood of being removed from an organisation's goals and needs, as well as the ways to support them
- Alignment between FM needs and the demand organisation's main business activities
- More efficient management of FM in general and delivery of facility services
- Consistent management practices from a methodology for developing a transparent, reproducible, measurable strategy
- Ability to see how much FM is improving its operational effectiveness and how it helps the company's main business
- Contribution to how much it helps the organisation save money and be more competitive
- Contribution to helping the organisation with sustainability and better use of limited resources

3.4 Why FM Organisations Do Not Consider Strategy

Some reasons for poor or no strategy are as follows (David, 2011):

- Insufficient knowledge or expertise in strategy
- Poor reward structures once operations are successful
- The facilities manager may spend all their resources responding to crises and dealing with emergencies, leaving no time for planning
- A view of it as a waste of time, because it doesn't translate into immediate results.
- A belief that it is too costly or time intensive.
- Lethargy – some individuals may not want to do the work required to create a strategy
- Satisfied with success – if a company is successful, people could feel there is no need to plan since everything is going well.
- Fear – unless a situation is urgent, there is little danger of loss if one does nothing. There is a chance of failure whenever something significant is done
- Overrated skill – as managers gain expertise, they could stop relying on formalised strategy. Overrating one's skill may be risky, and having a strategy is usually a sign of professionalism
- Prior negative experience – managers may have had a negative experience with strategy, mainly if the strategy was lengthy, complex, impractical, rigid, or done incorrectly.
- Focus on self – when someone has successfully used an old method to obtain position, privilege, or self-esteem, they frequently believe a new strategy is a problem
- Uncertainty about a manager's ability to learn new skills, aptitude for new methods, or ability to take on new positions
- A fundamental difference of opinion – some individuals could think a strategy is weak
- Suspicion – employees might not believe in top management

3.5 Factors Influencing FM Strategy in Africa

An FM strategy is influenced by many factors, as seen in Figure 3.1.

1 **Organisational**
- Maintenance priority
- Maintenance culture
- Funding
- Skills shortage
- Scale of FM operations
- Culture
- Management structures and support
- Condition of facilities
- Quality of building materials used
- Design type of facilities
- Quality of workmanship
- Use of facilities
- Reporting structures for faults
- Responsiveness towards faults
- Size of the organisation
- Training
- Level of contracting out of services
- Performance measurement
- Type of services provided

2 **Macro-Environmental**
- Political environment, such as the influence of tariffs and trade conditions and political stability
- Legislation

3 **Economic**
- The economic environment, which includes interest rates and unemployment rates
- Conditions in the market, like customers wants, needs, and aspirations, and the fact that the public sector has to do competitive testing

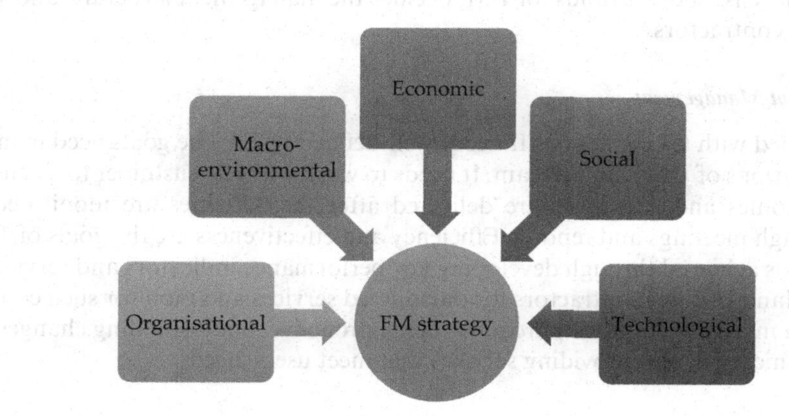

Figure 3.1 Factors influencing FM strategy are organisational, macro-environmental, economic, social, and technological.

Source: Author

4 **Social**
  - Attitudes of building users
  - The facilities manager's knowledge of users
  - Image of the organisation
5 **Technological**
  - IT
  - Innovation

3.6 FM's Role in Corporate Strategy

- Provide support for the overall strategy of the organisation
- Serve as a potential unique competency
- Provide internal uniformity of services to staff
- Be consistent with other functions, such as IT and HR
- Improve the service delivery performance of an organisation

3.7 FM Policy Arrangement

Once agreed upon, the FM strategy has to be included in an FM policy, which should:

- Focus employees on shared goals through full briefing sessions
- Change strategy into measurable objectives
- Ensure daily decisions and behaviours are in line with long-term goals
- Communicate to the in-house team, as well as outside providers and suppliers

3.8 Levels of Strategy in FM

3.8.1 Strategic Management

This is concerned with the long-term business plan and goals. The facilities manager performs market research, benchmarking, and customer management tasks at the strategic level. The manager obtains information about the competition and innovates, performs cost-benefit analysis, budgets funds for FM, decides the management structure, and supervises staff and contractors.

3.8.2 Operational Management

This is concerned with taking action. It needs well-defined goals. The goals need managers to define the actions of the facilities team. It needs to work with the customer to ensure that the right outcomes and standards are delivered after the activities are monitored and reviewed through meetings and reports. Efficiency and effectiveness are the goals of FM at this level. This is achieved through developing key performance indicators and service level agreements. Managers find contractors for outsourced services and monitor such contracts at this level. In management, the approach is being proactive, understanding changes, providing communication, and providing services that meet users' needs.

3.8.3 Tactical Management

At this level, the manager interprets goals and looks at what the FM unit needs to do and which activities would be done by different categories of FM staff. The activities carried out at this level include planning, developing actions to achieve formed plans, and resource allocation. Other tasks include budgeting and investment in facilities. Management supervises the team to achieve the company goal. It involves interpreting goals, working on achieving the developed goals, developing direction for facilities, and writing procedures for FM.

3.9 Defining Strategic Management

Strategic management "is the art and science of formulating, implementing, and evaluating cross-functional decisions that enable an organisation to achieve its objectives." This definition from David (2011) suggests that strategic management incorporates management, marketing, finance/accounting, production/operations, research and development, and information systems to achieve success for the organisation. *Strategic management* can also refer to *strategic planning*, which is often used in business, while the phrase *strategic management* is used in academia. Sometimes, *strategic management* refers to "strategy formulation, implementation, and evaluation, with *strategic planning* referring only to strategy formulation." The purpose of strategic management is "to use and create new and different opportunities for tomorrow; *long-range planning*, on the other hand, tries to improve for tomorrow the trends of today" (David, 2011).

Figure 3.2 is a framework for developing strategy in an organisation. There are three main stages in the development of a strategy, as defined by Johnson et al. (2020):

- Strategic analysis
- Developing strategic solutions
- Strategy implementation

3.10 Strategic Analysis Tools and Techniques

Strategic analysis is needed to understand an organisation's current property portfolio, real estate strategy, and currently managed services. It requires the following:

- The organisation's strategic plan highlights its goals, objectives, needs, and policies
- The property or real estate strategy describes its physical assets and their current use

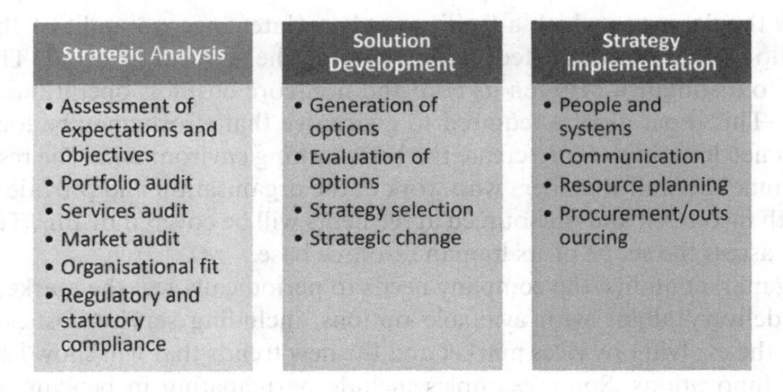

Figure 3.2 Johnson and Scholes' strategic management framework, modified by Atkin and Brooks (2015).

Source: Adapted from Atkin, B. & Brooks, A. (2015). *Total facility management*. Wiley Blackwell

Table 3.1 Strategic Analysis Actions and Tools

Development stage	*Tools or techniques*
Assessment of expectations and objectives	PEST
	SWOT
	Scenarios
	Quantitative analysis
	Scenario analysis
	Megatrends
Portfolio audit	Space analysis
	Estates register
	Maintenance plans
	Maintenance audit
	Risk assessment
Services review/audit	Current service level agreements/performance analysis
	Benchmarking
	Customer satisfaction levels
Resource audit	People training/skills audit
	Existing internal service provision audit
	Business process mapping or analysis
	Zero-based analysis
Market audit	External service providers' intelligence review
	Real estate availability
	Market trends
	Suppliers' competencies
	Suppliers' management analysis such as ISO 9000, ISO 1400
Organisational fit	Current management functions and expertise
	Stakeholders' determination and requirements
Regulatory and statutory compliance	Accident statistics and environmental violations analysis
	Claims history

Source: Atkin, B. & Brooks, A. (2015). *Total facility management*. Wiley Blackwell

- The human resources view includes permanent, temporary, agency, and contractor workers from a full resource view and has current processes and systems for services and facilities
- The budget forecasts use a complete cost analysis.[2]

To determine the degree to which a facility needs maintenance, an audit of the organisation's portfolio should be conducted, taking note of the services involved. The company must be able to distinguish between its core and non-core business operations in terms of service audit. This separation is required to guarantee that efforts may be focused where they are most needed – in order to create the best working environment. The resource audit needs to examine the staff members who work at the organisation and provide services. As expected, both insourced and outsourced agreements will be covered by this. The company also needs to assess the scope of its human resource base.

Regarding market audits, the company needs to periodically test the market to see how well service delivery aligns with available options, including service costs. This entails awareness of the evolving services market and the new trends that will show issues, opportunities, and innovations. Some examples include participating in benchmarking clubs, obtaining suggestive cost estimates from service providers, and comparing current pricing and service rates using public data. One may always decide whether a preferred alternative is the best, provided one is informed of the status of the services market. Regardless, market, space, and services audits done during the portfolio audit may already have some required data (Adewunmi, 2020).

3.10.1 Strategic Analysis Tools

In this section, some strategic analysis tools are discussed.

3.10.1.1 PEST Analysis Model

Political, economic, sociocultural, and technological factors are examined under the PEST model. Organisations may find the PEST model a valuable tool for studying market attractiveness. Applying this model "allows for a thorough analysis of the macro-environment within the sector".

Political: This refers to the nation's political stability and government policies.

Economic factors: Interest rates and credit availability are economic issues. The status of the economy can impact the financial performance of FM, service innovation, and planning.

Sociocultural: This refers to social and cultural aspects, such as language and social trends. Sub-factors, including lifestyle, societal norms, and personal preferences, are also considered here. This has to do with the characteristics and tendencies of the FM market.

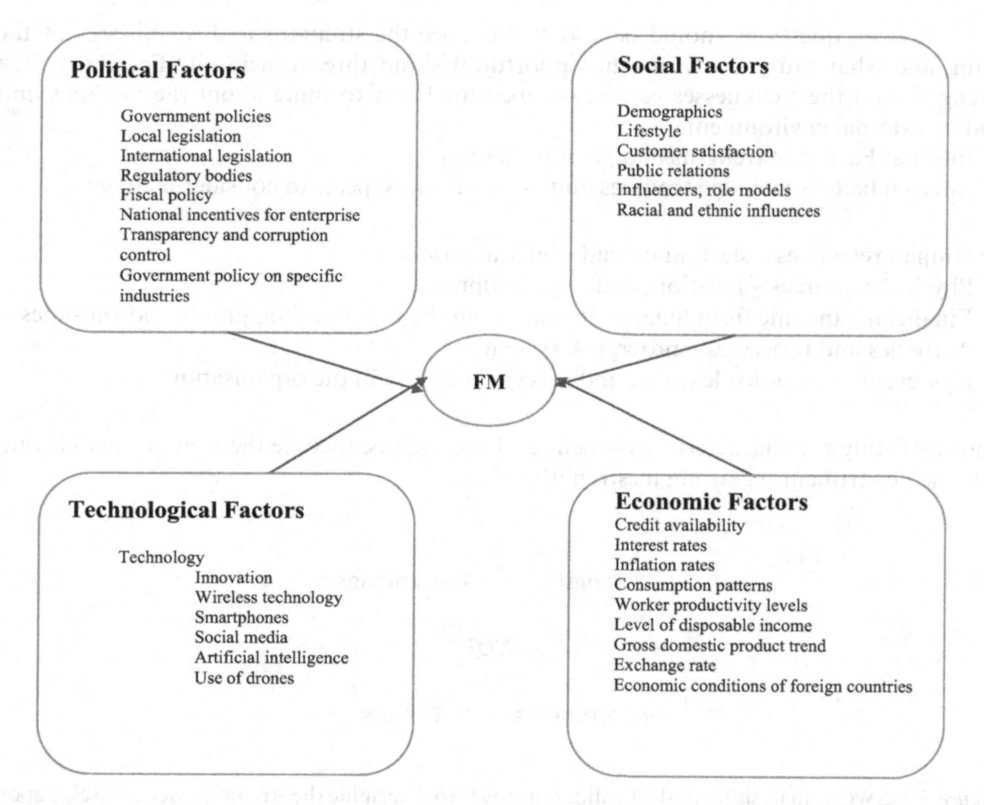

Figure 3.3 Political, economic, sociocultural, and technological factors are examined under the PEST model.

Source: Author

Technological: This "features factors that highlight the industry's technological development, such as engineering or scientific developments". This concerns automation, innovations, and technology that support FM.[3]

3.10.1.2 Strengths, Weaknesses, Opportunities, and Threats Analysis (SWOT)

Strengths, Weaknesses, Opportunities, and Threats analysis is also known as SWOT.

The benefits of SWOT analysis include:

- Examining potential new courses of action or solutions for issues
- Deciding on the best course of action for FM
- Understanding options may be helped by identifying probabilities for success and likely risks
- Determining likely areas for change
- Making changes to plans in the middle of them since a new opportunity could lead to new ways
- Creating a strategy for action
- Providing a straightforward method of organising data from surveys or research and discussing new projects or programmes

The following questions should be asked: What are the strengths and weaknesses of the team, and what can you say are the opportunities and threats facing it? For clarity, the strengths and the weaknesses can be grouped for brainstorming about the facilities unit and its external environment.

Internal Factors: Strengths and Weaknesses

Internal factors include resources and experiences. Aspects to consider include:

- Human resources – staff, unit heads, building users
- Physical resources – location, building, equipment
- Financial – income from tenants, income from the organisation, grants and subsidies
- Activities and processes – programs, systems
- Past events – steps for learning and success, integrity in the organisation

Start by listing the characteristics which can be strengths. Include the views of people outside the department regarding its strengths.

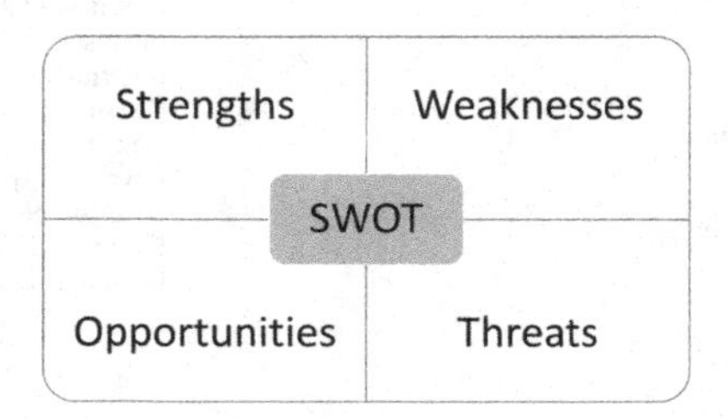

Figure 3.4 SWOT analysis helps the facilities manager to determine the strengths, weaknesses, opportunities of, and threats to, their operations.

Source: Author

External Factors: Opportunities and Threats
This includes the external environment of the facilities. They can be any of the following:

- Future patterns in FM or the culture
- The economy – local, national, or international
- Funding sources – foundations, donors, or legislatures
- Demographics – changes in the age, race, gender, and culture of the organisation or building users
- The physical environment
- Legislation (regarding legislation making the facilities' requirements difficult to meet)
- Local, national or international events

Conducting a SWOT analysis:
A SWOT analysis is often made during a planning or conference event, allowing people to think of ideas and analyse them for several hours. The process works best when everyone is involved and works together. The team shares what they have done when making the study.
How to do a SWOT analysis:
Assign a leader or group facilitator with good listening and assistance abilities. In big groups, elect a recorder to help the leader.

- Explain to the team the SWOT analysis objectives
- The team should work in smaller groups, depending on its makeup and the available time
- Have a recorder selected for each group
- Give each group 20 to 30 minutes to develop ideas for the project's advantages, disadvantages, opportunities, and dangers
- Encourage the groups to come up with suggestions
- After creating a list, narrowing it down to the top ten items might be helpful
- Call a group meeting at the scheduled time to discuss the outcomes. Obtain data from the groups
- Discuss the findings and document them depending on objectives and time frame
- Create a written SWOT analysis summary and distribute it to participants for future planning and execution.[4]

Sample SWOT analysis
The purpose of the SWOT analysis for Eskom, as indicated in Table 3.2, is to study the internal and external aspects of the company that will provide decision-making information by identifying and analysing an organisation's strengths and weaknesses as well as the opportunities available and threats to which they are exposed.

SWOT analysis identifies factors influencing the organisation's functioning and provides beneficial information in strategic planning. The SWOT analysis in Table 3.2 is from the perspective of FM, but could be fed into the SWOT analysis of the primary company.

3.10.1.3 Premises Policy

A business strategy is needed for a premises policy. It is a document that contains all the information about an organisation's standards and procedures. Future requirements should be considered, especially people, location, and information technology. It is a public document that may be changed, and senior management should support it since they can predict

Table 3.2 SWOT Analysis for Eskom – FM Component

SWOT ANALYSIS	
Internal Factors	*External Factors*
Strengths • Strong financial cash flow • String brand recognition • Economies of scales • Global player • Productive corporate culture • Strong management culture • Effective change management	Opportunities • High vacancy rate • Poor property management • Operating inefficiencies (FM) • Wide span of control • Overlap of decision making
Weaknesses	*Threats*
• Technology availability that reduces the cost of operations • Variety sourcing options, including outsourcing • Opportunity to redress environmental concerns • Scaling up FM to improve occupational health and safety	• Increasing demand for FM outsourcing which may impact the market price • Further deterioration of the infrastructure • Sick building syndrome and threat • Unstable economy • Poor political climate (unionised employees)

Source: Eskom

potential investment in the future. Building users are the internal clients who pay for services provided by facilities managers on the understanding that they will receive value for their money. A premises policy should show an organisation's present situation and future needs. It is often formed with a five-year time horizon in mind.

3.10.1.4 Skills Audit

A skills audit is a procedure for assessing and documenting the facilities team's abilities. The main goal of a skills audit is to determine the knowledge and skills that the organisation presently possesses and the knowledge and skills that the group needs. Skills audits are usually done to identify areas where training is required so that the team's skills and knowledge may be improved. However, other factors like reorganisation and arrangement, also need the completion of skills audits. A skills audit gathers more data than just the level of existing certifications. It initially determines the team's skill matrices and compares each person's current capabilities to the predetermined abilities needed for a given function. A skills gap analysis is the result of the skills audit procedure. The team will benefit from this knowledge by receiving the training and development required on account of the skills gaps. To verify that the right individual is in each role, the skills audit process will also produce data that may be used for internal employee selection.[5]

3.10.1.4.1 SKILLS OF A FACILITIES MANAGER

The core and optional competencies (based on requirements by RICS and IFMA) of facilities managers and other personal, interpersonal and professional skills useful to them are in Figure 3.5.[6]

Core competencies

- Customer service
- Project management
- Working with suppliers and specialists
- Maintenance management
- Corporate real estate management
- Procurement

Optional competencies

- Strategic facilities management
- People management
- Applying legislation, codes and standards
- Financial management
- Managing support services
- Risk management
- Building services
- Quality management
- Facilities operations administration
- Information management
- Sustainability
- Space management
- Management and business organisation
- Consultancy services

Personal, interpersonal and professional skills

- Ability to work with minimum supervision
- Ability to work in a team
- Oral and written communication
- Leadership
- Project management
- Understanding of professional ethics
- Planning
- Change management
- Analytical skills
- Systems thinking
- Commercial awareness
- Financial statement analysis
- Understanding the impact of FM in a global and societal context
- Research
- Local and international legislation
- IT skills

Figure 3.5 FM core and optional competencies and personal, interpersonal, and professional skills.

Source: Azasu, S., Adewunmi, Y., & Babatunde, O. (2018). South African stakeholder views of the competency requirements of facilities management graduates. *International Journal of Strategic Property Management*, *22*(6), 471–478.

Table 3.3 Skills Audit Matrix

Employee name	*Personnel rating*				*Evidence*
Competencies and skills	*None*	*Basic*	*Competent*	*Proficient*	

N/B: The competencies and skills highlighted in Figure 3.5 can be used to complete the column on competencies and skills in Table 3.3.

Source: Azasu, S., Adewunmi, Y., & Babatunde, O. (2018). South African stakeholder views of the competency requirements of facilities management graduates. *International Journal of Strategic Property Management*, *22*(6), 471–478.

3.10.1.4.2 METHODS FOR SKILLS AUDIT

Depending on the environment and strategy of the FM unit, a skills audit may be done using many methods. Analysing the environment and planning for the audit's goals are the first steps in implementing one. The background of the FM unit may be determined based on the amount of time available, arrangement challenges, the main goals of the skills audit, and the general socio-political environment. The FM strategy is the foundation for matching capabilities to the relevant department's current and future demands. This is important for the skills audit results to be valuable and consistent with the strategy.

The three main phases of a skills audit are as follows. The first step identifies each employee's abilities, the second stage identifies the skills needed, and the third step evaluates the findings and decides what skills must be developed. A training requirements analysis is often the process's output, allowing the facilities manager to provide information for recruiting and selection, performance management, and succession planning.

Identify what skills are needed:
To identify what skills are needed, the FM team should look at each job and write down what skills are needed now and in the future. The result is a skills matrix with descriptions of linked competencies. Definitions can be linked to levels of skill, like "basic," "competent," and "proficient."

Audit real skills:
The basic skills audit process includes a self-audit and a skills audit. The results are reported, including statistics, graphs, personal reports, and suggestions.

Identify what training is needed and plan how to get it:
Once the information from the skills audit has been gathered, it may be used to plan training and development and other HR activities. Actions decided upon are carried out after recommendations are discussed.

3.10.1.4.3 CURRENT STRATEGIES FOR SKILLS AUDIT IMPLEMENTATION

Skills audits may be done in many ways. Current approaches to skills audits include:

- Panel approach
- Consultant approach
- One-on-one approach

The panel approach, which includes knowledgeable people, subject matter experts, and HR staff, is the most reliable and impartial way to conduct a skills audit. The audit must be conducted honestly and transparently, with the facilities manager providing supporting documentation and presenting the conclusions from the self-audit.

3.10.1.4.4 SKILLS AUDIT RATING SCALE

The rating scale is helpful for statistical analysis of the skills audit findings. A person or panel rates a skill listing, and different computations understand and compile skills audit results. A definition and explanation are required for each rating.

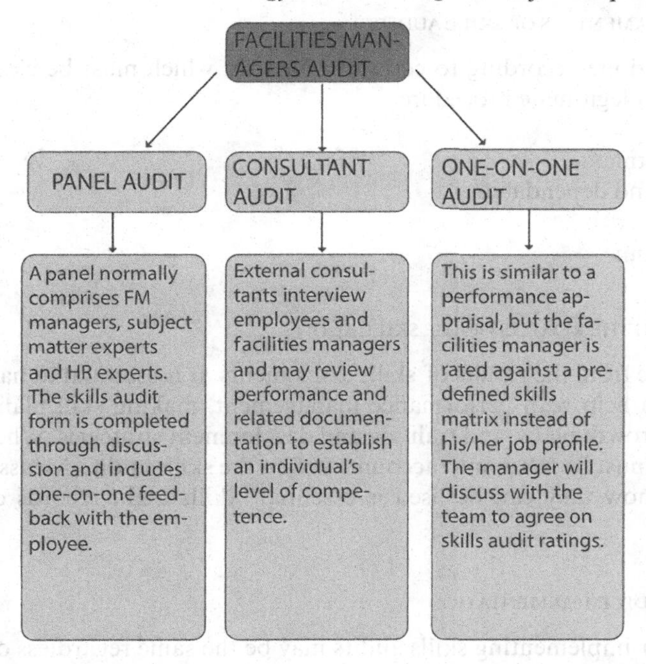

Figure 3.6 Strategies for skills audit include panel audits done through discussion and employee feedback, consultant audits done using performance and documentation, and one-on-one audits done using a predefined skills matrix.

Source: Author

Table 3.4 Example of a Rating Scale for a Skills Audit

Rating	*Description*	*Definition*
0	No evidence of competence.	Management requires a certain degree of competence, as well as the necessary training and abilities.
0.25	Some evidence of competence.	The manager may show partial competence and need work-related training and skills.
0.5	Evidence of competence, but needs more training.	Although the manager is competent, needs to get further training. The manager may not possess the degree of abilities required for the role.
0.75	Evidence of competence, but needs more practice with the skill.	The manager has received training and is qualified. This and complete competency might be enhanced with further job experience. The manager may possess the abilities required for the role.
1	Full evidence of competence.	The manager is skilled at the level suitable for the position.

Source: Author

3.10.1.4.5 THE FUNDAMENTALS OF SKILL AUDITS

Skills audits are done according to certain standards, which must be closely followed to provide a fair and legitimate procedure.

a Equitable treatment
b Authenticity and dependability
c Openness
d Helpful criticism

3.10.1.4.6 REPORTING THE RESULTS OF THE SKILL AUDIT

Reports are made from the results of skills assessments. It is essential to have these records because they can help with performance management, making skill plans for the workplace, personal growth plans, and training and development strategies. When keeping these records, privacy must be taken into account. Before the skills audit, discussing who can see the results and how they can be used is essential. Skills audit records can be made in many ways.

3.10.1.4.7 SKILLS AUDIT IMPLEMENTATION

The processes for implementing skills audits may be the same regardless of the team size. The main differences between skills audits in small and big teams will concern project budgets and timelines. Timelines and expenses are determined during project planning by defining the staff members involved, the number of positions involved, the goal of the skills audit, the methodology used during the skills audit, and whether or not external consultants will be hired. The first and most important phase in a skills audit for smaller teams should be determining what capabilities are needed and what competencies are already in place. This involves evaluating each person's ability level about recognised competencies and using the evaluation findings to plan skills development. The skills audit's primary goal is to ensure that knowledge helps implement FM strategy through skills development.

Effects on Costs

During the application of a skills audit, the following costs will arise:

- Training
- Time
- Administrative costs
- Information systems or software
- Communication
- The use of experts, if needed

The Problem of Not Doing a Skills Audit

The following can happen if a good skills audit method for planning for training purposes does not occur (Watson, 2004):

- Plans for training are not reliable
- Plans that are not tailored to the needs of FM

- Few or no plans for training and development are made by management and staff because they are not seen as offering value
- There is little or no match between training and growth and FM plans and goals

3.10.1.5 Services Audit and Review

This helps review current services to see where things stand currently (Adewunmi, 2020).

- Policies must include rules, guides, working processes, performance standards, quality standards, health and safety, staffing numbers, and finance approvals.
- Processes and processes will help fully understand how a company works, including budgets, procurements, purchase approvals, and payments.
- Service delivery audits on the property portfolio and service deliveries will improve customer relationships.

3.10.1.6 Balanced Scorecard

The broad measurements of an organisation can include FM goals and metrics, and the strategy can be obtained from the organisational strategies and vision. Facilities management has been involved in developing balanced scorecards. In FM, a balanced scorecard (BSC) is a strategic tool. It is needed for efficacy, efficiency, and maximising potential. Performance management techniques like balanced scorecards may improve organisational cultures, systems, and procedures. A BSC helps with the following tasks: defining mutually agreed-upon performance objectives, assigning and prioritising resources, telling managers to confirm or change current policies or directions to fulfil goals, and communicating performance outcomes in goal meetings. This shows the link between the strategy and the measuring system. A BSC can be measured along four dimensions.

3.10.1.6.1 CUSTOMER RELATIONS

This viewpoint shows the organisation's capacity to offer high-quality services, the efficiency of their delivery, and overall satisfaction and customer service. Customer-related performance metrics focus on developing customer value and how customers' needs are met. Facilities managers have many reasons to be concerned about this aspect, which shows the need for performance metrics that correctly represent crucial customer-oriented elements. The most important client criteria are often divided into time, quality, performance, and service.

3.10.1.6.2 INTERNAL PROCESSES

The business processes viewpoint focuses on examining the organisation's internal processes. While customer-based metrics are important, they may also be used to assess the internal work that the organisation has to do to satisfy consumers' expectations. Managers must thus focus on those vital internal company processes that allow them to meet client demands. Essential operations are examined closely to ensure positive results. Organisations should choose the procedures and skills necessary for excellence and define their metrics.

3.10.1.6.3 LEARNING AND GROWTH

Managers in FM companies are often judged on how well they do in the short term, which makes it hard for them to make investments that improve the skills of their employees, tools, and the way the organisation works. The main question with learning and growth problems is whether FM groups can continue to improve and give clients more value in the future. It will be difficult to see in the short term what the problem is with failing to improve employees, processes, and the organisation's skills over and over again. So, a process will only work if it has skilled, driven employees with access to correct and up-to-date information. This is even more important for companies where FM is undergoing big changes. Three things help a company learn and grow: its people, processes, and procedures. Long-term growth and change happen within this structure.

3.10.1.6.4 FINANCIAL

Financial success measures show if the organisation's plan, application, and effectiveness help the bottom line.[7]

Proposed BSC strategy for fictional company XX Facilities

Financial perspective: How do we look to our shareholders?
Corporate property management (CPM) aims to run XX Facilities in a way that supports business processes and helps the company reach its goals. The CPM will encourage people to be

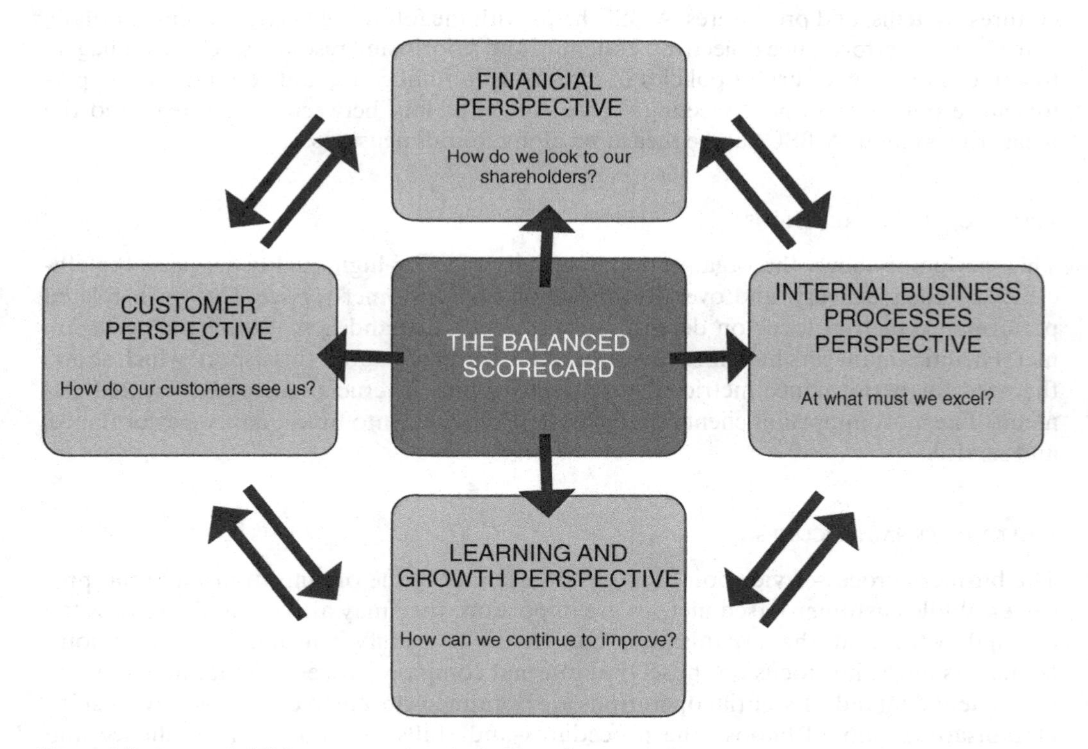

Figure 3.7 A sample FM strategy using the balanced scorecard.

Source: Amaratunga, D., & Baldry, D. (2003). A conceptual framework to measure facilities management performance. *Property Management*, *21*(2), 171–189.

responsible with money by lowering operating costs by focusing on the six strategy pillars of Cost Efficiency Leadership, Operating Excellence, Asset Management, Space Planning and Utilisation, and Energy Management. By focusing on these strategy pillars, the CPM will ensure that XX sites are built and run in a way that gets the most for the money and stays within the budget.

Customer perspective: How do our customers see us?
According to Kaplan and Norton (1992), the BSC requires managers to turn their customer service vision and goal statement into exact measurements that consider the things that mean most to customers. CPM will focus on finding ways to regularly meet service delivery standards and get a high degree of customer satisfaction. This is what the organisation wants to do with its business. Customer service is an ongoing process that starts with determining what the customer wants and coming up with answers that are fair and don't cost too much. People who rent want an open place to work that is also nice and efficient. The CPM needs to know how happy their customers are to make choices that hurt renters and bring about change in the business.

Along with the survey results, the business must be given a detailed action plan with due dates for dealing with the main issues. Setting up an online monitor for everyone in the company is one way to do this. To find out if the CPM meets business performance standards, performance must be benchmarked with that of similar organisations and SANS standards over and over again. This can be done with a set of success markers that SANS recognises.

Internal perspective: At what must we excel?
From the internal point of view, the CPM needs to resolve issues with internal policies and procedures that will improve measures that are based on customers. Every measure based on this internal view will help the CPM understand how the business is run and see if the products and services meet customer needs. These will show the CPM's most important skills and technical abilities, which will help XX's main business and operations.

Learning and growth perspective: How can we continue to improve?
Developmental concerns like creativity and skill development are dealt with from the learning and growth perspective. Additionally, it addresses the fundamental organisational tenets of people, systems, and managerial procedures. To acknowledge that both the environment and the demands of customers are changing, the CPM understands that innovation and constant learning are essential. All FM staff members will get personal and professional development support from the CPM.

The learning and growth enablers are separate to strategic goals. Their primary goal is to establish a structured organisational learning process that efficiently gathers best practices from within and outside the company and operational lessons. To enhance all facets of the organisation and be competitive, the CPM leaders and workers must consistently learn, grow, and propose creative ideas.

For the strategy implementation, the four elements of the BSC will link company operations with the vision, purpose, and strategy to stakeholder communication and performance monitoring against critical objectives. Fundamentally, the main goal of FM is to meet the company's immediate and long-term physical resource needs. The strategic objectives will be emphasised over the three-year term of the CPM strategy execution. The focus will be on the strategic goals in year 1. The many disciplines within the four BSC perspectives – which comprise the larger overall strategy – will implement the strategy. The CPM will coordinate daily tasks with the overall strategy. The initiatives recognised as affecting the overall plan will be prioritised. Finally, everyone on staff will be informed of the goals regularly. The proposed scorecard strategy (analysis and implementation) can be found in Tables 3.5 to 3.8.

Table 3.5 – Table 3.8 CORPORATE SCORECARD

FINANCIAL PERSPECTIVE = 30%	*Overall intent: Promote fiscal responsibility and sustainability*					
OBJECTIVES	*Measure*	*Units*	*Base*	*2019*	*2020*	*2021*
Reduce energy consumption	Energy conservation	kVA	New measure	10%	5%	5%
Reduce water consumption	Water conservation	KL	New measure	5%	3%	2%
Occupancy/Space cost	Annualised space costs	R/m²	R30/m2	R25/m2	R23/m2	R20/m2
Operational cost (Budget vs Actual)	Annualised cost savings (Rand)	%	New measure	10%	5%	5%
Capital Budget spend	% of CAPEX spent	%	New measure	80%	100%	100%
Waste recycled to total waste produced	Diversion rate	%	0	60%	70%	80%
Carbon footprint reduction	Reduction of greenhouse gas emissions	%	New measure	20%	15%	10%
Reduce the cost of moving employees	Churn costs	R/move/ workstation	New measure	10%	5%	5%

CUSTOMER PERSPECTIVE = 30%	*Overall intent: Improve customer relationships*					
OBJECTIVES	*Measure*	*Units*	*Base*	*2019*	*2020*	*2021*
Customer Satisfaction Survey	Customer satisfaction results	%	70%	80%	90%	95%
External FM Quality Assurance Audit results	Audit results	%	New measure	100%	NA	NA
Improve air quality – workplace comfort and productivity	IAQ results	%	New measure	90%	95%	95%
Introduce IT dashboard for critical FM indices	System in operation	%	New measure	100%		
Service quality levels	Response time	days	3 days	2 days	1 day	1 day
Accessibility for disabled	Level of accessibility	%	New measure	100%	NA	NA

INTERNAL PERSPECTIVE = 20%	*Overall intent: Efficient and effective processes*					
OBJECTIVES	*Measure*	*Units*	*Base*	*2019*	*2020*	*2021*
Provide a Health and Safety environment	NOSA Grading	5 star	New measure	4 star	5 star	5 star
Provide Health and Safety environment	No injuries or fatalities	Number	5	0	0	0
Business Continuity Plan	Finalised documented plan	%	New measure	100%	NA	NA
Complete the Procedure Manual for Asset Management	Finalised documented procedure	%	New measure	100%	NA	NA
Improve space planning and utilisation	New projects will be 10% of the recommended range	Number	New measure	80%	90%	95%
Introduce the use of Computerised Asset Management	CMMS implemented	number	New measure	80%	100%	NA
Improve space utilisation in the building	Internal audits and updates	Number	New measure	80%	90%	95%
Preventive vs corrective maintenance	Preventive maintenance ratio	%	60/40%	70/30%	80/20%	80/20%
Reduce unscheduled electrical and HVAC system outages	Number of outage/month	number	New measure	60	40	30

LEARNING AND GROWTH PERSPECTIVE = 20%	*Overall intent: Maximise human capital competencies*					
OBJECTIVES	*Measure*	*Units*	*Base*	*2019*	*2020*	*2021*
Employee Satisfaction Survey	Employee satisfaction results	%	New measure	80%	90%	95%
Average hours of training for Team Leaders and Managers	Number of hours/year	hrs	New measure	40hrs	40hrs	40hrs

(*Continued*)

Table 3.5 – Table 3.8 (Continued)

LEARNING AND GROWTH PERSPECTIVE = 20%	*Overall intent: Maximise human capital competencies*					
OBJECTIVES	*Measure*	*Units*	*Base*	*2019*	*2020*	*2021*
Average hours of training for front line employees	Number of hours/year	hrs	New measure	40hrs	40hrs	40hrs
Annual employee turnover	Turnover rate	< 14%	New measure	< 12%	< 10%	< 8%
Personal Development Plans for all FM staff	% staff who have been through PM process	%	New measure	100%	100%	100%
Positive changes in quality, cost, safety and delivery due to training	Number of initiatives/changes	Number	New measure	5	8	10
Develop a high performance team	Average days to fill a vacant position	days	New measure	60 days	60 days	60 days

Source: Anonymous.

Absa – A Kenyan Example of FM Strategy Using BSC

The mission is to provide the Nairobi branch of the Absa bank with facilities that will be adaptable to future technological changes, meet sustainability goals, and improve customer satisfaction.

The vision is to deliver state-of-the-art facilities that will deliver the best value in facilities management while not compromising the quality of services and ensuring business continuity.

The FM department is responsible for managing Absa's facilities by supporting core business operations and assisting the organisation in achieving its stated objectives. The department's aspirations are business continuity, customer satisfaction, operational efficiency, and agile and sustainable practices. Some of its operational initiatives are aligning practices with the industry and regulations, managing operational costs, and promoting continuous professional development. The performance measures are through changes in capital turnover, benchmarking of services that provide value, and cost of research and development.

The bank uses the balanced scorecard for strategy and, from a financial perspective, ensures that the organisation has the best FM service at a reasonable financial cost, which is crucial for the FM team to enhance the share value for the shareholders. The FM strategy also aims to meet the customer's goals from the customer's perspective. The internal perspective ensures that the technical aspect of operations functions in line with a set of criteria. The internal perspective made the FM department plan for which services to outsource, to provide in-house, and to provide through co-sourcing. The learning perspective deals with professional development and human resource matters such as recruitment, staff retention, empowerment, and inclusion. To facilitate learning, the facilities management team has a monitoring and evaluation team that monitors the strategy's performance. The learning perspective is implemented using the ADKAR model, which stands for awareness, desire, knowledge, ability, and reinforcement.

3.11 Strategic Choice Tools and Techniques

SWOT and PESTEL analysis are among the tools that can help you make a strategic choice. FMs use strategic choice tools to determine what their department will do in the coming years and beyond and the best way to get there. These methods help managers turn useful information into forms that can be used to make choices and take action. These tools can help you learn more about the business world, strategic issues, opportunities, and threats, which lowers the risk of making poor decisions. They can also help you set priorities in large, complex organisations and determine how critical different portfolios are, and they can also assist with presenting complex issues. In addition to their vital role, they may also be seen as valuable tools for communication.[8]

When generating options, it's essential to agree on the criteria used to judge them. There should be a clear split between putting together criteria and using them in options to ensure fairness. So, the factors should be easy to understand and include issues like end user satisfaction and how best value can be achieved. Options can be made by talking to stakeholders and inviting experts from outside the company. That means options should be thought about for filling in the gaps that have been found and ensuring that new solutions fit current and future needs. There should be a list of all the preferred choices in the facilities management options (Adewunmi, 2020).

Table 3.9 Strategy Choice Tools and Techniques

Development stage	*Tools or techniques*
Generation of options	Process mapping or business process engineering (BPR) Outsource modelling Modelling market offerings such as outsourcing or strategic partnerships
Evaluation of options	Cultural fit/gap analysis Risk analysis (physical, economic, financial, business, political) Cost-benefit analysis Stakeholder assessments and reviews Feasibility analysis Innovation and creativity potential Maintenance plans Life cycle cost appraisal
Strategy selection	SWOT (Strengths, Weaknesses, Opportunities, and Threats) analysis Financial modelling Organisational compatibility and synergies Decision trees Sensitivity analysis Optimisation model
Strategic change	Schein's "Unfreeze" change model

Source: Atkin, B. & Brooks, A. (2015). *Total facility management*. Wiley Blackwell

3.11.1 Process Mapping

This involves identifying, documenting, analysing, and developing a more efficient process. A process map is used to understand the organisation and enhance the efficiency of its operations.

According to Anjard (1998), a process map is "a visual tool for conceptualising FM work processes that show the relationships between inputs, outputs, and activities." A process map encourages fresh ideas about how work is done. It helps to discover important actions taken to achieve a result, the people who carry them out, and the places where these (important) issues continue frequently. Process mapping identifies the areas where changing procedures will most influence increasing quality. Teams become important in creating and using process maps. An effective process frequently involves many functions, each operating individually and independently.

The main mapping project phases include:

- Build teams and a framework
- Set up the process for mapping
- Get support from top management
- Learn how to map processes

Criteria for joining a team:

- Know the process well or at least one of the functions that helps the process
- Be trusted by the organisation and other team members
- Be able to speak for their function
- Know the "big picture" beyond their function
- Be creative
- Have much energy

- See the work of the team as a reward
- Be able to attend all team meetings

Process definition:

- Determine what the process is all about and what the goals are
- Make a project plan with due dates and outputs

Issues include:

- Difference between performance and what the customer wants
- Difference between performance and the ability to get ahead of the competition
- Need to make an opportunity or increase the gap in having a competitive edge
- Need to implement a significant change

Examples:

- Reducing current maintenance request turnaround time
- Reducing inventory costs
- Improving the quality of incoming parts from suppliers

Create a description of an important process:

- What started the process? Trigger events?
- From whom do the inputs come?
- Where does the process end? Final output?
- Who are the customers of the process? Internal or external? Both?

"IS" analysis:

- Discover how the current process works
- Search for opportunities for improvement

Goals:

- Ascertain what's going on with the current method and who's involved
- Discover how well the process is working now to redesign it for the future
- Identify problems that need to be resolved
- Find customers' perceptions and expectations
- Create a process map

Techniques:

- Developing the organisation (system) level view-relationship map, which includes:
  1 Services and information created
  2 Customer input – could be internal or external to the organisation
  3 Primary FM components or subsystems involved
  4 Inputs essential to FM
- Training on how to use map processes
- Creating process maps
- Identifying customers' perceptions and expectations and analysing them.

Ideal design:

- Create a new or better process that meets the goals and solves problems
- Develop a project plan
- Benchmark best practices
- Brainstorm new ideas
- Use the running tools
- Layout the "should be" process's macro-process phases
- Design sub-processes with micro-process phases
- Make a cross-functional process map
- Test the "should be" or ideal design
- Create detailed process specifications
- Create process measures
- List the steps that need to be taken, the deliverables that need to be made, and the dates that they need to be turned in
- Set due dates for all phases
- Set dates for the process sponsor and management committee to review the process

Implementation planning:

- Make the changes needed to put the "should be" process into action
- Get the management committee's go-ahead to put the plan into action
- Write down what was learned from the pilot programme
- Set up an official feedback loop
- Form a pilot execution team
- Keep track of the changes
- Create a team that focuses on processes[9]

The Process Profile Worksheet (Table 3.11) is one of the important worksheets for the mapping process. Each task should examine this form. The spreadsheet keeps track of essential data gathered regarding the procedure. It will be finished in stages when the process owners collect information.[10]

Table 3.10 Process Identification Worksheet and Process Description

Cleaning	
Trigger events	Process Name
Process description	
Process name	Process description

Source: Author

Table 3.11 Process Profile Worksheet – Cleaning Process for Reception

Process Number	*Process Owner*
Cleaning Budget	Ben Morris Segun Oni George Dlamini
Description	
The process of cleaning the reception and other open spaces	
Triggers	
Event beginning: Collection of cleaning materials from storage Event ending process: Clean and wipe the floors, touch surfaces daily Additional events: Complete cleaning, obtain cleaning materials, carry out the cleaning task	
Input – Items and Sources	
Disbursement to the service provider	
Process Units	Process Unit Owners
Check cleaning work Document cleaning work Monitor cleaning done	FM Supervisor Unit FM Supervisor
Objective(s)	Risks
Prompt and good quality cleaning	Slips and trips Infections Customer dissatisfaction
Essential controls/Risk mitigation	Measure of success
Separation of duties The person monitoring the cleaning work	Cleaning is done before, during, and after critical work hours. Quality of cleaning comparable to benchmarked site Absence of complaints from users

Source: Author

Managers are the best source of information for identifying the specific activities and choices that make up the process; the person mapping should be aware of each employee's duties. To obtain this, a workflow survey (see Table 3.12) must be sent to every employee involved in the process that is to be mapped.

One difference between a linear process map (see Figure 3.9) and a cross-functional process map (see Figure 3.8) is that the cross-functional process map shows how the process moves through the different functions. For example, it clarifies all the links between suppliers and customers. A process map can be made with varying amounts of information. The macro-level (the top level) of that process is where planning should start. This setting sets the system's reach. Usually, only a few decision points or files are added at this level. Around 5 to 20 processes are shown on this level of process map. After that, the process should be sped up to the mini-level so everyone can understand the basics. Several small processes can join together to make a single macro-process. The last step is the micro-level, which should have a naming system that refers to the map, map levels, and process specifications. Process numbers should be based on either the macro- or mini-level. The right amount of depth relies on the reason for mapping it. The best method starts at the top and gets smaller.

Table 3.12 Workflow Survey

Name:		**List of duties for the job**		
ID:				
Title:				
How long you've been in this job?				
Duty	**Where does the duty emanate?** (Colleague, e-mail, phone, boss)	**What are your duties?**	**What is the length of time for the process?**	**Where does the work go when it's done?** (A customer, someone in the FM department, or the e-mail)

Source: Author

Working alone or together, several functions are often needed for a process to go well. For this process to be shown correctly, the independent and dependent steps must be shown. Cross-functional process maps are the name for these kinds of maps. Service providers and customers can clearly understand each other's roles. The cross-functional map displays the steps in the process that each person completes, as well as the essential lines that separate the functions. The map also shows the order of the connections and what different processes do. If someone wants to, they can put time frames (cycle times) on the map. The map often goes from left to right in time, like a PERT chart (Jacka & Keller, 2009).

Table 3.13 shows process flow chart symbols for process mapping.

3.11.2 Outsourcing Models

Total Facilities Management (TFM) is the outsourcing of every service to a single service provider. The client will get a single bill that covers all services and has a relationship with the service provider, manager, or director. Not many organisations can supply services alone, so they must employ subcontractors to offer the full scope of services.

Single service providers are independent contracts for a service provided by a single contractor, like cleaning or catering.

Bundled services: Contracts are organised under bundled services. We can have two or three services provided by a single service provider. In a contract, a bundle occurs when, for example, in the aspect of maintenance, building fabric and grounds maintenance can be combined. Depending on the organisation, multiple groups of services might be formed.

In-house: When a service is provided internally by employees in an organisation, such service provision can be monitored using service level agreements. This option is preferred for security, business, or remote location concerns. The in-house team's organisational structure is determined by the services provided, the number of locations, and the sites. The in-house team should work towards the delivery of services whose quality matches that of external service providers. Sometimes, some technical aspects of the work done in the department can be outsourced for efficiency. An example is that of lift maintenance.

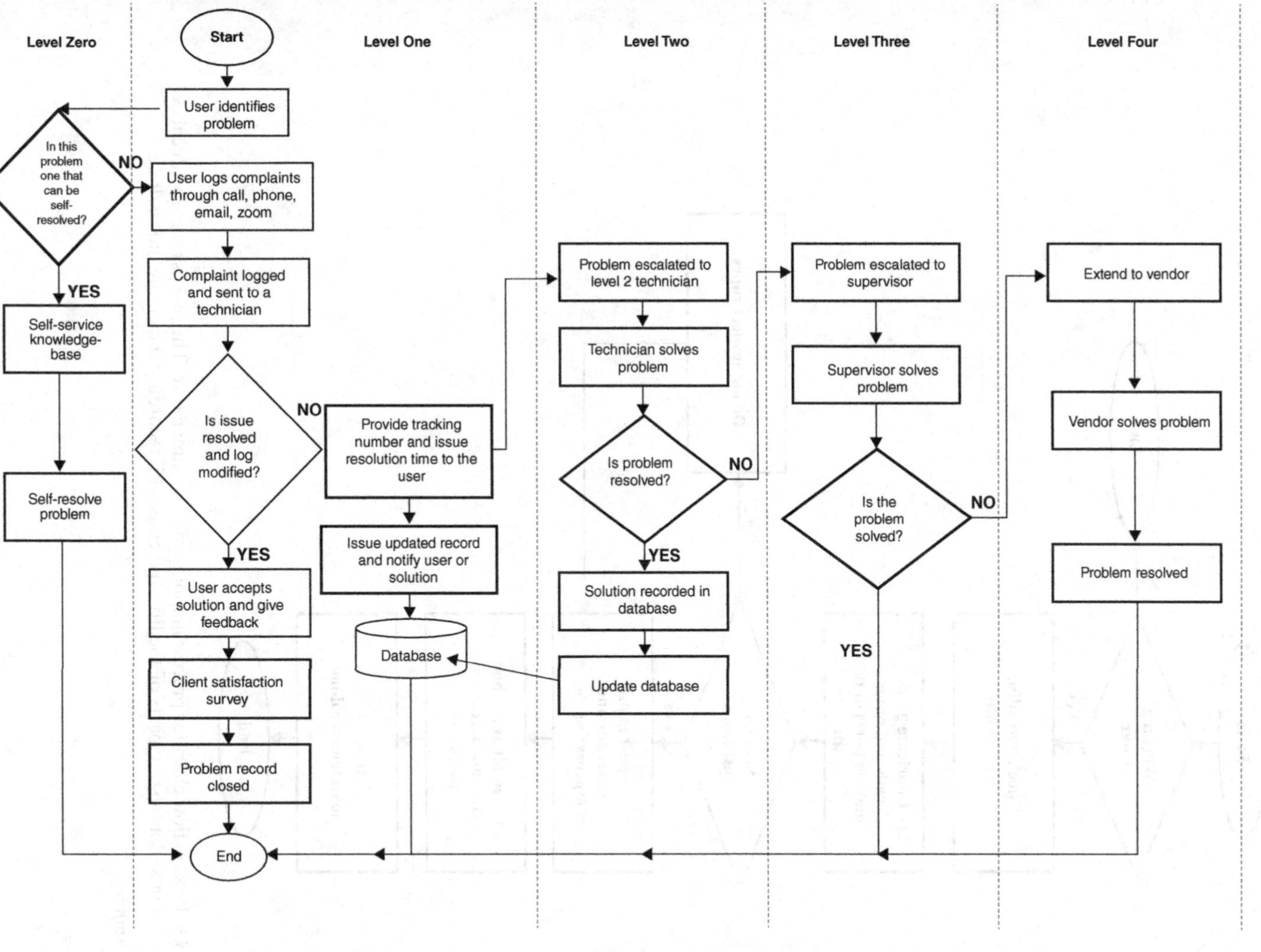

Figure 3.8 Flow chart for a maintenance help desk actions and choices. In the first level, the user lodges a complaint that can be self-resolved. If the service can't be self-resolved, it moves to level two and is sent to a technician who resolves the problem. If the problem remains unresolved, it moves to another technician and a supervisor in level three. In level four, a vendor will be contacted to assist with resolving the issue.

Source: Author

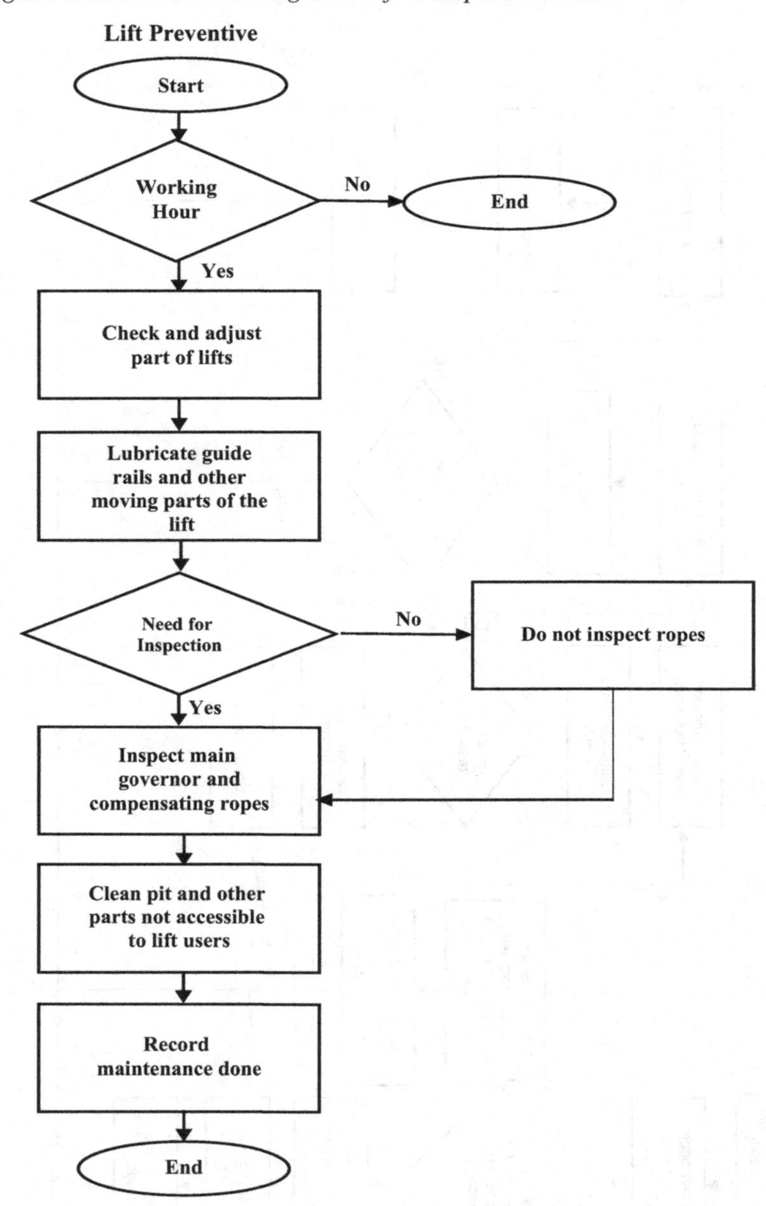

Figure 3.9 Process flow chart for process mapping of lift maintenance. The process starts with checking and adjusting parts of the lifts and ends with recording the maintenance done.

Source: Author

Table 3.13 Process Flow Chart Symbols

NO	*SYMBOL*	*DESCRIPTION*	*NO*	*SYMBOL*	*DESCRIPTION*
1		TERMINATOR	22		OFF-PAGE REFERENCE (INCOMING)
2		PROCESS	23		MULTI DOCUMENT
3		DECISION	24		SUMMING JUNCTION
4	YES NO	YES, NO	25		COLLATE
5		DATA	26		SORT
6		MANUAL OPERATION	27		OR
7		DOCUMENT	28		EXTRACT
8		PREDEFINED PROCESS	29		MERGE
9		STORED DATA	30		MAGNETIC DISC
10		INTERNAL STORAGE	31		DIRECT ACCESS
11		SEQUANTIAL DATA	32		DIRECT FLOW
12		DIRECT DATA	33		OFF PAGE REFERENCE (ARROW)
13		MANUAL INPUT	34		ON PAGE REFERENCE
14		CARD	35	TXT	ANNOTATIONS
15		PAPER TAPE	36		CONTROL TRANSFER
16		DISPLAY	37		CONDITION

(*Continued*)

Table 3.13 (Continued)

NO	SYMBOL	DESCRIPTION	NO	SYMBOL	DESCRIPTION
17	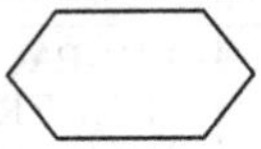	PREPARATION	38	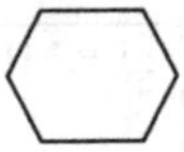	PREPARATION
18	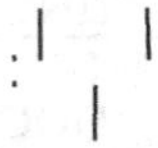	PARALLEL MODE			
19	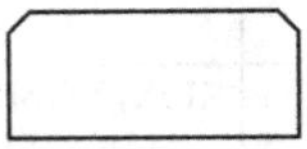	LOOP LIMIT			
20	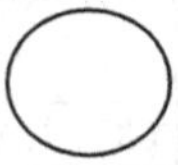	OFF-PAGE REFERENCE (CIRCLE)			
21		OFF-PAGE REFERENCE (OUTGOING)			

Source: conceptdraw.org

Integrated Facilities Management is outsourcing some services to a single service provider. The scope of services depends on the arrangement between the service provider and the client organisation. The client will get a single bill that covers the services and has a relationship with the service provider, manager, or director. Some organisations may need help with some of their FM services, so they employ subcontractors to offer such services.

The managing agent: An organisation acts as the managing agent when it oversees and coordinates services between the client and the contractor. Property management companies with managerial expertise and the financial ability to purchase cleaning, security, and maintenance contracts usually do this. Since there is a contractual arrangement between the client, contractor, managing agent, and client, it differs from TFM. The managing agent will be paid based on relationship management and service provider performance evaluation.

Example of Outsourcing for The Co-Operative, UK

Along with being one of the largest food groups in the UK, The Co-Operative Group also provides significant financial services, funeral services, pharmacy chains, and legal services. FM activities are essential to enhancing its brand values. The internal FM promotes the company's primary goods and pushes its "team members" to uphold ethical, sustainable, and corporate social responsibility principles in all facets of the company's operations, including recruiting, customer service, and purchasing.

In the past, The Co-Operative has been eager to maintain the independence of its services; however, it has shifted its attention to outsourcing operational tasks, which allows employees to concentrate more on long-term planning and strategic concerns (both business- and FM-related). The Co-Operative considered the following strategies for outsourcing:

- Choosing partners with care and ensuring that outsourcing is suitable
- Working hard internally to assemble the proper team and framework

- Establishing a supplier management role inside the team and implementing a supplier management framework with outside help rather than letting the contractor do what they want
- Recruiting and training the appropriate individuals
- Installing the right controls
- Prioritising proper infrastructure and lowering risk after that
- Determining what is measurable in terms of output
- Attracting suppliers who strive for inclusivity across all facets of society and operate within the communities where operations are located
- Complying with the conditions of trade and paying suppliers on schedule
- Assessing value based on cost, risk, and quality considerations, not just cost alone.

3.11.3 Stakeholder Expectations

Stakeholders are individuals or groups an FM organisation depends on to fulfil its objectives. External stakeholders include clients, suppliers, service providers, shareholders, investors, senior managers, regulatory bodies, and trade unions. Few people inside an FM organisation have the authority to influence the strategy. Influence is likely to occur because members of stakeholder groups, which might include departments, locations, or various levels of the chain of command, share expectations. Depending on the plan, different stakeholder groups will be available, and people may belong to more than one stakeholder group.

External stakeholders can be categorised into three types regarding the nature of their relationship with the organisation and how they can affect the success or failure of a strategy:

- Stakeholders from the *"market" environment*, suppliers, and shareholders should be included. These people have a financial stake in the organisation and impact how value is created
- Stakeholders from the *social/political environment*. The social part of the strategy will be affected by entities like lawmakers, regulators, and government agencies
- Stakeholders in the *technological environment*, like early adopters, standards agencies, and tech owners, will impact how new technologies are used and industry standards are adopted

Stakeholder mapping shows what stakeholders want and how much power they have. It also helps people understand the political goals. It shows how important it is for each stakeholder group to express its views about the organisation's objectives and chosen strategies, and stakeholders need the power to do so.

3.11.3.1 Power/Interest Matrix

Figure 3.10 shows the power/interest matrix. It explains the political environment in which a particular method in FM might succeed. This is achieved by categorising stakeholders according to their influence and how likely they are to support or oppose a specific goal of FM. The matrix shows how facilities managers relate to various stakeholder groups in the four quadrants. Essential stakeholders (Segment D) are the supervisors or head of departments supervising the project and must accept the strategies. The Segment C stakeholder concerns are the most difficult (the top management belongs to this group). Even though these stakeholders may often be very passive, things might go wrong if their interest in the project is not retained, and they quickly move to Segment D, influencing how the strategy is implemented. Managers may need to increase the degree of interest of significant stakeholders (such as top management) to perform their expected roles more effectively. This might also be about how effectively information and briefing can help managers do their duties. Segment A involves staff or contractors that must be supervised closely. Facilities managers must provide information to stakeholders in this group as they are senior executives working in another department, not the FM department. These stakeholders may play a critical role in influencing the behaviour of other influential stakeholders, such as through lobbying.[11]

Sample stakeholder analysis (hypothetical)

A stakeholder analysis was done in 2021 for a university renovation project. Four main categories of stakeholders were identified for the renovation project. Assumptions made are that the project was Tender Based. The Client, not the Contractor, appointed a Project Manager, and the Contractor must appoint a Site Supervisor.

End Users:	**Internal**: Faculty Staff, Cleaners, and Lecturers **External**: Students, Guest lecturers, and Visitors
Steering Committee:	Senior Faculty Staff is responsible for governing the project and ensuring it is delivered according to their planning and vision.
Professional People:	Architects, Engineers, Quantity Surveyors, Project Managers, Site Managers
Contractors:	The Main Contractor and Subcontractors working for them.

Level of Influence:	
1. Steering Committee	High
2. Professional People	High
3. Project Manager	Moderate
4. Contractor	Low
5. End Users	Low

(The steering committee needs to ensure the end users' needs are accounted for)

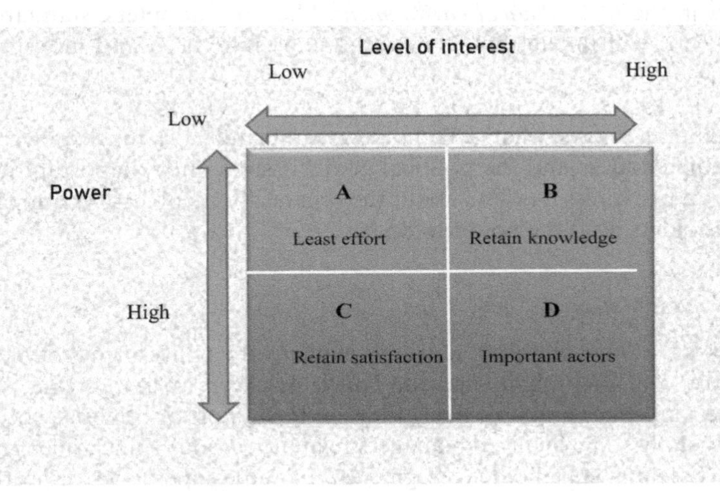

Figure 3.10 Stakeholder mapping: The power/interest matrix. The matrix shows how facilities managers relate to various stakeholder groups in the four quadrants.

Source: Author

Table 3.14 Stakeholder Needs and Expectations

Stakeholder	*Expectations/Needs*	*Measures*
End Users	• Receive communication regarding the aspects of the project that will impact them • Have a good, productive, and healthy environment for learning when the project is complete • Have physical needs satisfied • Find out the project benefits once it's done • Be a part of the planning process so that their ideas and wants are considered • Identify what the processes are	• To get the public involved by using questionnaires, information booklets, etc. • To have in-depth conversations with the end users to discover what they want and need from the new building
Steering Committee	• To be regularly informed about the progress of the project • To ensure that the idea is carried out in the final result • To communicate with the project team with all details • To ensure the project team chooses the right things	• Having regular meetings with the project team • Ensuring that communication is clear
Professional People/Project Team	• To see that the client's idea for the project is implemented • To confirm the project is on schedule and meets all the rules and regulations • Ensure the project is cost-effective • Have a plan of how a project will be carried out and a clear way for communicating the project • Ensure that the planning and development approach works	• To have regular meetings to talk about each person's role • Clear communication between professionals in the same field
Contractors	• Ensure that the planning methods are put into action • To make sure that daily building work is done • Ensure the building work is done on time and according to plan • Ensure that the building supplies needed are on site • Ensure that all employees have the right skills to do the job • Ensure the building is safe for workers and staff to do their jobs • Ensure the project's supply chain is in order • Knowledge of project expectations and process	• Ensure the right skills for the project are used • Have daily meetings so that the project team understand their roles

Source: Anonymous

Table 3.15 Action Points / Strategic Choices

Stakeholder	*Action*
Senior Faculty Staff	• Create a project plan and goal • Select competent professionals to handle the project • Organise meetings to ensure that professionals have an understanding of the project
Project Manager	• Needs to learn about the project. • Hold meetings with stakeholders who have an interest in the matter • Make a schedule for the construction work • Work towards timely delivery of the project • Ensure that there is communication with the contractors • Ensure the project aligns with the client's (the faculty) vision
Contractor	• Possess knowledge of project expectations
Architects	• Understand and plan for the needs of the clients • Present to clients available options on how to reach their project goals • Conduct meetings with the engineer to assist with implementing the designs
Engineer	• Ensure that the plans that were made follow the rules • Meet with the architect and the contractor to discuss how they want certain parts of the build to be conducted • Advise on how to do certain things in the building • Ensure the project follows the regulations
Quantity Surveyor	• Ensure the project stays within its budget • Meet with the architect about what materials to use • Meet with contractors and suppliers to make sure you have the best materials • Ensure the build is cost-effective
Site Manager	• Conduct health and safety meetings every day • Weekly meetings with the project manager to understand what is expected regarding planning and deadlines • Ensure that difficulties are reported to the project manager • Meet with the staff daily to discuss the project • Supervise the daily activities of the building
Lecturer	Take part in discussions about what facilities are needed and possible modifications, such as HVAC, lighting, projector facilities, sound, and the acoustics of the rooms
Students	Take part in meetings and surveys to ensure they are a part of the design process. Ensure the suggested plan meets users' wants. Needs include feeling safe, privacy, a visually appealing environment, thermal properties (like HVAC and heating), and a clean and well-kept environment

Source: Anonymous

3.12 Strategy Implementation

The implementation programme will need the following:

1 Executive ownership across the core and non-core businesses
2 Operational targets agreed for internal customers and teams
3 Clear management accountabilities to be identified
4 Correct resource alignment plans
5 Organisational structure and reporting systems between client and FM supplier
6 Informed/intelligent client where outsourcing occurs

Table 3.16 Implementation

Development stage	*Tools or techniques*
People and systems	Change management
	Personnel development and training
	Business process re-engineering (BPR)
Communication	Organisation intranet
	Newsletters, noticeboards, memoranda
	Instant messaging and corporate social networking
	Workshops and seminars
Resource planning	Planning, scheduling, and control
	Resource levelling/optimisation
Procurement/outsourcing	Service provider selection
	Market testing
	Performance benchmarking

Source: Atkin, B. & Brooks, A. (2015). *Total facility management*. Wiley Blackwell

7 Training programmes
8 Change management
9 Project management that includes project teams, timetables, milestones, risk management plans, transition or mobilisation plans, budgets, and spending rates

In terms of people and processes, the organisation must enhance staff members' competencies and skill sets to ensure they are thoroughly knowledgeable about the definition and application of facilities management. These goals will be met, and capacity will be increased with the help of individual mentorship, education, and training. Ensuring that tasks are closely monitored and controlled will also assist in ensuring that they change as planned. This is not a one-time event; there must be a continuous culture of improvement, with regular performance reviews and appropriate corrective action. Effective communication between the organisation, as a knowledgeable customer, and the service providers ensures that implementation is understood and considered. It is crucial to include all stakeholders involved in the discussion concerning structure and organisation to prevent problems. Planning and regulating the efficient and effective use of resources is known as resource planning (Adewunmi, 2020). Strategy implementation is usually for 3–5 years with a baseline period where milestones are measured for each year (example can be found in Tables 3.5 to 3.8).

3.13 Strategic Implementation Tools

Some of the strategic implementation tools are discussed below.

3.13.1 Organisational Structure and FM

The facilities manager may need to change the organisation structure once the FM strategy has been established. The FM department may take various forms in an organisation, including finance, procurement, HR, operations, customer services, logistics, property management, and building services. The nature of the organisation and the business sector will determine the right location in the overall business structure.

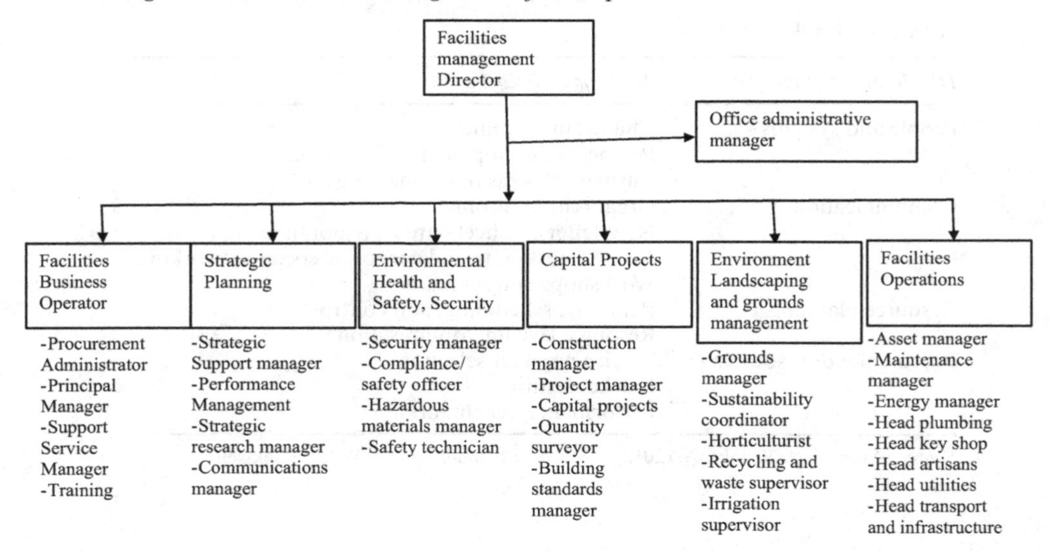

Figure 3.11 Example of an organisational structure of the FM department of a higher education institute. The structure is at three levels. The head is the facilities director who has six heads of units reporting to them. At level three there are various professionals, artisans, and technicians that report to the head of each unit.

Source: Author

3.13.2 Facilities Management Change Management

Employee involvement in facilities management is important, as is the increase of participatory facilities management. In addition, customer services are becoming more important for employee and organisation support. Facilities management needs reception and agent services in this regard.

3.13.2.1 Implementing Change

Change can be implemented by involving employees in the move. During transition times, there may be a tendency to believe that the facilities management team is forcing change on the workforce rather than that it is necessary for the organisation. The workers should communicate with one another honestly and transparently.

3.13.2.2 The Project Team and Preparing the Stage

Six elements are useful for effective implementation relating to learning events that can form the basis for working with the FM team and users concerning project implementation (Pritchard, 2007).

- "Initial contact stage" – these requirements include creating a good FM team to carry out the organisation's strategic goals and distributing information about the change to users. Certain project team members may not be well-organised; this communication should occur early in the project. Other users may also view these individuals as project supporters

- "Main focus" or connecting the concepts to actual events. For example, the project team members know their responsibilities, the path forward, and what is expected of them
- The project needs to maintain drive throughout the "making it happen" stage: remembering that open communication allows for a two-way flow, the team and the end users should be involved in the project while ensuring openness to listen to recommendations and comments
- The "review stage" relates to the outcomes phase to show how it has helped them enhance their working procedures
- "Communication" fills all six original parts
- The "measurement" phase is focused on analysing how the team collaborated, any problems that came up, and measuring the project's effectiveness (time, cost, and quality)

3.13.2.3 Alternative Workplace Strategies and Space Utilisation

Workplace design modifications and on-site and off-site working methods are examples of alternative workplace strategies (AWS). Examples include flexible work schedules (such as staying longer in the office), hot desks, and telecommuting (a mix of home and office-based working). An organisation might save 30% of its office spending by implementing AWS. AWS must consider three essential resources – people, technology, and space. When developing AWS, it is important to keep the employee in mind, since the workplace may affect attitudes, job satisfaction, and morale. The working environment and cultural norms will change due to the implementation of an AWS, reducing confidence and satisfaction. In Africa, AWS includes flexible work arrangements where employees have the choice to vary the times they arrive at work and leave work; there is also the compressed work week option where the staff work less than five days a week or short-term employment approaches; and the third option is that of telework, which allows employees to work from different locations other than their workplaces.[12]

3.13.2.4 Communication

Discovering what is happening is the first step in the change process. The next step is denial, which means saying it won't happen. The third step is resistance, which means employees are worried and need personalised communication to understand how this affects them. And finally, there is acceptance. The communication plan should help employees get used to the planned change by making them feel informed and schooled about it. Laframboise et al. (2002) added ideas for employee comments to stress how important it was for staff members to be involved.

3.13.2.5 Change Management Theory

According to Lakomski (2001), "Organisations are frequently balancing the pressures for change with the need for stability; this becomes destroyed when the desire for change is bigger than the resistance to it, creating a change in the organisation's stability". An organisation can successfully implement a change by changing current behaviours and patterns, moving them to the desired state, and then changing them.

Employees will experience fear when a company changes how space is used. Since people's behaviour may be changed to fit the culture, there may be great resistance to change. Employees are exposed to any unexpected effects this change to the usual working method may have. This emotion is expected in change programmes since the first shock is followed by denial and rage. People are attached to the past unless they accept the nature of change and its causes, the transitional phase where the new culture develops until it supports the new vision. Maintaining a balance between change and stability is important when dealing with cultural change resistance, which will limit change. Cultural change will likely happen when conditions such as low organisational performance or competitive marketplaces exist.

3.13.2.6 Communication Methods

Communication with the people the change is meant to help is critical to its success. The team in charge of facilities should ensure the right tool is used for the job. Questionnaires, focus groups, project walls, project rooms, project websites, electronic message boards, post-occupancy evaluations, idea boxes, democracy, and panels are all communication tools.[13]

Example

Discovery Building, Sandton, Johannesburg

The decision to combine all the premises into No. 1 Discovery Place was based on Discovery's commitment to keeping its people together and to creating open spaces that align with its values. Discovery's global headquarters in Sandton was developed in a joint venture by two of South Africa's leading property companies, Growthpoint Properties Limited and Zenprop Property Holdings. Discovery is the main tenant, having entered into a 15-year lease arrangement. Discovery participated in the design of the building. Employee relocation started in October 2017, while the Change Management effort started in March 2017.

Change Management Practices

Stakeholder Management: This is the systematic characterisation, evaluation, planning, and execution of actions involving stakeholders. Senior management, middle management, staff, change leaders, middle management, move champions, HR, IT, PA communities, and the building committee were all participants.

Communication: This was done by creating a solid multi-channel communication strategy specific to each placement phase. Presentations, meetings, seminars, buzz sessions, monthly updates, walkthroughs, forums, and communications were used to raise awareness.

Risks and Issues: Involves identifying, monitoring, and managing resistance, risks, and problems influencing project execution. For the 1DP Project, two processes were used to identify risks: The Space Planning team held RFI (Request for Information) sessions with the Move Champions from each department, and the Change Management team held workshops on Business Impacts, Risks, and Requirements with the Change Leads from

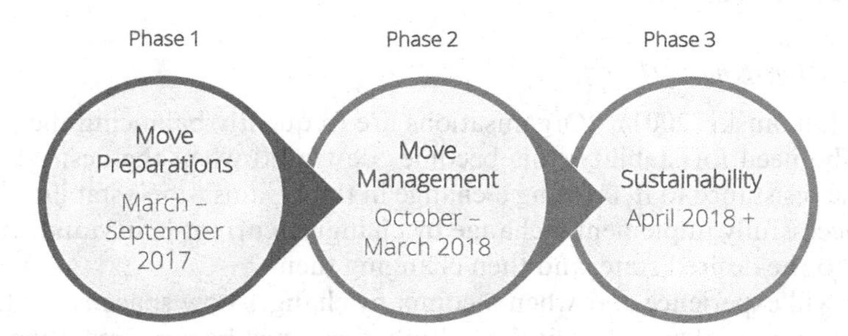

Figure 3.12 Three-phase change management strategy.

Source: Kruger, E. (2019). Change Management for Discovery Building, Discovery Place, Sandton. Presentation

each Department. Each Departmental Head reviewed, revised, and approved these materials. All identified risks and requirements were recorded in a risk and condition log to be monitored. The relevant project stream, which would mitigate the Risk and Requirement, was used to categorise and separate this log. Each project received the Risk and Requirements logs to be addressed and mitigated. The company officially highlighted ADHOC risks, which were recorded and delivered to the right Project Streams. The Change Management team scheduled meetings with each Project Stream to address and resolve the risks and requirements brought up by the business. The Move Champions from each group received updates. A company Readiness Tracker was shared with the company 12 weeks before the relocation as part of the Change Management process. This tracker listed all active risks and chose Day 2 for their acceptance.

Data Management: Identify organisational enablers, measure them, and take action to ensure they are in place. Multiple checkpoints, procedures, and already existing ones were used to get this. Research, predictive modelling, in-depth knowledge, and push for significant change were used. It involved wellness surveys, sending out engagement surveys regularly for time-efficient measurement, tracking, and quick action, receiving unplanned feedback to help the employee experience, and using targeted surveys linked to milestones (GF Survey) to produce insightful data.[14]

Knowing the project's key players is essential to making a good change management plan. It's also necessary to have a good communication plan for a move so that employees are ready to receive information about the changes being considered. Starting with an overview of the project, individuals who will be touched and where they can learn more about the project should be in the plan. The communication plan should also include a summary of the project goal created by the top management. This can be shared with employees in a meeting. During the whole project, there must be constant communication. Once everyone is on board with the project, employees can fill out online surveys to see how satisfied they are with it and find ways to improve it.

Table 3.17 Change Management Delivery Areas

Delivery area	*Measurement*	*Results*
All employees are sufficiently aware, informed, and able to move	Business readiness assessment	Achieved (BR assessment all green)
Efficient and effective move to 1DP	Minimal disruption to staff and other stakeholders	Achieved (Post-occupancy survey feedback)
Accurate risk management	All people, process, and technology risks identified and mitigating action in place.	Achieved (Risk metric audited by Group Risk)
Enhanced organisational engagement	Improved engagement metric into space as an enabler of performance	2% average increase in results (Health and Bank not relocated at measurement time)
Project teamwork and morale	At least 80% optimism and confidence	Achieved (89% measured in morale survey)
Stakeholder satisfaction	Client survey indicating at least 80% satisfaction with key move metrics	Achieved (89% measured in morale survey)

Source: Kruger, E. (2019). Change Management for Discovery Building, Discovery Place, Sandton. Presentation

3.14 Conclusion

This chapter discussed the concept of strategy, the level of strategy, and the definition of strategic management. Strategies appear are at strategic, operational, and tactical levels. The chapter includes the benefits of strategic FM, why FMs need to consider having a strategy, and guidelines for effective strategy development in FM. It also contains factors influencing FM strategy in Africa. Some of the relevant factors are lack of maintenance culture, funding, skills shortage, quality of building materials used, quality of workmanship, responsiveness towards faults, training, performance measurement, political environments such as the Influence of tariffs and trade conditions and political stability and legislation, attitudes of building users and FM knowledge of users and level of use of IT and innovation.

The chapter also discussed FM's role in corporate strategy. It covered strategic management framework in FM, strategic choice tools and techniques, SWOT analysis, premises policy, skills audit, balanced scorecards, strategic choice tools and analysis, process mapping, outsourcing models, and stakeholder analysis. It included sections on strategic implementation, organisational structure, and change management.

3.15 Guidelines

- Develop a communication and orientation plan for maintenance for all stakeholders
- Have a clear goal for the organisation and let the FM strategy align with it
- FM should work with top management
- Understand the future direction of the organisation
- Invest time in planning for facilities
- Focus on creating value from what is done
- Promote and have quality control processes in place
- Determine the facilities' budget and life cycle costs over short and long periods to have accurate information
- All stakeholders should understand all the steps involved in the process
- Evaluate the goals of the organisation
- Involving all stakeholders in the process
- Optimise the use of space
- Facilities staff should be trained regularly and involved in continuous improvement
- Staff and contractors should have performance measures that support the goals and action steps
- Have good rewards and penalty structures for staff and contractors
- Developing a strategy with processes that support the goals of the organisation
- Developing and implementing a strategy using the right strategic tools
- Staff and management should be aware of it and also learn from the process
- It should be simple
- It should be reviewed
- It should use data to monitor the progress and check if the goals are met
- Understanding trends in the industry and political environment will help with a vision for the future
- Understand and comply with FM standards, codes, and legislation.
- Introduce innovation through technology

3.16 Questions

1. Do you understand your organisation's mission?
2. Explain a strategic FM plan
3. Provide the reporting structure of any FM department
4. What are the consequences of not having a strategic FM plan?
5. Create mission statements for your FM department
6. List seven characteristics of a mission statement
7. List eight benefits of having a clear mission statement
8. Evaluate the premises policy in your organisation
9. Using the *Strategy* framework, map key issues relating to the strategic position, strategic choices and strategy implementation for FM in an organisation you are familiar or work with
10. Which scenario would your organisation most desire, and which would they most fear? What are the implications of these situations for companies outside the sector of your organisation?
11. For an organisation of your choice, perform an FM PESTEL analysis and identify key opportunities and threats.
12. identify the resources and capabilities of an organisation with which you are familiar,
13. Prepare a SWOT analysis for an organisation of your choice. Please explain why you have chosen each factor included in the analysis, in particular, their relationship to other analyses and what are the conclusions you arrive at from your analysis, and how would these inform an evaluation of strategy
14. What is the FM strategy model used in your organisation?
15. Review your current FM strategy for the core business
16. Identify the factors that can influence facilities management strategy in your organisation
17. Using the stakeholder mapping power/attention matrix, identify and map out the stakeholders for any aspect of FM, for example, space management, for an organisation of your choice about current strategies; different future strategies of your choice. What are the implications of your analysis for the organisation's strategy?
18. Draw the FM team function as it fits your organisation as a whole
19. Draw a process map for an essential service within the FM department in your organisation
20. Draw out how you will use change management strategy to implement a new project in your organisation

Notes

1. David, F.R. (2011). *Strategic management: Concepts and cases*. Prentice Hall.
2. Atkin, B., & Brooks, A. (2015). *Total facility management*. Wiley Blackwell.
3. Adewunmi, Y. (2020). Political and social influences on facilities management in South Africa. *The Bulletin of Facilities Management Special Edition*, *1*(2).
4. Renault. (2022). What is SWOT Analysis? What is included in SWOT Analysis of Renault? www.embapro.com.
5. Watson, H. (2004). *Skills audits* (The Skills Framework). Fasset: Finance and Accounting Services Sector Education and Training Authority.
6. Azasu, S., Adewunmi, Y., & Babatunde, O. (2018). South African stakeholder views of the competency requirements of facilities management graduates. *International Journal of Strategic Property Management*, *22*(6), 471–478.

7 Amaratunga, D., & Baldry, D. (2003). A conceptual framework to measure facilities management performance. *Property Management*, *21*(2), 171–189.
8 Kalkan, A., & Bozkurt, Ö.Ç. (2013). The choice and use of strategic planning tools and techniques in Turkish SMEs according to attitudes of executives. *Procedia – Social and Behavioral Sciences*, *99*, 1016–1025.
9 Anjard, R. (1998). Process mapping: A valuable tool for construction management and other professionals. *Facilities*, *16*(3/4), 79–81.
10 Jacka, J.M., & Keller, P.J. (2009). *Business process mapping: Workbook*. Wiley.
11 Johnson, G., Whittington, R., Regnér, P., Angwin, D., & Scholes, K. (2020). *Exploring strategy*. Pearson.
12 Atiku, S.O., Jeremiah, A., & Boateng, F. (2020). Perceptions of flexible work arrangements in selected African countries during the coronavirus pandemic. *South African Journal of Business Management*, *51*(1), 10.
13 Laframboise, D., Nelson, R.L. & Schmaltz, J. (2002). Managing resistance to change in workplace accommodation projects. *Journal of Facilities Management, 1*(4), 306–321.
14 Kruger, E. (2019). Change Management for Discovery Building, Discovery Place, Sandton. Presentation.

Bibliography

Adewunmi, Y. (2020a). Political and social influences on facilities management in South Africa. *The Bulletin of Facilities Management Special Edition*, *1*(2).

Adewunmi, Y. (2020b). Developing strategies for FM. University of the Witwatersrand Lecture Notes.

Alexander, K. (1994). A strategy for facilities management. *Facilities*, *12*(11), 6–10.

Amaratunga, D., & Baldry, D. (2003). A conceptual framework to measure facilities management performance. *Property Management*, *21*(2), 171–189.

Anjard, R. (1998). Process mapping: A valuable tool for construction management and other professionals. *Facilities*, *16*(3/4), 79–81.

Atiku, S.O., Jeremiah, A., & Boateng, F. (2020). Perceptions of flexible work arrangements in selected African countries during the coronavirus pandemic. *South African Journal of Business Management*, *51*(1), 10.

Atkin, B. & Brooks, A. (2015). *Total facility management*. Wiley Blackwell.

Azasu, S., Adewunmi, Y., & Babatunde, O. (2018). South African stakeholder views of the competency requirements of facilities management graduates. *International Journal of Strategic Property Management*, *22*(6), 471–478.

Barrett, P. (2000). Achieving strategic facilities management through strong relationships. *Facilities*, *18*(10/11/12), 421–426.

David, F.R. (2011). *Strategic management: Concepts and cases*. Prentice Hall.

Finch, E. (2012). Facilities change management in context. *Facilities Change Management*, 1–16.

Gilleard, J.D. and Rees, D.R. (1998). Alternative workplace strategies in Hong Kong. *Facilities*, *16*(5/6), 133–137.

Jacka, J.M., & Keller, P.J. (2009). *Business process mapping: Workbook*. Wiley.

Johnson, G., Whittington, R., Regnér, P., Angwin, D., & Scholes, K. (2020). *Exploring strategy*. Pearson.

Kalkan, A., & Bozkurt, Ö.Ç. (2013). The choice and use of strategic planning tools and techniques in Turkish SMEs according to attitudes of executives. *Procedia – Social and Behavioral Sciences*, *99*, 1016–1025.

Kaplan, R.S. and Norton, D.P. (1992). *The balanced scorecard – Measures that drive performance*, *Harvard Business Review*, *70*(1), 71–79.

Kruger, E. (2019). Change Management for Discovery Building, Discovery Place, Sandton. Presentation.

Laframboise, D., Nelson, R.L. & Schmaltz, J. (2002). Managing resistance to change in workplace accommodation projects. *Journal of Facilities Management*, *1*(4), 306–321.

Lakomski, G. (2001). Organisational change, leadership and learning: Culture as a cognitive process. *International Journal of Educational Management*, *15*(2), 68–77.

Pritchard, N. (2007). Efficient and effective implementation of people-related projects. *Industrial and Commercial Training*, *39*(4), 218–221.
Renault. (2022). What is SWOT Analysis? What is included in SWOT Analysis of Renault? www.embapro.com
Watson, H. (2004). *Skills audits* (The Skills Framework). Fasset: Finance and Accounting Services Sector Education and Training Authority.

4 Customer Relationship Management in the Management of the Workplace

4.0 Introduction

We start by looking at the meaning of customer relationship management (CRM). CRM is the strategy of the core business in an organisation that combines how an organisation operates internally and relates externally to produce and provide value to specific customers for profit. It is backed up by reliable customer data and made possible by information technology.[1] CRM is the strategic process of choosing the customers that an organisation may service and guiding relations with customers.[2] The goal is to enhance the organisation's present and future customer value. CRM in FM happens at the strategic management analysis stage and is a customer-focused activity that involves assessing customer needs, objectives, expectations, and behaviour, and controlling those aspects to influence organisational performance (Hoots, 2005).

The historical evolution of CRM shows that it has been used since the early 1990s. Many efforts have been made to define CRM in the past, but there is no agreement on what CRM entails because it is an undeveloped practice. Customer relationship marketing is another name for CRM. CRM has been associated with software programmes that automate enterprises' marketing, selling, and customer care processes through information technology (IT) organisations. When Tom Siebel launched Siebel Systems in 1993, it was the first time such companies used the term CRM. Some believe that CRM is a systematic method for creating and rewarding customer networks and that technology may or may not have a role (Buttle, 2009). In FM, the goal is to meet the needs of users of facilities who are the customers. CRM is important in FM and is one aspect of FM that has been under-researched and documented in Africa. Many approaches to building support have not been geared towards meeting users' needs. The essence of this chapter is to introduce readers to CRM in general and FM and its models.

4.1 Customer Relationship Management in FM

In FM, CRM refers to obtaining thorough information about customer wants, expectations, and behaviour, and controlling those aspects to influence organisational performance. Instead of being a cost centre for facilities services, CRM is a customer-intensive business role in facilities management. On the management side, it suggests that the FM affects the customer's opinion of service success actively rather than inactively. The manager's main goal with CRM is to move customer service from the inside-out. Outside-in, customer-driven CRM suggests that operations should come first, followed by a study of the customer. The term "outside-in" refers to designing facilities and systems around the demands of the users (Figure 4.1). Facilities managers should manage customers'

DOI: 10.1201/9781032656663-5

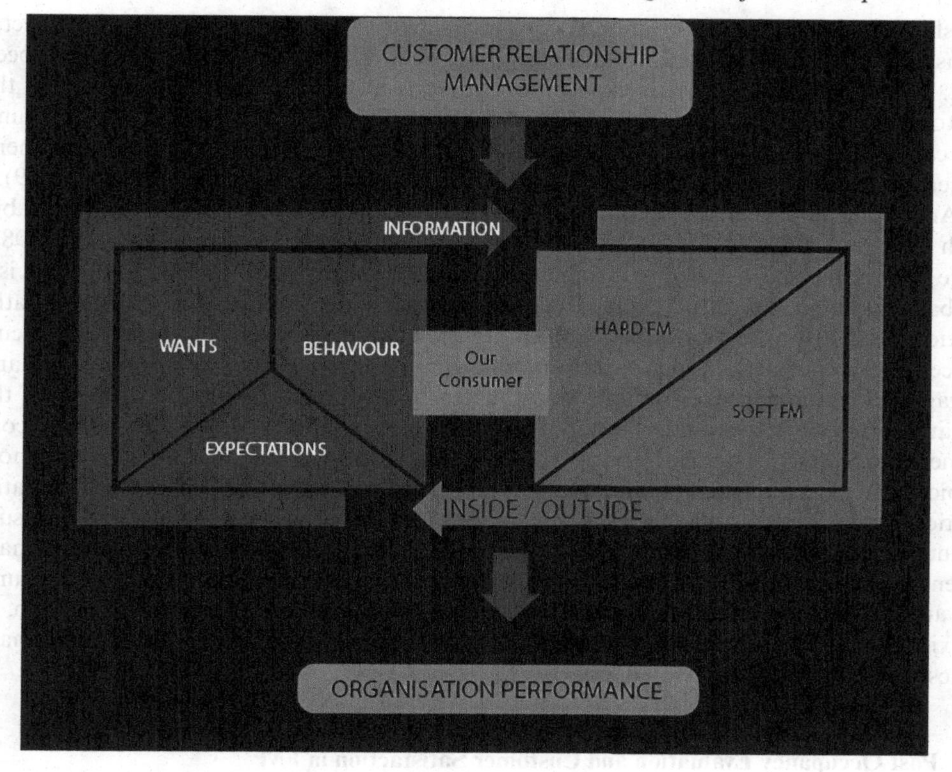

Figure 4.1 Customer relationship management in FM.
Source: Author

expectations in addition to meeting their needs. When there are discrepancies in client satisfaction and service performance, providing successful services becomes a significant difficulty.[3]

4.2 Customer Satisfaction in FM

4.2.1 Customer Satisfaction

Buttle (2009) has defined customer satisfaction as follows:

> Customer satisfaction is the customer's fulfilment response to a customer experience, or some part thereof, a pleasurable fulfilment response. Dissatisfaction is an unpleasurable fulfilment response. The "experience" part of the definition suggests that the satisfaction evaluation can be directed at any or all elements of the customer's experience, including product, service, process and any other components of the customer experience.

Customer satisfaction is a complex idea associated with a buying decision through intellectual judgement for a choice that has been made (Selnes, 1993). Comparing a customer's impression of an experience, or a part of one, with their expectations and the expectations-disconfirmation model of customer satisfaction is the most popular method of measuring satisfaction. According to this model, consumers are content if they believe their expectations have been satisfied and unsatisfied if they believe their expectations have not been met. When perception exceeds expectation, there is a positive disconfirmation, and the customer may be pleasantly surprised or even happy. This approach assumes that consumers can evaluate performance and have expectations. One should note that consumers' uncertain expectations are sometimes met, yet the client is still dissatisfied (Buttle, 2009).

According to earlier research, client happiness and service quality are connected but with different theoretical views (Sureshchandar et al., 2002). Parasuraman et al. (1988) state that "satisfaction is tied to a single transaction, whereas perceived service quality is a global evaluation, or attitude, referring to the superiority of the service." Customer satisfaction results from service quality, indicating the link between service quality and occurrences after a transaction, such as attitude changes. Because FM service is personalised and engaging for each consumer, it may be estimated by customer satisfaction. Instead of the organisation's objective quality criteria, customer satisfaction is determined by how consumers subjectively rate the services they receive. Customer satisfaction helps promote choices and determine how service quality and behavioural intention are related. The satisfaction assessment results from customers' evaluations of service quality and pleasure influence behaviour after a purchase. Customer satisfaction has been linked to purchase intentions, customer retention, and the likelihood of repeat transactions. Happy consumers are likelier to remain loyal, but unsatisfied customers tend to forego patronage. So, to maximise decision-making and draw in clients, FM suppliers should include customer happiness in evaluating the performance of FM.[4]

4.3 Post Occupancy Evaluation and Customer Satisfaction in FM

Post Occupancy Evaluation (POE) evaluates how well buildings match users'/consumers' needs and identifies ways to improve the design and performance of buildings and ensure they are fit for purpose. It involves systematically evaluating opinions about buildings in use from the viewpoint of the people who use them (customers). Opinions of the customers/users are sought about issues such as the services, condition of the building, environmental conditions, privacy, spatial arrangement, information technology, communication, social interaction, safety, and location. An example of part of a section of a POE survey can be seen in the feedback form in Figure 4.2.

There may be three types of POE. The type depends on the availability of finance, time, manpower and the required outcome. The general approach to each type includes planning, conducting the study and interpreting the results. The three types are in Figure 2.2. The data collection methods include questionnaires, individual and group interviews, behavioural mapping, technical assessment tools and mathematical models. Indicative and investigative methods, which are simple, are usually used.

In Africa, a reactive approach to maintenance is usually adopted, which results in many buildings being in a poor state. A more customer-focused approach is needed, and many users and occupiers of these buildings, especially the public ones, are unsatisfied. There also needs to be more orientation for those working in the built environment area on the benefits of adopting POEs, as well as more awareness among users of the importance of their feedback during the evaluation process. In addition, many facilities managers are not involved in building projects right from conception. Although they are professionals familiar with the need for contact with users, they may not contribute to the design of such

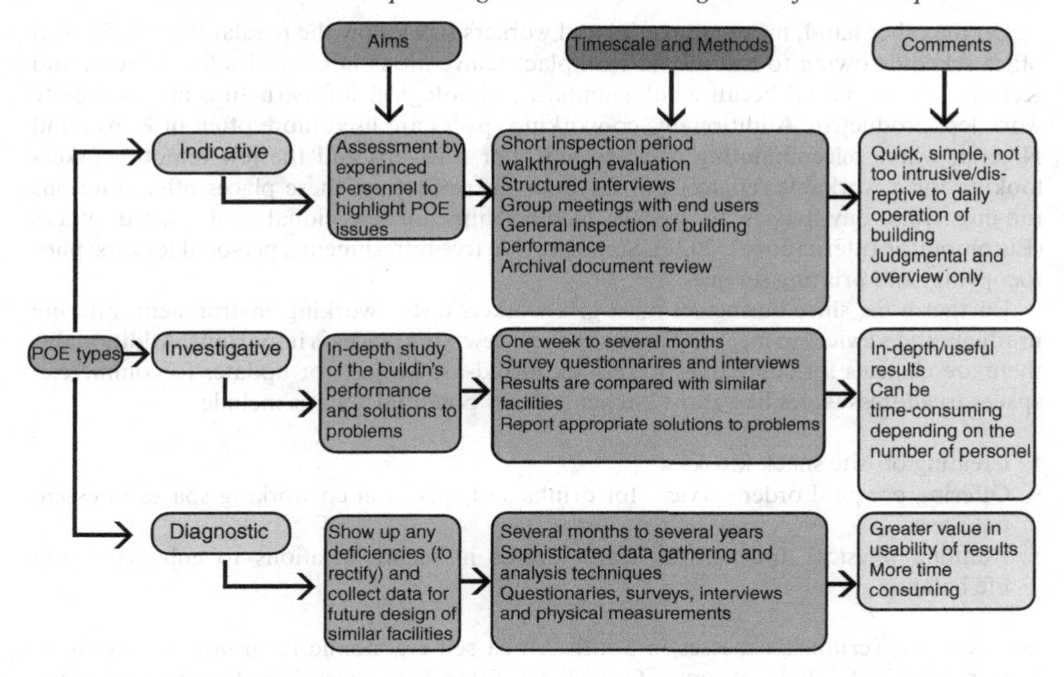

Figure 4.2 Types of POE based on the goals, timescales, and methods used.
Source: Adapted from Adewunmi, Y., Omirin, M., Famuyiwa, F., & Farinloye, O. (2011). Post-occupancy evaluation of postgraduate hostel facilities. *Facilities*, *29*(3/4), 149–168

buildings. Thus, suggestions still need to be made to ensure clients and users get the best from their buildings. Further, more enforcement of regulations is needed to improve building conditions, and many managers, especially in the public sector, need access to sufficient funds to manage the buildings under their portfolio.[5]

4.4 Workplace Trends in Africa

The onset of the COVID-19 coronavirus led to the re-evaluation and redesign of practices, products, properties, and interpersonal connections. New workplaces have impacted changes in Africa. Previously, expensive data plans and low internet penetration (with only 25% of African households having access), meant remote work in Africa had not succeeded. There was also a belief that workers who worked from home would need to be more productive. Due to the pandemic, people began working in shifts, on rotation, and for shorter periods, and companies were forced to provide their employees with internet and mobile assistance in their homes so they could continue working. Implementing the new working style was easy for those who already supported it in some way, for instance, working from home one day a week (Euromonitor International, 2021).

In South Africa, the availability of technology and a more laid-back work environment made remote work possible. There are even small villages known as "zoom towns" in South Africa, which provide calmer lives, more roomy houses, and peaceful outdoor areas for millennial remote workers. These areas may be closer to nature preserves or beaches while also within easy driving distance of urban centres.

On the other hand, most companies and workers still follow the regular 9-to-5, physical office schedule owing to traditional workplace conventions in countries like Nigeria and Kenya. This is crucial because sub-standard technological infrastructure makes remote work less productive. Additionally, co-working spaces are used more often in Kenya and Nigeria, where telecommuting is expanding. For start-ups and the few remote workers looking for a workable replacement for poor infrastructure, these places offer solutions ranging from day passes to weekly memberships for individual and shared offices (Euromonitor International, 2021). Some provide free refreshments, personal lockers, photocopying, and printing services.

Furthermore, since businesses must give workers a safe working environment, offering products and services to meet customers at their new workplaces is important. Additionally, there are chances for companies to rethink mixed-use projects or updates to commercial spaces to address issues like ghost kitchens. Other potential actions include:

- Creating on-site snack kiosks
- Offering pre-paid order services for drinks and snacks in co-working spaces in essential areas
- Running physical and mental health events in secure locations to enhance work-life balance

For example, certain businesses in South Africa sell ergonomic furniture. According to Euromonitor's "Global Consumer Trends" (GCTs) in 2021, several products and services – such as weekly taxi subscriptions, pay-per-use conference rooms, and equipment rentals – supported shift-based or remote/home working.[6]

4.5 Consequences of Customer Changes on FM

Managers can identify new and unobvious client preferences by studying their behaviour patterns and sampling customers on social media and review sites. Since the FM sector is user- or customer-focused, value addition may be seen when managers coordinate their plans to satisfy client needs. The use of CRM is essential in facilities management. Facilities managers' interactions with clients can take several forms because many different types of clients and parties are involved. As such, the manager must be familiar with clients from various viewpoints. Managers must be aware of the demands of their customers, keep an eye on their feedback, and act on it while considering the organisation's objectives and resources. Managers must take on an interactive role to deal with consumers and occasionally be willing to give up authority. Managers must be creative and aware of cultural trends to build positive customer interactions. Managers must be diplomatic, honest, and transparent to develop consumer relations. They can adopt techniques that prevent them from avoiding clients.

4.5.1 Implications of the Customer Change on FM in Africa

The African market is characterised by a young, dynamic, and sophisticated population, primarily in the upper-middle class, and as such, managers on the continent must develop strategies to meet their requirements. Managers should realise that they have a taste for luxury goods and have moved on from traditional parts of the culture. Consumers on the continent are also technology savvy, though there is little data access in some locations. Managing customer relations using technology may be easy to embrace on the continent. A lot is happening in the retail market, as shown by developments in many cities, which provide opportunities for facilities managers on the continent. The increase in exchange rates compared to the dollar presents many challenges for customer relationship management. This drives the cost of service provision and makes it unpredictable. This makes customer service provision, especially on the supply side, difficult to manage, which will affect customer relationships as prices fluctuate compared to the proposed agreed prices. This is because many materials used for service delivery are imported in African countries.

Poor infrastructure can influence logistics and cause delays in service provision. Some locations have problems with good service provision because of their location, causing variations in the price, which can be an issue in customer relations management. COVID-19 has also brought many changes in the workplace, as many people have to work from home. The sudden change in working arrangements has brought many dynamic changes to customer relations management since people now work in new formats. Also, spaces have to be redesigned sometimes, and new types of spaces are created and integrated with the home space. The manager still needs to manage such areas to keep the organisation running. COVID-19 has also brought about heavy reliance on technology and the need to cut costs. Hence, managers need to rely more on data and cost-effective technology to manage their teams.

4.6 Types of Customers

4.6.1 Client

This represents the main customer or the person or group with the most authority over the FM organisation. The client is employed in the FM organisation's service. The client has a relationship with the FM, is involved in decision-making, and pays for its service. Examples of clients include a board of directors, top management, and service providers; for a good relationship between clients and the FM, communication is needed between clients and the FM. Clients are usually interested in the cost of services provided, want value for money, and want compliance with regulations to ensure that the facilities are safe and adverse consequences are avoided.

4.6.2 Customer

Customers receive information, counsel, services, or goods from the facilities manager. The customer can be internal or external to the FM organisation. Internal customers are people paying facilities managers for their services. Internal clients work in the same organisation as the facilities manager and are at the receiving end of the service provision of the facilities manager. Examples of customers are the FM department's team, other department employees, department supervisors, trustees, sponsors, business clients, and visiting employees from other locations. Customers are usually interested in the cost of work performed, compliance with regulatory requirements, concerns about the quality of work done by the facilities manager, that the business continues to function even in the event of an unforeseen eventuality, communication with the facilities manager regarding the service provided, documentation of the work done, and its reports.

4.6.3 Stakeholders

A person or body with an interest in a company, service, or project is known as a stakeholder. A stakeholder might be an individual or organisation that receives services. Examples include contractors, inspectors hired by investors or trustees, and suppliers since they participate in an activity, receive compensation, and are interested in it. Other departments and teams from the same organisation might also be considered stakeholders because they help set up or provide support services or engage in a project using their skills. When they team up with the facilities team, they require assistance. Examples of stakeholders include the organisation from which the manager receives its resources, such as the community, trade unions, project groups, and investors, who may be included in this category of interested parties. People working to deliver, monitor, sponsor, report results, communicate, or coordinate on behalf of others or impacted by the actions done in providing a service may also be considered stakeholders. Stakeholders connected to or involved in service-providing project work and contractual work may or may not participate.

The stakeholders will be interested in financial considerations about payment, compliance with statutory requirements, and regulations; the contractor will need assurance that the company will not put them in any hazardous conditions while working and will, thus, uphold their commitments. The facilities manager must assess the contractor's performance, will be concerned about the safety of the work performed, and will inform the designated staff should anything be brought to his notice. Furthermore, the contractor will need a permit to work and request a confirmation of the necessary work using checklists, while the facilities manager will need daily reports of work done.

4.7 Relationship Management

4.7.1 Meaning of Relationship

Interaction throughout time is necessary for a relationship. Most of us would not classify a situation as a relationship if it consisted simply of a single transaction, such as purchasing a motorised cleaner from an uncommon retailer. A dyadic relationship is a partnership between two people that builds into a relationship through time. Each incident is time-bound and has a name, a beginning, and a finish. A facilities manager builds good relationships with stakeholders to have their satisfaction, trust, and cooperation. Facilities managers may encounter various situations, including buying cleaning supplies, calling suppliers, negotiating contract conditions with contractors, handling complaints, and resolving billing issues. A series of exchanges make up each incident in turn. Action and reaction to that action make up the relationship. Each character will interact with and behave toward the others in each incident. A range of communication behaviours, including speech, actions, and body language, are included in each incident. Communication is two-way and needs feedback, which can be used to meet the users' and stakeholders' needs. There must be an emotional link or connection for an attachment relationship (Buttle, 2009).

A relationship begins only when the partner transitions from independence to dependence or interdependence. It is not a relationship when a customer purchases a plumbing tool from a hardware store. Suppose a client prefers an office space because they like the ambience and its services; this is more of a relationship. There is reliance but no interdependence (between the client and the office). In a facilities organisation's procurement, the facilities personnel may view them as harsh and transactional, while their suppliers may believe they have established a relationship. A relationship is a social concept that develops when two people act by that perception. Furthermore, relations can be one-sided or shared, with either one or both persons believing they are connected. Efforts should be made to resolve conflicts if they happen in time for good relationships with stakeholders.

4.7.2 Changes Within Relationships

Time changes relationships. Parties get closer or further apart; there are more or fewer contacts. They can vary significantly regarding the quantity and diversity of incidents and their dealings because they evolve. According to Dwyer, customer-supplier interactions can progress through five phases: Awareness, exploration, expansion, commitment, and dissolution.

The moment one side is made aware of the other as a potential partner for an exchange is known as awareness. The inquiry and testing phase, or exploration, is when the parties examine each other's performance and capabilities. Some trial buying is done, and if the trial is unsuccessful, the partnership may be ended with little cost. There are five subprocesses in the exploration phase: Attraction, communication and bargaining, power development and exercise, norm construction, and expectation development. A rise in dependency describes the period of expansion. As more transactions occur, trust begins to grow. Increased adaptability and a shared understanding of responsibilities and objectives make the commitment phase.

Not all relationships go to the commitment stage, and many end before it. A partner could decide to end the relationship due to a trust violation. Bilateral or unilateral relationship dissolution is also possible. It is a bilateral termination when both parties agree to end the relationship. When one party ends the relationship independently, this is known as a unilateral termination (Buttle, 2009). Clients may end partnerships for various reasons, including changes in the company that make their focus change, making it difficult for some managers to align with these changes. Changes can come from regular service failures in service delivery. Suppliers may choose to end their partnerships because they failed to help reach supply volume targets, so cost-to-serve can be one way to solve the issue and keep the connection going. This relationship development pattern highlights commitment and trust as qualities of well-developed partnerships. Failures in service delivery could result from the FM department's inability to analyse customer demands based on the timing and duration and how they fit into the organisation. Sometimes, facilities managers may need to be more organised regarding their work scheduling, which can cause a loss of resources. Service failure can also come from inadequate time, skills, and equipment to meet users' needs. Therefore, facilities managers should quickly adapt to change and get the necessary tools, technology, and training to avoid such failures. Some skills needed in CRM include communication, problem-solving, negotiation, FM knowledge, budgeting, technical, analytical, and project management.

4.7.3 Trust

Trust has a goal. These emotions are focused despite the possibility of a generalised feeling of comfort and confidence. A party may have faith in another party's:

- Honesty – a belief that one party acts in the other's best interests
- Benevolence – an idea that the other party's statement is trustworthy or believable.
- Competence – the conviction that the other person possesses the skills necessary to perform the requested actions

Building trust is a relationship-building investment that will pay off in the long run. Trust develops as partners exchange experiences, analyse one another's intentions, and evaluate them. Risk and uncertainty are diminished when they get to know one another better. Both parties are driven to invest in a relationship when there is mutual trust. When trust is lacking, the relationship suffers, and conflict and uncertainty grow.

The nature of trust also changes during relationships (Buttle, 2009):

- Calculus-based trust is present in the early phases of a relationship. One person says, "I trust you because of what I am getting or anticipate getting out of the connection."
- Knowledge-based trust is based on the interaction history and understanding of each party, which enables one side to predict how the other will behave
- Identification-based trust develops when two people thoroughly understand and can stand in for one another during social dealings. This appears in the latter phases of relationship growth

A lack of trust weakens the link between facilities managers and clients. Clients' demands can be demanding and challenging to satisfy in FM. It, therefore, can be necessary to communicate using diverse methods to clients and put in systems that encourage facilities managers to promote customer satisfaction. The facilities manager needs to thoroughly understand the client's goals, needs, and objectives to align processes in FM to meet such needs. Also, systems that are transparent and fair help manage customer relationships. Another dimension of trust is the position of the facilities manager. At the strategic level, it is easier to get trust from management regarding FM initiatives. A facilities manager should have the skill to present a business case for a project to the organisation and demonstrate the project's benefits to the company. The manager should be able to analyse the facilities, conduct space planning through space vacancy rates and usage, and understand the demand for space and facilities to project future customer needs. Other capabilities include using reports and dashboards for communication. Benchmarking can also be used as a business case to determine and help facilities meet industry standards and show the best-fit service provision among alternatives. There is a need to possess the skills to show how facilities projects can add value to the organisation.

4.7.4 Commitment

Commitment is needed for partnerships to be effective and last over the long term. Relationship commitment was described by Morgan and Hunt (1994) as:

> [A]n exchange partner believing that an ongoing relationship with another is so important as to permit the maximum effort to maintain it; that is, the committed party believes the relationship is worth working on to ensure that it endures indefinitely.

Trust, similar ideals, and the belief that a partner would be hard to leave all contribute to commitment. To safeguard the financial interests of the partnership, commitment pushes partners to work together. Committed relationships reject quick fixes in favour of more steady, long-term advantages involving their present partners. When clients have a choice, they only choose trustworthy partners since commitment involves compassion, which leaves them open to opportunism (Buttle, 2009).

The investments one partner makes in the other serve as proof of commitment. If the other reacts, the connection develops, and the parties grow more dedicated to one another. One side invests in the potential relationship. Time, money, and neglecting existing or alternative relationships can all be considered investments. The amount invested in a relationship represents the costs associated with ending it. Since some relationship investments may be lost, highly committed partnerships have high termination costs. Moving to an alternative may also result in significant cost savings. Facilities managers should commit to processes that align with the customers' goals and objectives. There should also be performance measures and key performance indicators that cater for the customer.

4.7.5 Relationship Quality

Understanding commitment and trust leads to the belief that some relationships may be of higher quality than others. Trust and commitment are the key characteristics of a high-quality relationship. However, additional features include relationship satisfaction, shared objectives, and cooperative rules. Relationship happiness and commitment are not the same (Buttle, 2009). Facilities managers' commitment to the supplier results from investments made in the relationship, and such investments are made if the committed party is satisfied with the supplier's transactional history. Investments are made in fulfilling relationships. Mutual goals occur when two parties share goals through cooperation and ongoing relationships. Cooperative norms arise when relational partners collaborate effectively and dependably to solve issues. Facilities managers need to know the quality of their customer connections because CRM initiatives frequently aim to promote close and value-driven relations.

4.8 Factors That Promote Relationships Between Facilities Managers and Customers

Meeting customer demands and achieving organisational goals are the important incentives behind facilities managers' desire to develop client relations. They will get more remarkable outcomes when they manage their client base to find, attract, please, and keep consumers. These are the main goals of several CRM strategies. The size of the client base grows when customer retention rates are improved. A solid customer base helps an organisation's performance and increases the average customer retention rate, which reduces management costs by preventing the need to replace lost clients. Managers can better understand customers' needs and expectations as their relationship grows. Suppliers can also better recognise and meet client requirements since they know what they can improve.

4.9 Factors That Limit Relationships Between Facilities Managers and Customers

Among the causes are:

- **Loss of control**: Under an outsourcing agreement, the facilities manager may give up control of the organisation's resources. The contractor may have specific requirements and resources for the tasks to be carried out.
- **Exit costs**: Not every relationship lasts, and ending a relationship is not always simple or economical. When a relationship ends, investments occasionally are not repaid. Investments in relationships can range from negligible to large.

- **Use of resources**: Relationships need the investment of resources like people, time, and money. Facilities managers must choose if it is preferable to dedicate resources to various FM components. Resources may turn into lost expenses after they have been committed. Lost costs are previous expenses that cannot be recovered. Since they cannot be regained, these are usually not considered when determining whether to stay in a relationship.
- **Opportunity cost**: When service providers commit resources to one client, they cannot be given to another, which results in opportunity costs. Having relationships has significant opportunity costs. Even though supplier B appears to be a better offering, you could have to forgo an association with them if you commit resources to supplier A. When a facilities manager agrees to contract maintenance services with an existing customer, they may lose the opportunity to work with other potential maintenance service providers.

Facilities managers may not want relationships with contracting organisations for the following reasons:

- **Cost**: Managers may be concerned about possibly higher costs if they include a management fee and the overhead profit the suppliers charge.
- **Quality**: Since suppliers' standards are not under the employees' control, they may result in worse quality due to their use.
- **Risk**: The supplier might not be willing to take on responsibilities unless the whole sum is paid to cover risks, preventing any risk transfer.
- **Goal**: Facilities managers may be concerned that all new services and resources would incur an additional fee or that the agreed-upon expertise may not exist.
- **Roles**: Managers may be reluctant to work with suppliers since there is an opportunity for responsibility to be spread through excessive subcontracting and parallel contractual responsibilities, reducing the need to hold the supplier directly accountable. It may be challenging for managers to apply control over the contract.
- **Innovation**: Facilities managers need help understanding that innovation offered at the beginning of a contract is not sustained and is based on old procedures rather than providing answers to client concerns.
- **Investment**: If there is no investment, the provider may perform only to the level necessary to support the contract and provide the required quality of service.
- **Information**: Facilities managers do not want to deal with data systems that are difficult and expensive to organise compared to returns. All data reporting comes at an additional cost, and the data collected rarely applies to planning. Data should be changed to meet the goals and performance standards of the contract.
- **Rigidity of the supplier**: Here facilities managers may have concerns about dealing with the client account. Suppliers making all changes based only on cost might be another problem.
- **Customer side**: Employees in the FM department receive on-the-job training and have access to all other available training. A limited customer perspective may be worse for suppliers if the supplier is not part of the client organisation.

4.10 Communication in Customer Relationship Management

4.10.1 Communication

Communication is "the social process by which people in a specific context construct meaning using symbolic behaviour".

Communication has the following specific characteristics:[7]

a It is a social process
b It is continuous
c It relates to human beings
d It is contextual, meaning it may differ in different situations
e It is symbolic by nature
f It shares thoughts and ideas, which produce a response

Communication advantages in customer management include having happy clients, a motivated workforce, and staff from other departments. Additionally, it boosts customer reputation and encourages creative thinking while creating plans to deal with clients' reluctance to change.

Failure to effectively manage client relationships might result from poor communication. Customers, such as users and service providers, may hold the facilities manager responsible, or circumstances may prove the opposite, resulting in disputes. The facilities manager plans roles, communicates clearly, and informs all parties about their obligations. Lack of clear communication from the facilities manager might lead to service providers performing incorrect tasks, leading to issues. The facilities manager manages workers, clients, and suppliers no matter where they are located. The facilities manager may occasionally need to work virtually with other team members. The manager requires a solid understanding of communication's main concepts, a flexible collection of valuable tools, and practical application expertise.

4.10.2 Barriers to Communication

Communication barriers affect the clarity, accuracy, and effectiveness of the message. One of the oldest categorisations of barriers is the following:[8]

1 Acceptable barriers: People interpret the same communication differently – acceptable issues with message content, meaning, and the importance of sending and receiving.
2 Organisational barriers: These barriers arise due to issues with members' functional specialisation of duties, power, authority, status connections, and ownership of information.
3 Social obstacles: These barriers also come from relationships, values, and attitudes of the people involved in the communication process.
4 Individual barriers: This issue comes from individual differences in thinking and acting abilities, including physical ailments or impairments. It is also a result of a person's ability to receive and transfer information, including deficient hearing, reading, and unfavourable psychological situations.
5 Location and economic hurdles: Time, geography, and the impact of time on message reception can all be sources of communication difficulties.
6 Channel and media barriers: The channel and medium used in the communication process impact its efficiency and accuracy. Problems related to transmitting a message fall under this category effectively. (For example, it is preferable to communicate verbally rather than in writing when sending a message.)
7 Technology barriers: These are obstacles from communication technology improvements. The technology produces more information than the recipient can handle. The advances in media further increase the boundaries from technological processes.

4.10.3 Overcoming Communication Barriers

The following are some of the strategies for resolving communication problems (Ras & Das, 2009):

- **Monitoring of the communication process**: Monitoring the message at all times during the communication process can help reduce the impact of these obstacles. To increase communication, the facilities manager must collaborate with other parties. Since he oversees people, the facilities manager is a leader whose actions will affect how he interacts with others, especially those on his team.
- **Communication**: Efforts must be made to educate people about the organisation's view purposely. The facilities manager should use the right methods depending on the situation. Technical vocabulary should be avoided when communicating with people outside the FM Department or the organisation to prevent disagreement and promote collaboration.
- **Feedback**: The right feedback should reduce the customer's wrong impression. The disadvantages of the customers' careful perception should be discussed in order to reduce obstacles. For example, comments from customer surveys should be communicated to them to improve on the positive relationship with consumers.
- **Correctness**: To increase openness and efficacy, messages sent between parties should be accurate. The message's precision to prevent disputes between managers and clients or between managers and suppliers; the message to be delivered should be precise, clear, and practical.
- **Encouraging strong relationships**: To prevent conflicts, it is essential to promote strong relationships between the organisation and its employees. This will also make it easier to communicate effectively. Innovative FM methods may be hindered by disagreement or a lack of tolerance for other people's viewpoints.
- **Administration in the organisation structure**: Departmental bosses should have the same perceptions as their subordinates and vice versa to improve communication.
- **Flat structure**: The design should be clear and simple. Complex structures make reporting more difficult and harmful if someone within the organisation possesses inaccurate information. The status gap will be reduced with effective organisational structure change. A status impact may occur when one individual is significantly higher on the ladder than another. FM in many organisations tends to be fragmented, which in many departments may result in overlapping functions and make it difficult to communicate information. FMs should encourage departments to adopt flatter, more integrated organisational structures.
- **Specialisation of labour**: To avoid information overload and delays in the distribution of information, there should be an appropriate division of labour among the parties in FM.
- **Organisation policy**: Organisations should design policies to benefit all members. It needs to be adaptable and straightforward.

4.11 Factors That Improve Customer Feedback Instrument Design

Introduction: Start with a brief introduction that describes the goal of the questionnaire, what will be done with the data obtained, and how to use it.

Feedback information: This should cover the information necessary to avoid wasting time gathering and analysing data. If you ask too many questions, people can stop answering you. Additionally, it is crucial to ask all the necessary questions, which should be written with the survey's objectives in mind.

Straightforward questions: To adequately reply to instructions and questions, the person receiving them must understand them: the respondents' backgrounds and language influence how questions are worded. Additionally, questions that will make participants feel anything should not be asked. For respondents to feel more at ease answering the questions, they must be assured that confidentiality will be maintained.

Consistent structure: The layout needs to be clear, or the sequence of questions logical to avoid confusion and irritation by those surveyed for better responses to the questions asked.

Professionalism: The tool should be interesting, and the exercise should offer rewards, making it easy to convince the individual to use the instrument. The instrument should have a professional appearance using templates and common word-processing tools.

Types of feedback: Feedback instruments should be distributed using the most convenient and cost-effective means. Some, such as internet portals and emails, can be distributed physically or electronically.

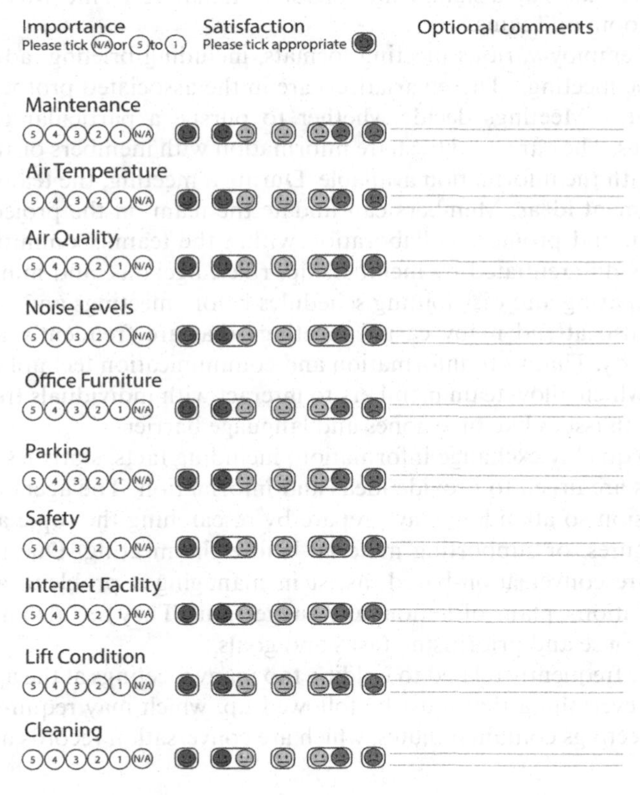

Figure 4.3 An example of a customer feedback form.

Source: Author

4.12 Communication Methods in FM

This section discusses the circumstances in which facilities managers must interact with clients to provide services, satisfy needs and expectations, assess satisfaction levels, and evaluate the effectiveness of services. Service-related information is gathered and supplied to customers. Communication methods is part of the stakeholder communication plan, which should be for each type of customer and goal (Mukadam et al., 2023). This section discusses some of the communication approaches' characteristics, benefits, and disadvantages (DDCE, Utkal University, Bhubaneswar, n.d).

4.12.1 Meetings

Meetings with the FM team and facilities users help with communication. Facilities meetings provide opportunities to send out important information and communicate company goals. It helps the FM unit to be more efficient and effective. A meeting is a planned gathering that enables individuals with similar interests to interact, debate potential solutions, exchange, and come to choices. All sides are expected to contribute and provide ideas for queries, conversations, and involvement, making it a two-way communication technique. Attending meetings takes up a significant portion of a manager's life. Meetings can also be online through Zoom or Teams.

Organisations employ various meeting formats, including briefing, advisory, consultative, and executive meetings. These variations are in the associated protocols and communication techniques. Meetings decide whether to pursue a particular course of action among alternatives. They are used to share information with members of the FM team and solve problems with the information available. During a meeting, the team can brainstorm on how to implement ideas. Members can update the team on the projects they are currently working on and promote collaboration within the team. Committee meetings are formal gatherings differentiated by membership, rules, agendas, and minutes. The procedure includes preparing and distributing schedules before meetings and taking exact minutes. Managers also attend many casual meetings that are frequently a component of project team activity. Thanks to information and communication technology, we also have "virtual" teams, which allow team members to interact with individuals from different cultures and cope with issues like time zones and language barriers.

Meetings help quickly exchange information, including facts, statistics, data, and opinions. Participants are urged to provide ideas and information. The agenda should include topics for discussion so attendees may prepare by researching the topic and contributing relevant facts, figures, or supporting material before the meeting. Due to their two-way character, they are conversation-based, assist in managing a problem, and seek to find agreement on solutions, plans of action, and suggestions. They are also helpful in obtaining a prompt response and prioritising tasks and goals.

Limitations are frequently related to holding too many meetings at the appropriate times. This will impact everything that must be followed up, which may require calling another meeting. Most meetings contain minutes, which are conversation records and specifics.

4.12.2 Presentations

A presentation is a planned activity in which the main speaker will give the audience information that has been carefully prepared. It may take the shape of a PowerPoint presentation with prepared slides. It could be a lecture-style communication using a whiteboard or flipchart setup, or it might look like a meeting where everyone is seated around a table. At the same time, the speaker presents written notes. A presentation could demonstrate how to perform or report on the progress of a task.

Managers use presentations, from electronic webcasts and presentation dashboards to casual face-to-face briefings. For presentations, audiovisual aids appropriate for the style and demands of the audience should be used, and people giving the presentation should be aware of their technological strengths and limitations and how these may affect communication techniques.

The venue arrangement is important for a manager to deliver a good presentation, as that would influence the layout and colours used in the slides. The venue's arrangements should include the rooms, tools such as the audio-visual equipment, and refreshments, as they significantly impact the presentation. The four main concerns throughout presentation delivery are establishing an attractive introduction, showing transitions and varying speed between essential ideas, covering up the presentation, and addressing questions, interruptions, and conversation. The manager should practise the presentation ahead of the delivery.

Presentations have the potential to be educational, draw the audience's attention to information that they may need to be aware of, or express decisions and actions. Given that this is a one-way communication mechanism rather than a meeting, there is minimal chance for dialogue.

4.12.3 Questionnaires

They are a very efficient way to find out what clients want and expect, and they can also show how effectively services are doing. Discussions are more casual than surveys; the information gathered (online or paper) can be analysed afterwards for trends and particular issues.

Forms are sometimes used in surveys, which can help gather data and information but are less effective at exploring underlying issues. Open-ended inquiries may be used to discover different topics since they enable people to express their ideas and perspectives. Due to their directness and observability, the results of closed-ended questions may be easily analysed using statistical analysis displayed in graphs. Trends are easy to determine from the data collected. Limitations include the need for a specially designed questionnaire to help get the required data.

4.12.4 Interviews

Interviews are regular two-way communication that can help in forming shared meanings. Interviewers may use their personal experience as a resource to ask questions. Questions could include open-ended and closed-ended questions, just like surveys. Closed questions gather information on facts and similar statistics, whereas open questions are used to learn about people's perspectives, opinions, and ideas. Unlike questionnaires, interviews can be used to obtain reliable and rich information from a fewer number of people.

An FM organisation may conduct interviews for various purposes, including research, hiring, promoting, appraising, coaching, and mentoring new hires. They are also used to resolve disputes and disciplinary issues. With practise, listening is a skill that may be enhanced. Focusing on what is being said, avoiding bias, expressing attention and encouragement, being aware of sentiments and emotions, clarifying concerns, and summarising conversations are all essential interviewing skills.

The interviewer should understand the differences between open, closed, probing, leading, and hypothetical questions for correct use during interviews. Interviews should be recorded so the information obtained from the exercise is properly documented. The interviewer should not ask too many questions and learn to engage the interviewees during the interviews. Interviewers should know how interactions and replies might change while passing on their questions. It involves dedicating time to thinking about and responding to the interviewer's perceived requirements; interview candidates can reduce the expected burden of interviews.

Interviews often focus on what the interviewer wants to hear rather than serving as a debate forum. They are more professional than informal chat and less engaging than business meetings. Everyone who participates in an interview offers a wide range of information unless the same questions are asked, making it challenging to identify trends and common elements. Interviews must be arranged and agreed upon because they require time to conduct.

4.12.5 Focus Group Discussions

This involves interviewing small groups of people with common interests. For example, the facilities manager could have a focus group with office building users to ascertain building concerns for maintenance and future design purposes. Handling numerous people's views can be challenging if a group gives input, so the moderator should be trained. A focus group needs someone to speak on behalf of a group during group discussions. The problem in speaking with represented individuals is that the chosen representatives' views may sometimes differ from the general views of the group; also, because a venue has to be arranged with refreshments, it can be more expensive to conduct than surveys. Compiling information from this exercise is also more difficult than surveys.

4.12.6 Staff Suggestion Programmes

The FM department can use this to promote creativity, engagement, and exchange of ideas and perspectives. Staff can use this method to share ideas for improving the unit's activities. The manager should make some of the adjustments so that staff members will notice them. Problems can come from the need to carefully evaluate the information, mainly if the suggested methods are based on opinions. The need for confidentiality can be a challenge since those who use such strategies do not want others to think poorly about them or their reputation or position. Additionally, if the management group does not implement the ideas, staff members will be discouraged from taking part in the future.

4.12.7 Suggestion Boxes

These are boxes where staff and facility users submit completed forms that contain suggestions and complaints without disclosing their identity. They are located near the coffee maker, in the cafeteria, at the front desk, or in other public areas. There could be a problem with implementing the suggestions and providing feedback to the participants.

4.12.8 Complaints Logs

These show how concerns reported to a facilities manager are recorded and how to note the specifics and timing of an action. They should be accessible at all times and need information for practical purposes, such as trends, frequencies, the timing of an action, the person in charge of the activity, and the person to whom the action is transferred if attention is needed. Also, a complaint log will be noted when resolving the problem, and the client is notified.

The person in charge must frequently check records and examine the information inside them to show the details needed for management reasons. Using electronic recording, copying, emailing, or computer reporting is unnecessary for manual techniques like keeping a paper-based log. So, an electronic format should employ a compliance log wherever possible. Since information is documented, the problem is that it is unlikely that the users will do this of their own accord.

4.12.9 Inspections

Facilities managers carry out a physical inspection of buildings to ensure that they are safe and function properly. Nowadays, drones are also used to inspect buildings. The inspector will want to find out what is happening and search for specific items they anticipate finding, such as service delivery performance, employee morale, artistic quality, data they may need to gather, copies of reports, letters, and complaints. Inspections can be carried out by an internal or external inspector, a team leader, or a member of management acting on behalf of the FM. The purpose is to get information, check if the service meets expectations, achieves performance goals, and leaves consumers pleased. The weaknesses include the length of time required to complete this task alone. With inspections, some parts of the buildings, such as the roof and high parts, can be difficult to access; it can be subjective depending on the inspector's skill. The use of an inspection checklist is therefore recommended.

4.12.10 Informal Meetings

A casual conversation will enable consumers to be more truthful and transparent about how they feel, so they may share difficulties with a degree of confidentiality that a more formal approach does not allow; a questionnaire may be the first step in assisting with this. During informal conversations, the manager can pick up on soft issues. Soft issues relate to a customer's feelings, including their level of inconvenience and any anxieties or concerns they may have.

It is essential to practise communication skills like active listening, and the service provider should let customers talk sensitively about what they have to offer. The service user must understand that whatever is shared will only be beneficial to avoid any potential disadvantages of this strategy, such as the concern that confidentiality may not exist.

4.12.11 Reports

These are carefully produced documents that will have one of two main functions:

- A factual report which conveys factual information without recommendations or conclusions
- A leading report intended to bring the reader to accept conclusions, recommendations, and change.

Reports are documents created to give the necessary information authoritatively and are made with the reader in mind. They should be well-researched and organised, with sections and subheadings that each focus on a different aspect of the subject at a time. They must be presented attractively, with parts that are well put out and show professionalism while informing the reader of the findings or reporting on them.

If the report is a leading rather than a factual one, it will also provide recommendations for the reader regarding the findings or conclusions. It should be targeted at the appropriate technical level and have depth and information for readers as the audience should already be recognised.

During the preparation stage, one of the most important responsibilities is finding or creating a detailed specification (or "brief"), including topics like the kind of audience, the intended goal, context, and sources. When drafting, authors should think about logical flow, clarifying the structure, and utilising figures and tables to summarise information. Writing itself may be a learning experience. Due to their communication with the subject matter, authors frequently discover new connections and may need to be adaptable to address new concepts. The final stage calls for proofreading and a fair viewpoint. In multi-authored reports, it can be especially difficult for authors and editors to check to maintain consistency in language and writing style.

4.12.12 Briefing Documents

These documents provide facts, information, and directions like a report. They should be short, concise, readable, and reliable. They need the reader to apply it to their condition or circumstances, are educational, help them come to conclusions, and do not contain suggestions. A briefing paper should contain the information a wider audience requires and may be distributed to a broader audience than a report. Before drafting a document like a report, the author should be aware of the audience or reader. They feature an opening that states the document's purpose and should be reasonably organised with topic headings and subheadings.

4.12.13 Briefings

A briefing conversation and a briefing paper are similar in that they provide the same details, circumstances, and information, but verbally instead of in writing. Best practice advises capturing the basic outline so it may be sent out, thereby recording information, even if a written record does not fully back it up.

4.12.14 Memos

A memo is not a document type but rather an instrument to interact semi-formally with a colleague or a limited set of recognised colleagues. A memorandum is a more formal document to create and distribute an official message for a specific corporate audience. The author of the note above should be aware of this as it is a document that other users must send information to address a problem that will engage others. It sends information; it should not contain emotional statements, highly personal ideas, demands, or threats. Email may send many less formal communications, which can be written in simple, ordinary language; memos should only use standard language. Since they are official papers, they look like letters. The intended audience, however, frequently consists of co-workers or business partners. Memos can be personally delivered or prepared on official corporate stationery and sent through email. The limits are that it can only be used within an organisation and not externally and does not provide detailed information.

4.12.15 Newsletters

A newsletter can be in hard copy or electronic copy and contains information regarding activities in an organisation or industry sent to members, customers, employees, or other subscribers. Newsletters generally contain one main topic of interest and are instructive and less formal than a report or briefing paper. They target a large but interested audience with current information that is of general interest. Since it might be hard to identify everyone who could want this, it is typically sent to a larger audience so that the appropriate individuals can receive the information. Regardless of a reader's position, background, or level of expertise, the newsletter will suit them all. For electronic newsletters, the issue can be from them sometimes getting into spam emails. It may not be sent at the right time, and too many could be sent out.

4.12.16 Letters

These professional communications papers carry various suggestions to others or the organisation. They are documents frequently created on pre-printed business stationery and might be an official notification or request. The receiver will receive a tangible copy of the document that may be kept on file or in a record.

In an organisation, letters play a crucial role in communication and are linked to several ways of communication, including email. They are useful in communication because they help with honesty as people write alone and are aware that records of their actions are kept. They revolve around three tasks: "Setting the scene, communicating the main idea, and outlining any necessary action." In the case of standard letters, when a courteous, clear, and concise approach is typically most suitable, letters should include the recipient and the aim of the communication. With the help of database and text-handling technology, personalised letters that are targeted and written may be produced in large volumes.

Electronic mail (email) and text messaging, which are text-based and conversational mediums, are appropriate communication tools. They are the most appropriate means to send information to external parties – even if they are not mainly a method to interact with them – for instance about ending a contract. It also involves requesting information from outside organisations and notifying them of changes. The constraints are that the letter writer must think about what information needs to be transmitted and create the document to send it.

4.12.17 Notice Boards

These are one-way communication strategies to post content that potential onlookers will view. Small posters with messages are used and may be designed to be visually appealing. We also have digital display boards with developments in technology. This approach effectively communicates crucial information to encourage engagement. The issue with notice boards may come from user abuse if they are not monitored. There could be problems with the cost and technical management of digital display boards. Also, it may not always work in locations prone to load-shedding since it is powered by electricity.

4.12.18 Internet

The internet is a commonly used and effective form of communication. Now, most information is accessible online. However, not everything has a corporate function, making matters more complicated; some individuals find it difficult to distinguish between official and social purposes. For instance, Facebook was created to stay in touch with friends but is also used to communicate in commercial scenarios. Now custom-designed websites link to social media handles such as LinkedIn, X (formerly known as Twitter), and Instagram accounts. Online viewers can access an organisation's news, technical specifications, operating procedures, readily available data and statistics, images, diagrams, and textbooks through websites and social media handles. However, the internet provides solutions via blogging, filling out forms, sending emails, and making reservations online. Managers use the channel to encourage customers to contribute messages as well.

Managers should consider how they and their companies use the Internet for commercial objectives. There could be differing views on the best way to use the platform for the stated purpose such as sending information to clients and customers and allowing customers to send or volunteer information to them.

4.13 Frequency of Communication

Developing a plan on what to communicate and the best way to achieve this will be possible if the frequency of communication is known. Examples of communication frequencies include:

- A once-off arrangement, for example, when preparing a site map for contractors
- A repeat communication for weekly records such as the performance report of a cleaning team
- A monthly progress report
- A quarterly report may be for an activity such as customer satisfaction data analysis of a new office service
- If an annual communication was made, the findings after an annual review before a service level agreement renewal

Other frequencies include:

- As and when required
- At the end of a holiday period
- Fortnightly
- Bi-monthly
- Hourly

4.14 Data Collection, Customer Databases, and Data Analysis

4.14.1 Data and Information

Data is a collection of unprocessed, raw information. They are usually numerical. On the other hand, information consists of facts and the results, interpretations, or conclusions generated from the points to manage the issue. For instance, customer complaints sent to a help desk may contain statistics such as the overall volume, regularity, and response time. Information may include, for example, the number of complaints received, the number of complaints compared to last month's number, the timing of the protests, recurring themes in the complaints, such as equipment failures, and the response time to customer reports. Information is used to find faults and enhance response timing. Both internal and external sources of information can provide data.

Table 4.1 Data and Information Sources

Forms of internal customer data	*Forms of external customer data*
• Utility costs • Environmental costs • Indoor air quality of buildings • Occupancy levels in buildings • Number of maintenance activities undertaken daily, weekly, quarterly • Staff hours • Staff sickness records • Space usage • Hours of operation • Demographic details of employees • Help desk contact details, number of recorded issues, number of issues dealt with each day • Energy usage data from metre readings • Average number of visits monthly to a building • H&S incidents, number of entries in the first aid book • Type of inspections undertaken • Amount of work undertaken against pre-arranged KPIs • Frequency of project activities occurring in the building • Number of times churn is undertaken (moving offices and desks) • Catering requests • Cleaning performance by staff	• Profit margin of the contractors • Profile of the contractors • Hours of operation of contractors • Scope of service provided • Performance of contract • Number of external visitors to the site over a given time • Compliance with regulation • Frequency of contractor site inspections done • Deliveries made by contractors • Number of contractors responding to tendering requests • Access to funding for contractors • Number of staff employed by the contractors • Frequency of handling or integrating legal updates into working procedures • Training of service providers • Activities of competitors • Cost savings from efficient procurement activities • Price of supplies

Source: Author

4.14.2 Analysis

Data analysis may provide reports with resulting conclusions, prepare reliable information to be presented to a group of people, and offer solutions to management problems. Information from a space occupancy exercise may be used to plan for space in an organisation. The study results can be used for planning and improvement reasons. Decision-making processes can be guided by information from data analysis.

4.14.3 Sources of Satisfaction and User Data

These could come from the following sources:

- Self and research, such as data obtained from metre readings, internet research, software, running a database report, and fact sheets from the utility company
- Information from customers and service users on satisfaction levels and opinions

4.14.4 Voluntary Feedback

Managers need to encourage customers and service users to provide feedback on what they think about the services received from the FM department. The facilities department does not have to wait for service users to volunteer information; they must request such information. Managers must develop strategies that make users more responsive to providing such information. Such strategies should be interactive as much as possible and must allow respondents to express their honest feedback and genuine concerns freely. There should be ways to provide information in an orderly manner.

Mechanisms for this include:

- Complaints logging
- Staff suggestion systems
- Help desk arrangement
- Meetings
- Questionnaires
- Interviews
- Focus groups
- Inspections, audits, and site visits
- Discussing soft issues through casual conversation
- Web pages

After gathering and analysing user data, managers can draw conclusions based on their findings. Table 4.2 shows the performance criteria for a building, based on which users can complain about a building and its services.

Table 4.2 Performance Criteria to Obtain Information from Building Users

1	Level of cleanliness
2	Adequacy of natural lighting
3	Control of artificial lighting
4	Adequacy of lighting levels in the corridors
5	Overall perception of lighting quality
6	Room temperature during summer/dry season
7	Room temperature during winter/rainy season
8	Perception of temperature in a building
9	Air quality within the building
10	Air quality in the corridors
11	Control of natural ventilation
12	Overall perceptions of indoor air quality
13	Noise from outside the building
14	Overall perceptions of noise in the building
15	Overall comfort level in building
16	Faulty electrical systems
17	Faulty plumbing systems
18	Faulty HVAC systems
19	Faulty fire systems
20	Faults with the operation of the lift/elevator
21	Furniture arrangement
22	Amount of space
23	Conversation privacy in the building
24	Visual privacy in the building
25	Conveniences
26	Interior design of the building
27	Telephone system
28	Overall satisfaction
29	Car parking
30	Fire safety
31	Security level
32	Internet facilities

Source: Author

Service user issues or problems reported by customers and facilities managers must be considered in order to implement the CRM strategy properly. Table 4.3 lists some of the issues:

4.14.5 Help Desk

For this section, we use the examples of the North-West University (NWU) in Potchefstroom, South Africa, and Singapore Management University, which operates through the Integrated Work Management System. Planon software is a world-class building management software for logging in and tracking maintenance requests. It has the following embedded in the system: Space reservation, move requests, infrastructure applications, capital project management, and OHS incident logging and space management. The implementation was done by Vetasi, an international consultancy company that deals with enterprise asset management solutions, IT service management, and integrated workplace management systems.

Case Studies

North-West University

This university has many campuses spread over two provinces. The university had about 65,000 students, over 1,000 buildings, and 40,000 spaces.

Table 4.3 Customer Issues

Type of Issues	*Gaps created by issues*
Human resources	• The team is poorly allocated around the building • A higher skill set is required to work in some areas, but staff undertaking this may not have the necessary skills • Lack of training of the FM team in CRM • Lack of understanding of the company goals will influence cutting corners • Staff undertaking the work may be poorly trained • When staff are not enough, it can take more time to finish work • Some staff get bored working in the same areas all the time (lack of variety) • Time management is not well organised or planned, and work is conducted too quickly or not thoroughly enough • Demand for perfection by customers/users • Getting the job done right the first time • Customers do not complain • FM workers do not know what level of quality customers expect • Communication skills of the FM, such as language barriers • Critical customers always get better service • Lack of dispute-handling procedure • Customers do not appreciate a good job • Corruption can influence the quality of service delivery • Protests can influence the quality of service delivery
Equipment	• Outdated equipment with no budget for replacement • Equipment is old, out of date, and needs modernising • Facilities do not adequately support operations • Poorly trained operatives operate the equipment • Poor quality equipment is used • Equipment is suitable for some parts of the building and not other parts • Noise created by equipment use is less distracting for those working on other building floors.

(*Continued*)

Table 4.3 (Continued)

Type of Issues	*Gaps created by issues*
Materials and supplies	• Inadequate material supplies can cause conflict between FM and the contractor • No proper record-keeping system • Inadequate IT resources
Cost	• There is a limited budget • The building has been partially refurbished in office areas on some floors, leading to easier / more efficient cleaning • Not enough people to do the job because the budgets are defined and inflexible • The organisation runs a no-overtime allowance policy, so staff only work to a given extent • The budget cannot support service success • Inability to convince a boss to fund a project
Time	• FM workers seem to get switched from job to job. • FM workers are not able to readily meet all facility needs • Things are changing too quickly to keep up • There is not enough time to do everything needed • Outsourcing can cause delays in response to customer demands • Protests can delay service delivery
Processes	• It is quicker to do specific tasks in some parts of the building than in others, but tasks are shared equally, so not enough attention is given to those areas where it is easier to do the task • The complexity of some parts of the building's tasks requires more attention • Expectations of people working in the Head office are higher than in other company offices • There is a poor sequencing of cleaning activities undertaken in some places. • Meaningful work is undertaken at the wrong times of the day • Different rooms require different cleaning processes, but the same processes are utilised for all areas • Some areas may be restricted-access areas, and only specific staff are allowed • Tightly packed furniture may affect the ease of work

Source: Author

As part of the Facilities Strategic Optimisation Project (FSOP), the NWU had a clear objective of creating and improving the governance, delivery model, processes, and procedures associated with the facilities functional area, taking into account the DHET 17 Elements of the Macro Infrastructure Framework (MIF) to deliver a service in support of NWU's business and digital strategies. NWU realised a clear need for an Integrated Workplace Management Solution (IWMS) tool, which would assist in achieving the above objective.

After thorough Request for Information (RFI) and Request for Proposal (RFP) phases, Vetasi was identified as the vendor of choice due to their vast experience in providing market-leading solutions within the facilities and asset management industry, their deep understanding of the university's requirements within the higher education sector, and their vast experience in implementing the Planon IWMS solution which satisfied 12 of the 17 DHET Elements of the MIF by itself. This decision was supported by Vetasi's unique ability to leverage international experience whilst ensuring local delivery.

Figure 4.4 North-West University Campus.
Source: Vetasi

Project Scope
The IWMS functional scope for NWU included Asset and Maintenance Management, Space and Workplace Services Management (specifically Reservation Management), Capital Project Management, and Contract Management.

The implementation strategy adopted was at three levels:

Level 1
Preventive maintenance was implemented in the three campuses in over 1,300 buildings, with the automatic routing of various types of orders enabled by more than 500,000 rules. The integration of Planon helped the real-time transfer of maintenance cost information between the systems, which reduced administrative efforts.

Level 2
Roll out of planned maintenance, condition assessments, lease and reservation management, and budgeting.

Level 3
Roll out of capital project management to help with the design and construction of projects.

There was also system integration where a separate system integration project stream was managed and spanned over all three business releases. Each business release had its own unique and detailed integration requirements.

Results and Lessons Learned
The system provides a platform (desktop and mobile) for the university staff to create service requests for any facility-related issue that requires attention from the facilities

management department. The service request is automatically routed to the correct responsible party based on the information provided in the service request. The work order process enables the allocation of technical resources to attend to the problem, and the requestor can monitor the progress of the work at any time. The system provides the ability to manage planned maintenance more effectively based on the predefined maintenance schedules. Maintenance backlogs are visible and can be acted on more effectively.

The phased approach helped to prioritise key needs and receive value as soon as possible, which helped the implementation team to get knowledge when the project advanced. One of the major success factors of the project was that emphasis was placed on change management, where functional area experts and change champions from the user base were used to promote the system at NWU. A well-formulated training strategy supported this. This exercise proved incredibly effective as the number of user uncertainties was identified and addressed ahead of the go-live date, an eye-opening result for Vetasi, who witnessed the importance of effective change management on a large-scale project.

Key contributors to the success were the internal meticulous planning and readiness the university performed before the implementation, which allowed for a structured, efficient, and effective rollout. A common challenge usually faced with large-scale technological implementations lies in the organisation failing to fully grasp their needs and expect the tool to provide solutions miraculously; however, this was never the case at the NWU. The university spent time to ensure that the Planon tool could provide value towards pain points identified during the Gap analysis.

All three levels relied on a network of system integrations to transport information between systems. In total, 20 integration points across 9 systems guaranteed no data duplication. The average time to define, create, test, and deploy a full integration was three weeks, demonstrating the excellent collaboration between NWU and Vetasi.

A large emphasis was placed on up-skilling in-house staff to be competent system administrators. This ensured the system could be operated, maintained and improved with limited Vetasi support. Vetasi encourages all Planon customers to take ownership of their system to maximise value. NWU has fully taken ownership of the system, which has set a new standard. The project was successful because of the NWU team's ability, desire to learn, and dedication. NWU witnessed significant benefits, including cost savings and operational efficiency improvements.

Singapore Management University

Singapore Management University (SMU) is internationally recognised for its world-class research and distinguished teaching. Established in 2000, SMU's mission is to generate leading-edge research with global impact and produce broad-based, creative, entrepreneurial leaders for the knowledge-based economy. It is known to be a pioneer in Singapore for its interactive and technologically enabled seminar-style teaching in small class sizes. Planon's IWMS provided SMU with an efficient, real-time solution to overcome workplace and facility challenges.

The SMU workplace and facilities management offices needed a software solution to manage their work processes. They found that, for example, AutoCAD drawings were being misplaced or not updated as they had no central database. They were looking for several important things: An integrated system with a good user interface and a bidirectional AutoCAD facility. The students use Planon on their smartphones to find the location of specific rooms or even search for a professor. Student satisfaction is very important to SMU, and using these apps is important for them.

Figure 4.5 Singapore Management University Campus.

Source: Vetasi

Results and Lessons Learned

The university can now effectively manage service request workflows. Also, online management dashboards are in place to access service requests, fault trend reports and KPIs. The software was implemented in less than three months, providing SMU with an immediate Return on Investment (ROI). SMU could take advantage of over 30 years of experience and many predefined workflows, which meant no need for customisation. In addition, SMU and Planon set up a local implementation team at SMU. This meant that all of the expertise was available on-site as required.

4.15 Added Value for Facilities Management Customers

Added value in facilities management is about making workplaces better, meeting users needs, closing service delivery gaps, and meeting facilities' sustainability goals (Jensen et al., 2013). Value can be created using service related methods, such as improving service quality and service level agreements.

4.15.1 Service Quality

Quality means meeting the requirements of norms and requirements (Crosby, 1979). According to Grönroos (1990), service may be considered an intangible activity that a service provider provides to clients to address their issues, and it has nothing to do with ownership.

SERVQUAL and SERVPERF are two methods for evaluating the quality of a service. In their Gap model, Parasuraman et al. (1988) created SERVQUAL. This model refers to the "gap" between consumers' expectations and their views of the service provided as service quality. The expectation is the belief in anticipated performance and the formed opinion of a person's experienced service. It gives managers a systematic way to measure and monitor service quality; SERVQUAL brings value and focuses on understanding client expectations and creating internal policies that compare with company operations (Buttle, 2009).

Under the Gap model, SERVQUAL is used to measure service quality. Parasuraman et al. (1985) created the Gap model based on gap analysis. This model states that the following five gaps are the leading causes of service quality issues (Zeithaml et al., 1990):

1. Gap 1 is the difference between customer expectations and management's perceptions of those expectations
2. Gap 2 is the difference between management's perceptions of customers' expectations and service quality specifications
3. Gap 3 is the difference between service quality specifications and delivery
4. Gap 4 is the difference between service delivery and external customer communications about service delivery
5. Gap 5 is the difference between customers' expectations and perceived service

Gap 5 is influenced by Gaps 1–4, which should be analysed to identify any corrective actions to diminish or eliminate Gap 5.

With regard to Gap 5, Parasuraman et al. (1985) defined service quality as a function of the discrepancies between customers' expectations for the service execution before the service encounter and customers' views of the service they received: Perception-minus-Expectation. Service dimensions and associated features are used to determine service quality. There were initially ten dimensions (tangibles, dependability, responsiveness, communication, credibility, security, competence, courtesy, understanding/knowing consumers, and access). Then, using 22 characteristics, these 10 qualities were divided into 5 general dimensions. They were tangibles (the appearance of physical facilities, equipment, and staff), reliability (the ability to deliver the promised service dependably and accurately), responsiveness (the desire to help customers and provide prompt service), assurance (the expertise and courtesies of employees and their capacity to inspire trust and confidence), and empathy (the provision of individual care and attention to customers). The 5 dimensions include a total of 22 service qualities.

The attributes used in a sample FM SERVQUAL instrument are shown in Table 4.4.

Cronin and Taylor (1992) created the performance-only metric SERVPERF. SERVPERF only evaluates SERVQUAL's perception components by using the exact dimensions and characteristics as SERVQUAL, which reduces the number of attributes in the questionnaires from 44 to 22.

4.15.2 Service Level Agreements

A service level agreement (SLA) is a legal contract between a service provider and a client that outlines each party's duties concerning the services rendered and the performance requirements to be met. An SLA is a practical method for managing distinct bundles of tasks or services, and SLAs are negotiated between providers and clients. An SLA tries to lay out the conditions for handling a service that the Customer and Supplier have agreed. It specifies the service goals and regulates both parties' expectations. It may be unofficial and simple or formal and a part of a contract; it must expressly state the service, the performance criteria expected, and the manner of payment.

Table 4.4 Example of a SERVQUAL Instrument for Meeting Rooms and Conferencing Facilities

Dimension		*Attribute*	*Quality Attribute for Meeting Rooms and Conferencing Facilities*
Tangible	1	Up-to-date equipment	Spaciousness of rooms
	2	Visually appealing physical facilities	Attractive reception
	3	Neat-appearing employees	Exhibition facilities
	4	Visually appealing materials	Visually appealing brochures and stationery
			Adequacy of teleconferencing equipment
			Adequacy of video conferencing equipment
			Adequacy of electronic equipment
			Appealing interior and exterior décor
			Functional stage
			Adequate facilities for the disabled
			Women friendly space
			Cleanliness of the space
			Clear directions for using the space
			Hygienic toilet facilities
			Comfortable furniture
Reliability	5	Availability of staff	Availability of the staff to perform service
	6	Dependability	Timely performance of service
	7	Timeliness	Trained and knowledgeable staff
	8	Correct performance of service the first time	Staff with good communication skills
	9	Maintenance of error-free records	The ability of staff to create the right impression at first
			Maintenance of venue error-free booking system
			Accurate knowledge of the pricing of venues
			Accuracy of the menus provided
Responsiveness	10	Providing users with information of service time	Quick booking of venues
	11	Providing prompt service to users	Willingness of staff to assist promptly
	12	Willingness to assist users	Promptness in food service
	13	Responsiveness to customers requests	Available staff to provide the service
Assurance	14	Courteous staff	Courteousness, friendliness, and integrity of staff
	15	Ability of staff to make customers feel safe	
	16	Ability of staff to ensure customer confidence	
	17	Employees of the company know how to respond to customers' enquiries	

(*Continued*)

Table 4.4 (Continued)

Dimension	*Attribute*	*Quality Attribute for Meeting Rooms and Conferencing Facilities*
Empathy	18 Understanding the users' needs	Paying attention to detail in responding to the users
	19 The company's operating hours are convenient for all its customers	Availability of the spaces
		Understanding users' needs
	20 Employees of the company give customers personal attention	Listening carefully to complaints about venues
		Ability to staff to solve problems
	21 The company has the customer's best interest at heart	Having the customers' best interest in mind
	22 Dealing with customers with care	Making the users feel relaxed when using the venue
		A variety of basic services are provided in the venue
		Choice and variety of menu
		Provision of the latest technology in the venue

Source: Author

In an SLA, service objectives should use the SMART criteria:

Specific – target a specific area for consideration
Measurable – quantify indicators of success
Achievable – agreed upon, attainable, acceptable
Realistic – relevant, reasonable, within available resources
Time-related – specify when the result(s) is achieved

Service level agreements use parameters, such as the frequency and calibre of reporting and the parties overseeing the performance levels, to link monitoring the scope of services to the service objectives through Key Performance Indicators (KPIs). The KPIs or objective performance measurements may partially or wholly determine the cost of the service. Customers benefit from SLAs because they are less unclear about the services, their standards, and pricing. The limits and responsibilities of the customer and provider are made clear by a good SLA.[9]

4.15.3 Value from People

For many organisations, people are a significant source of value. Through customer relations management, the manager is responsible for providing value, reducing complaints, and rewarding customer loyalty at a lower cost of service delivery. The manager must keep track of client information and know their needs, and he is in charge of overseeing the client relationship. This would require managing the customer's objectives, responding to their questions and complaints, and ensuring they are satisfied. In addition to having a thorough awareness of what the manager's organisation can provide, the manager's skills may include effective communication, analytical thinking, problem-solving, and customer expertise.

4.15.4 Value from Place

Physical proof comprises the observable buildings or locations businesses use to demonstrate their worth to clients. The aspects of the physical property included are rent, property taxes, insurance, depreciation, service fees, and projects. Controlling costs, running expenses, upkeep, cleaning and housekeeping, electricity, water, sewerage, waste disposal, landscaping, and service fees are all possible design considerations. Organisations should inventory the physical evidence that influences how people perceive value. The objectives and values of the organisation are communicated to clients and users through premises.[10] For instance, hospitals should show their clients excellent hygiene and care through clean uniforms and facilities, and well-kept grounds. Corporate organisations should have colours and interior design that reflect their brand image to the public.

4.16 Applicability of Implementation Models to FM

There are many models for CRM. Some of the important ones will be discussed in this section (Buttle, 2009). These models are implemented in the corporate world and can also be adapted to FM. FM departments can also use the processes and skills used in these models to improve CRM.

4.16.1 The IDIC Model

The IDIC model by Peppers and Rogers suggested that companies should take four actions to build closer relationships with customers:

- **Identify** who customers are and build a deep understanding of them
- **Differentiate** customers to identify which customers have the most value now and which offer the most for the future
- **Interact** with customers to ensure that there is an understanding of customer expectations and their relationships with other suppliers or brands
- **Customise** the offer and communications to ensure that the expectations of customers are met

4.16.2 The QC Model

The QC model, as seen in Figure 4.6, is similar to the creation of a consultancy; however, the word "relationship" is not present in the model. The model outlines several tasks companies must compete in to attract and keep clients, and the model shows people going through procedures and using technology to support those operations (Buttle, 2009).

4.16.3 The CRM Value Chain

Francis Buttle created this model, which consists of five primary stages and four supporting circumstances that ultimately lead to increased customer profitability. The preliminary stages are customer portfolio analysis, customer intimacy, network development, value proposition development, and managing the customer lifecycle (see Figure 4.7). This is to ensure that an organisation develops and delivers value propositions that attract and keep profitable customers with the help of its network of partners, suppliers, and employees.

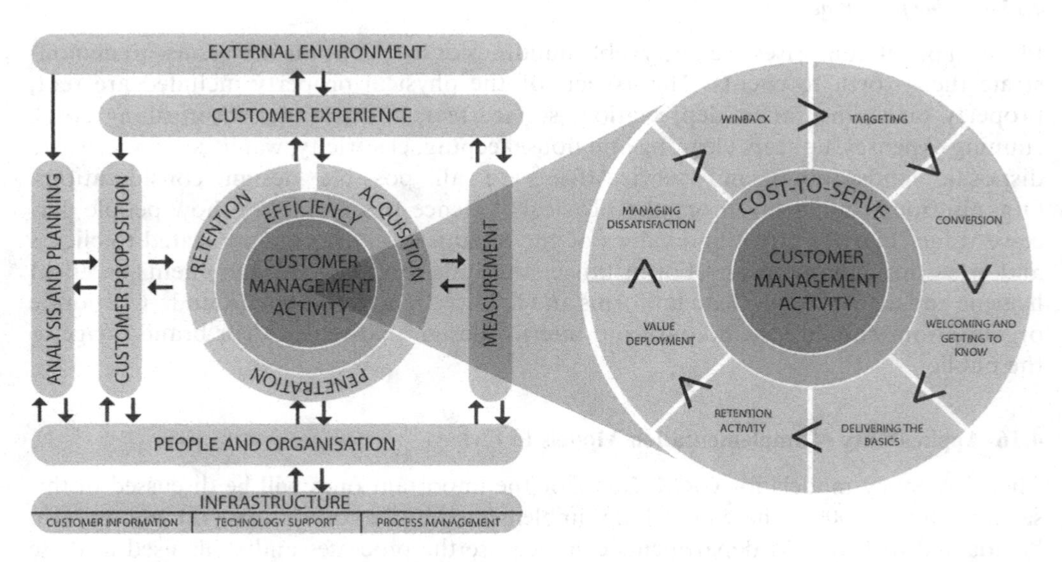

Figure 4.6 The QC Customer Management Model.

Source: Buttle, F. (2009). *Customer relationship management: Concepts and technology*. Butterworth-Heinemann

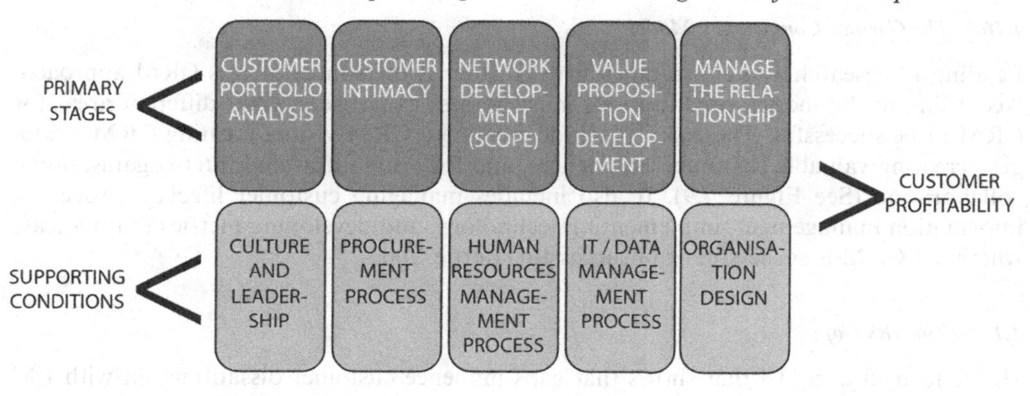

Figure 4.7 The CRM value chain.

Source: Buttle, F. (2009). *Customer relationship management: Concepts and technology*. Butterworth-Heinemann

The supporting factors, including leadership and culture, data and IT, people and processes, aid in the effective and efficient operation of the CRM strategy (Buttle, 2009).

4.16.4 Payne's Five Process Model

Adrian Payne created this model. Strategy formulation, value generation, multichannel integration, performance assessment, and information management processes are the five primary processes in CRM (see Figure 4.8). The first two examples are analytical CRM, where the information management process is an example of operational CRM, and the multichannel integration process is an example of strategic CRM (Buttle, 2009).

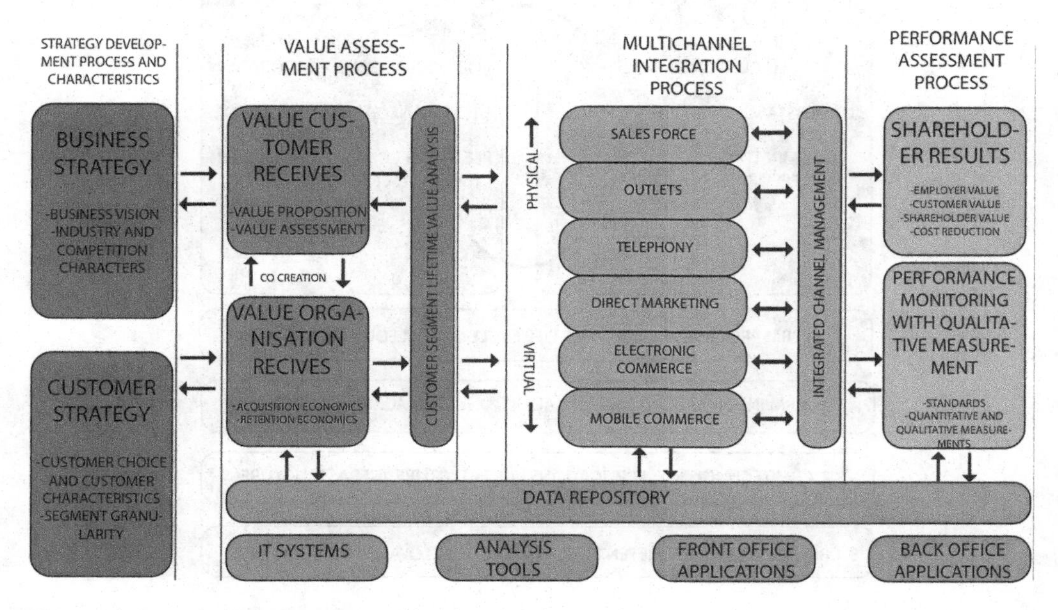

Figure 4.8 Payne's CRM strategy and implementation model.

Source: Buttle, F. (2009). *Customer relationship management: Concepts and technology*. Butterworth-Heinemann

4.16.5 The Gartner Competency Model

Leading IT research and consultancy firm Gartner is the source of this CRM approach. According to the model, organisations must possess expertise in eight different areas for CRM to be successful. These skills include creating a CRM vision, creating CRM strategies, creating valuable customer experiences, and fostering intra- and inter-organisational collaboration (See Figure 4.9). It also includes managing customer lifecycle processes, information management, implementing technology, and developing metrics that indicate whether a CRM is successful or unsuccessful (Buttle, 2009).

4.16.6 The 3Rs Gap

Hoots formed a model that shows that gaps influence customer dissatisfaction with FM service delivery in the "3 Rs": Resources, responsiveness, and respect. The cornerstone of a successful FM CRM is effectively addressing gaps in the "3 Rs". Time, money, people (see Figure 4.10), facilities, materials, supplies, and equipment can all be considered resources. Response gaps can occur when bureaucratic processes clash with customers' flexible needs, the FM department's structure conflicts with organisational requirements, and present conditions obstruct future needs planning. Respect gaps may be caused by problems such as

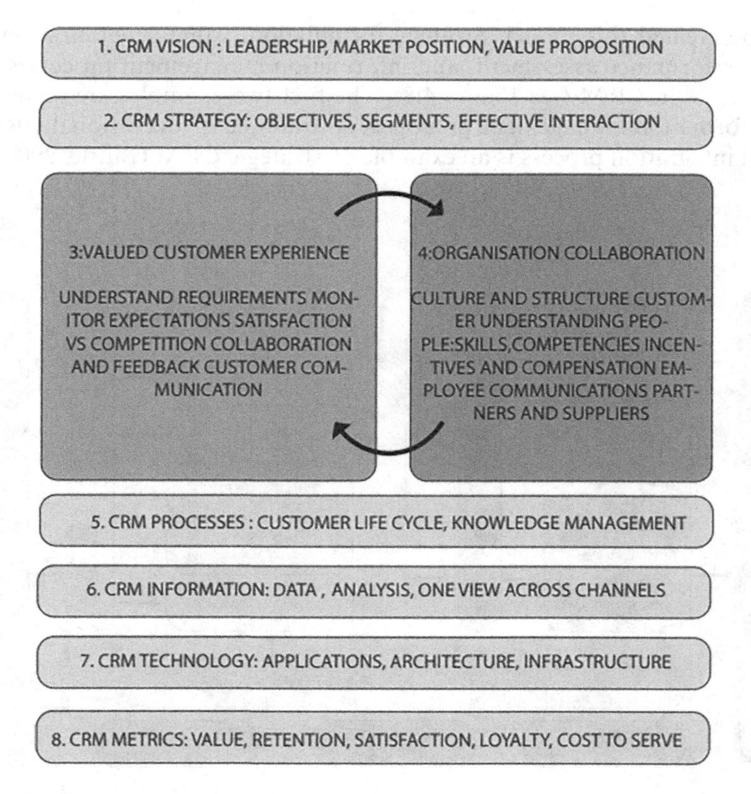

Figure 4.9 Gartner's competency model.

Source: Buttle, F. (2009). *Customer relationship management: Concepts and technology*. Butterworth-Heinemann

Table 4.5 Applicability of CRM Models to FM

Model	*The steps*	*Comments*
IDIC model	Customer identification, differentiation, interaction, customisation	The facilities manager performs this role in many facilities
QC model	Considers customer retention, looks at the external environment, measures customer satisfaction, and is involved in value development, customer knowledge, win-back and targeting	The facility manager performs all these activities. Customers are knowledgeable, and the manager determines customers' needs to meet and exceed them. Communication of quality is done. They also know how to add financial value. to organisation's decisions
CRM value chain	Customer portfolio analysis, customer intimacy, network development, value proposition development, and managing the customer lifecycle	Facilities managers manage resources using the right technology, people and time. They manage customer relationships and add value to the company's decision-making. They also use the right networks, such as the help desk, to access information on customer needs or service requests from the organisations
Payne's process model	Strategy formulation, value generation, multichannel integration, performance assessment	The CRM model in FM looks at how to create financial and quality value for decision-making and services delivered. The multi-channel integration in FM does not consider channels such as sales or outputs but facilities. The other channels, such as telephony and mobile, are used in CRM in FM. In FM, there are specific measures and KPIs for measuring success. The Help Desk is also a good data management system used for CRM
Gartner's competency model	Skills include creating a CRM vision, creating CRM strategies, creating valuable customer experiences, and fostering intra- and inter-organisational collaboration. It also includes managing customer lifecycle processes, information management, implementing technology, and developing metrics	All the skills outlined in this model are within the competencies of the facilities manager for managing facilities

Source: Author

FM quality not meeting customer quality expectations, customer views of the FM department, and consumer perceptions of the manager's perspective on customers.

4.17 Implementation of FM CRM

The CRM may be implemented utilising Hoots' gap closers or Deming's wheel. Gap closers address the customer's resource, responsiveness, and respect gaps. Deming's wheel put out

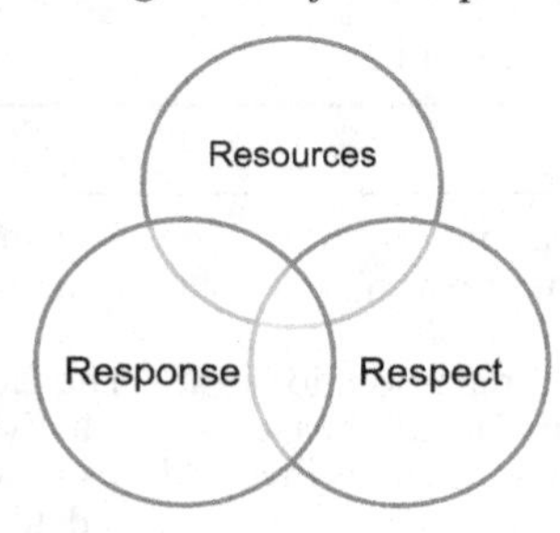

Figure 4.10 The 3Rs CRM model.

Source: Hoots, M. (2005). Customer relationship management for facility managers. *Journal of Facilities Management*, *3*(4), 346–361

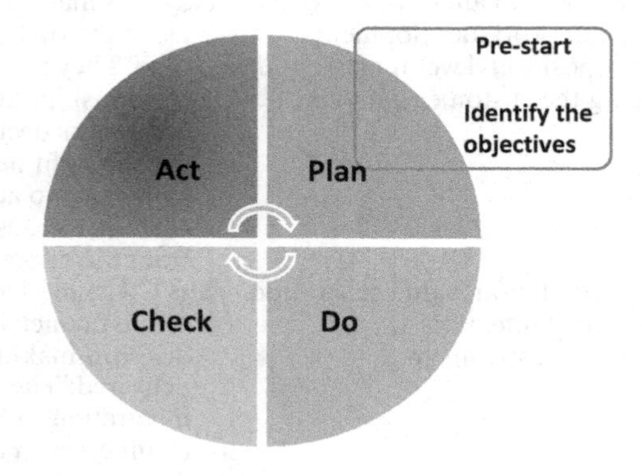

Figure 4.11 Deming's wheel.

Source: Author

a framework for some acts that offer a solution for a course of action. Each activity must have objectives, be planned to be completed at the most advantageous moment, and then be evaluated to see whether it was effective. Managers take action after observing the activity. Figure 4.11 is a diagram of the action's progression:

P – Devise a Plan to make objectives work

D – Initiate (Do) the plan and carry out all of the critical activities and duties

C – Monitor (Check) the activities and duties and assess if the plan is still working towards its intended targets

A – Act upon findings and make changes where necessary

Plan again before implementing an enhanced second plan, monitor this, and act upon new findings. Deming's wheel is also known as the PDCA cycle.

Table 4.6 shows some service user issues and solutions for proper implementation of the CRM strategy.

4.18 Conclusion

In FM, CRM refers to obtaining thorough information about customer wants, expectations, and behaviour and controlling those aspects to influence organisational performance. Because FM service is personalised and engaging for each consumer, customer satisfaction

Table 4.6 Gap Closers

Type of Issues	*Gaps closers*
Human resources	• Reallocate the number of staff, so there is adequate cover in the areas which matter most • Train the staff • Communicate goals to staff • Create some personal development planning • Improve team organisation • Exercise leadership and motivational skills • Re-define job descriptions, responsibilities and allocated tasks • Give the staff some variety; allow them to work in different areas of the building • Delegate some of the organisation to supervisors • Engage with stakeholders • Understand the company's goal • Teach the staff how to improve their time management skills and processes • Utilise a dedicated project manager for complex works • Raise a business case to ask for more staff or resource • Allocate a staff member to communicate more often with the members of the teams and act as a spokesperson • Set performance ratings for FM staff • Get funds to buy new equipment or lease new equipment • Say things in pictures • Get the job done the first time • Eliminate outmoded thinking about the quality/cost/time dilemma • Reward getting the job done the first time correctly • Do not ask customers how well things are done • Stop fighting the inevitable and embrace the necessary change • Set up a 24/7 help desk • Reward workers who honour customer quality demands • Ensure transparency • Put dispute-handling procedures in place • Support organisation needs and be fair to all
Equipment	• Review the type of equipment being used for each job • Put the better equipment in areas that need a greater level of attention • Improve the maintenance regime • Invest in better equipment if what we have is unsuitable • Train more operators to use the equipment • Train those who already use it to do so in an improved way • Keep maintenance records • Put records where occupants can see them (e.g. on the back of doors in the canteen, toilets, reception) • Demonstrate to crucial occupants that equipment is maintained appropriately to be as efficient as possible and so that it will not distract workers
Materials and supplies	• Get the right amount of supplies for the work • Ensure there is record-keeping and that there are reports presented • Use IT to save time and get complaints resolved

(*Continued*)

Table 4.6 (Continued)

Type of Issues	*Gaps closers*
Cost	• Get to lease if a purchase cannot be made. • Review the Service Level Agreement to ensure doing all proposed services • Increase activities if customers are paying for something they do not get • Demonstrate to occupants that there is a training budget and to improve the services provided • Demonstrate to occupants that there is an equipment maintenance and refurbishment budget and to improve the services provided • Demonstrate where purchases are to show where investment took place • Show that the materials used are good • Offer a better service if the organisation is willing to pay for a higher standard of service • Share the cost
Time	• Closely align tasks with the organisational structure • Set core competencies and then staff/equip to accomplish them • Set performance measures in SLAs • Have a business continuity plan
Processes	• Review when work should take place • Spend more time on problematic areas • Do not try to work in critical areas at the wrong times • Review the service level agreement (SLA) to see what was agreed • Enhance the SLA to include new or upgraded specification • Increase the visibility of staff so occupants can see the efforts made • Encourage staff to report specifics using the helpdesk or other chosen feedback mechanism • Create a report of achievements and circulate this to key occupants • Be familiar with space planning constraints

Source: Author

may be used to estimate it. For POEs, although FMs are familiar with the need for contact with users, they cannot contribute to the design of such buildings.

The consequences of changes in customer demands on FM are that managers must be creative and aware of cultural trends to build positive customer interactions. Managers must be diplomatic, honest, and transparent to develop consumer relations. They can adopt techniques that prevent them from avoiding clients. The implications of changes in customer demand for FM in Africa could be from the increase in exchange rates compared to the dollar, which presents many challenges for customer relationship management. This drives the cost of service provision and makes the cost of service provision unpredictable. Poor infrastructure can influence logistics and cause delays in service provision. Some locations have problems with good service provision because of their location, causing variations in the price and can be an issue in customer relations management. COVID-19 has also brought many changes in the workplace, as many people have to work from home. The sudden change in working arrangements has brought many dynamic changes to customer relations management since people now work in new formats.

In relationship management, failures in service delivery could result from the FM department's inability to analyse customer demands based on the timing and duration and how they fit into the organisation. Sometimes, facilities managers may need to be more organised regarding their work scheduling, which can cause a loss of resources. Service failure can also come from inadequate time, skills, and equipment to meet users' needs.

Also, systems that are transparent and fair would help manage customer relationships. Another dimension of trust is that of the position of the facilities manager. At the strategic level, it is easier to get trust from management regarding FM initiatives. Facilities managers as managers can better understand customers' needs and expectations as their relationship grows. Suppliers can also better recognise and meet client requirements since they know what they can improve.

Communication advantages in customer management include having happy clients, a motivated workforce, and staff from other departments. Additionally, it boosts customer reputation and encourages him to think creatively while creating plans to deal with clients' reluctance to change. Some of the communication approaches include meetings, briefings, presentations, questions, focus groups, memos and interviews.

Data about customers can be collected from internal and external sources. Data analysis may provide reports with resulting conclusions, prepare reliable information to be presented to a group of people, and offer solutions to management problems. Customer-related problems include human resources, equipment, materials and supplies, cost, time and processes. This chapter also discussed the applicability of different CRM models to FM and the gap closers to customer-related problems in FM.

4.19 Guidelines

- Facilities managers should quickly adapt to change and get the necessary tools
- Technology and training should be used to avoid such failures
- Some skills needed in CRM include communication, problem-solving, negotiation, FM knowledge, budgeting, technical, analytical, and project management skills
- The manager should be able to analyse the facilities, conduct space planning through space vacancy rates and usage
- Understand the demand for space and facilities to project future customer needs
- The manager should use reports and dashboards for communication
- Benchmarking to determine and help facilities meet industry standards can also be used as a business case to show the best-fit service provision among alternatives
- There is a need to possess the skills to show how facilities projects can add value to the organisation
- There should also be performance measures and key performance indicators that cater for the customer
- Get organisational support
- Use change management to drive success
- Refer to Table 2.9 as well

4.20 Questions

1 What is customer relations management in facilities management?
2 How has customer behaviour changed an organisation's approach to managing buildings?
3 Identify a service in the context of managing facilities in an organisation, client, user, and stakeholders involved in providing the service
4 For any new support service, identify the levels of interest of those involved in the service provision
5 Provide examples of a range of information that interested parties would want (at various times and given in various ways) regarding services provided

6 Try to recall situations with complicated relationships between the facilities manager, the client, and the management. Was the situation well managed, or could it have been better? Provide reasons
7 What are the barriers to communication, and how can these barriers be overcome?
8 Provide different channels of communication and their strengths and weaknesses
9 Provide four examples of when to communicate specific things. Ensure each example shows some of the different arrangements and frequencies
10 Describe how the help desk works and how it is set up?
11 How can feedback from customers be used to improve service delivery? Explain using examples
12 Identify the causes of the issues affecting customer satisfaction and what practical actions to address the causes, where possible
13 What are the models of customer relations management used in an FM organisation, and how are they implemented?

Notes

1 Buttle, F. (2009). *Customer relationship management: Concepts and technology*. Butterworth-Heinemann.
2 Kumar, V., & Reinartz, W. J. (2006). *Customer relationship management: A database approach*. Wiley.
3 Hoots, M. (2005). Customer relationship management for facility managers. *Journal of Facilities Management*, *3*(4), 346–361.
4 Hui, E.C., Zhang, P.H., & Zheng, X. (2013). Facilities management service and customer satisfaction in shopping mall sector. *Facilities*, *31*(5/6), 194–07.
5 Adewunmi, Y., Omirin, M., Famuyiwa, F., & Farinloye, O. (2011). Post-occupancy evaluation of postgraduate hostel facilities. *Facilities*, *29*(3/4), 149–168.
6 Euromonitor International. (2021). Global consumer trends in Sub-Saharan Africa: https://www.euromonitor.com/article/global-consumer-trends-in-sub-saharan-africa.
7 Blundel, R. (2004). *Effective business communication: Principles and practice for the information age*. Prentice Hall.
8 Ras, N. and Das, R. (2009). *Communication skills*. Himalaya Publishing House.
9 Adewunmi, Y. (2021). Measuring performance using KPIs. University of Witwatersrand Lecture Notes.
10 Williams, B. (2003). Facilities economics in the UK, United Kingdom: International Property and Facilities Information Ltd

Bibliography

Adewunmi, Y., Omirin, M., Famuyiwa, F., & Farinloye, O. (2011). Post-occupancy evaluation of postgraduate hostel facilities. *Facilities*, *29*(3/4), 149–168.
Adewunmi, Y. (2021). Measuring performance using KPIs. University of Witwatersrand Lecture Notes.
Blundel, R. (2004). *Effective business communication: Principles and practice for the information age*. Prentice Hall.
Buttle, F. (2009). *Customer relationship management: Concepts and technology*. Butterworth-Heinemann.
Cronin Jr, J.J., & Taylor, S.A. (1992). Measuring service quality: A reexamination and extension. *Journal of Marketing*, *56*(3), 55–68.
Crosby, Philip B. (1979). *Quality is free: The art of making quality certain*. American Library.
DDCE, Utkal University, Bhubaneswar (n.d.) Business Communication, accessed April, 2024 at https://ddceutkal.ac.in/Syllabus/MA_English/Paper_21.pdf
Dwyer, F.R., Dahlstrom, R., & DiNovo, T. (1995). Buyer-seller Relationships – Theoretical perspectives. In K.K. Möller & D.T. Wilson (Eds.), *Business marketing: An interaction and network perspective*, 71–109. Springer Dordrecht.

Euromonitor International. (2021). Global consumer trends in Sub-Saharan Africa: https://www.euromonitor.com/article/global-consumer-trends-in-sub-saharan-africa

Grönroos, C. (1990). *Service management and marketing: Managing the moments of truth in service competition*. Lexington Books.

Hoots, M. (2005). Customer relationship management for facility managers. *Journal of Facilities MManagement*, *3*(4), 346–361.

Hui, E.C., Zhang, P.H., & Zheng, X. (2013). Facilities management service and customer satisfaction in shopping mall sector. *Facilities*, *31*(5/6), 194–207.

Jensen, P. A., Sarasoja, A. L., Van der Voordt, T., & Coenen, C. (2013). How can facilities management add value to organisations as well as to society? In *Proceedings of the 19th CIB World Building Congress*.

Kumar, V., & Reinartz, W.J. (2006). *Customer relationship management: A database approach*. Wiley.

Mukadam et al. (2023). What is the best way to develop communication plan for Facility Management (FM), Identify your stakeholders, define your communication goals, accessed January 2024, https://www.linkedin.com/advice/0/what-best-way-develop-stakeholder-communication-yvxwe

Morgan, R.M. & Hunt, S.D. (1994). The commitment-trust theory of relationship marketing. *Journal of Marketing*, July.

Parasuraman, A., Zeithaml, V.A., & Berry, L.L. (1985). A conceptual model of service quality and its implications for future research. *Journal of Marketing*, *49*(4), 41–45.

Parasuraman, A., Zeithaml, V., & Berry, L.L. (1988). SERVQUAL: A multiple item scale for measuring consumer perceptions of service quality. *Journal of Retailing*, *64*, 12–40.

Parasuraman, A., Zeithaml, V.A., & Berry, L.L. (1994). Alternative scales for measuring service quality: A comparative assessment based on psychometric and diagnostic criteria. *Journal of Retailing*, *70*(3), 201–230.

Peppers, D., & Rogers, M. (2004). *Managing customer relationships: a strategic framework*. John Wiley & Sons.

Ras, N. and Das, R. (2009). *Communication skills*. Himalaya Publishing House.

Selnes, F. (1993). An examination of the effect of product performance on brand reputation, satisfaction and loyalty. *European Journal of Marketing*, *27*(9), 19–35.

Sureshchandar, G.S., Rajendran, C., & Anantharaman, R.N. (2002). The relationship between service quality and customer satisfaction – A factor specific approach. *Journal of Services Marketing*, *16*(4), 363–379.

Zeithaml, V., Parasuraman, A., & Berry, L.L. (1990). *Delivering quality service: Balancing customer perceptions and expectations*. Simon & Schuster.

5 Benchmarking in Facilities Management

5.0 Introduction

Benchmarking involves identifying a point of reference (or benchmark) against which comparative performance can be determined. The benchmark can be internal to an organisation or external concerning competitors or best practices.[1] Benchmarking changes in perception or action are ways of comparison and improvement. The practice helps find areas of weakness and sets realistic goals or targets.[2]

Benchmarking is a strategic tool, and in facilities management, it is "a process of comparing a product … – indeed, any activity or object – with other samples of a peer group, to identify 'best buy' or 'best practice' and target oneself to emulate it" (Williams, 2000). Benchmarking in Africa is needed as local organisations want to compete internationally and meet global best practices. However, benchmarking has not been successfully implemented in some organisations because of a lack of understanding of the process.[3] The information obtained from benchmarking can be used to make decisions in strategic management.[4] Some of the other strategic benefits of benchmarking in FM are that it helps determine how to make the most significant use of facility resources, it improves FM practices and supports the organisation's position in the marketplace, and it helps organisations show top management performance relative to the others in the industry. This chapter focuses on benchmarking in facilities management.

5.1 Definitions of Benchmarking

According to Camp (1989), "Benchmarking is the search for the best industry practices that will lead to exceptional performance through implementing these best practices". Anand and Kodali (2008) defined it as:

> a continuous analysis of strategies, functions, processes, products or services, performances compared within or between best-in-class organisations by obtaining information through appropriate data collection methods to assess an organisation's current standards and thereby carry out self-improvement by implementing changes to scale or exceed those standards.

A successful application of benchmarking includes comparing it with previous implementation experience, choosing the project to focus on and then finding partner organisations, working well with others from various disciplines, getting support from top management, and committing to the budget. This means that one cannot depend on information that is not original or fails to match a particular trend.[5]

DOI: 10.1201/9781032656663-6

Benchmarking should happen more and more at the input and process stages. Organisations work in a world that changes quickly and must use information sources to compare progress. When following best practices, they shouldn't go against how the company is set up, managed, or its culture. Large companies use benchmarking to ensure they get the correct information and find the best ways to do things. It should be able to assist and have an organised strategy. Any plan for using comparisons requires consideration about how to gather the correct information as efficiently as possible (Adewunmi, 2014).

5.2 Origins of Benchmarking Globally

It is said that Japanese businesses began doing comparison studies in the 1950s by going to Western competitor's factories to learn about their procedures and technology. In the 1950s, cost/sales and investment rates were the first tools used to calculate corporate performance. This made organisations consider their pros and cons and see how they measured up with their competitors in the market. However, this method did not provide choices for more improvements.

In the 1960s and 1970s, the development of computer technology brought about the more extensive use of benchmarking. In the 1980s, benchmarking became popular in the US for encouraging continuous improvement. The Malcolm Baldrige Quality Award was another factor. By the late 1980s, benchmarking was being used in the UK.

To keep up with the growing competition, Xerox Manufacturing Operations chose in 1979 to look at how much it cost to make one copy machine compared to its primary competitors. This decision was the start of a comparison of possibly helpful management tools. They found that product prices in the US were much higher than those at Fuji-Xerox in Japan. What was even stranger was that Xerox's biggest Japanese competitors were selling copy machines for the same price it cost Xerox to make theirs in the US. Xerox found a way to beat its competitors in terms of speed. The company started a competition testing methods to try to close this performance gap. To do this, guidelines or standards were made for product features, selling prices, manufacturing costs, and cycle times. The organisation's work was compared to that of its primary competitors. The findings showed that Xerox usually did not perform as well as its closest competitors. However, some elements provided Xerox hope, and it is important to remember that Xerox's strategy strongly emphasised process concerns rather than results or finished goods.

The benchmarking process may be used to understand any organisation, not just competitors. To do this, business practices are compared to those of organisations that are leaders or innovators in a specific business function by focusing on standard criteria, such as manufacturing, marketing, engineering, and finance. Thus, Xerox examined the products that an organisation made as well as the design, production, marketing, and service of those products. Benchmarking, therefore, includes two parts in its generic form: particular metrics or standards of assessment and procedures. Metrics are benchmarked to see performance gaps, followed by benchmarking practices or processes to increase understanding and identify practices/processes. Both of these parts are necessary for benchmarking to be effective. Only the practices or procedures on which the measurements are based will explain why a performance gap exists, since it is hard to understand why it exists only from the numbers.

Since 1989, when Camp released his first textbook on the subject, benchmarking has been applied to many significant industries, including telecommunications, information and communications technology (ICT), health, automotive, and aviation, as well as fields like education. It has also been widely used in manufacturing, especially by US and Japanese businesses (Adewunmi, 2014).

At the beginning of the 1990s, 65% of the Fortune 1000 companies used benchmarking to get ahead in management. In France, benchmarking became so popular that half of the French 1000 companies used it daily, and 80% of those companies saw it as a powerful way to bring about change. In 1999, more than 70% of managers worldwide used four management tools: strategy planning, goal and vision statements, comparisons, and measuring customer happiness. Benchmarking is recognised as a global quality improvement method (Adewunmi, 2014).

5.3 Types of Benchmarking

There are four stages of benchmarking in most fields. These are competitive benchmarking, process benchmarking, strategy benchmarking, and global benchmarking. Benchmarking has changed over time. It used to be all about constantly and systematically evaluating goods and services. It is more about identifying, learning, and implementing the best practices available (Adewunmi, 2014).

The classifications of benchmarking below are also used in facilities management (see Figure 4.2). Although the basic benchmarking process is the same, there are also some differences, each with advantages and disadvantages in line with the focus of the benchmarking. The three basic types are:

1 Internal benchmarking
2 Competitive benchmarking
3 Generic (functional) benchmarking

Internal benchmarking examines how similar organisational methods are used in various departments or places. The problem with this method is that the results will only show the company preferences, and they cannot demonstrate or close any performance gaps with competitors. There are, however, possible benefits if the organisation is diverse, especially if the business has branches in other countries. Getting information for internal benchmarking should be easy, which is a noticeable gain. This kind of comparison takes time because competitors may be busy growing their market share. The organisation in question is also busy measuring its performance simultaneously.

Competitive benchmarking focuses on direct competitors, ideally ones with the same customers. The disadvantage is that it might be problematic to gather information. One way to get around this is for competitors to join the process, which is beneficial. Because of this, benchmarking is thought to be better for bigger organisations than for smaller ones. Based on similar technologies and procedures, the knowledge gathered will be helpful. Competitive comparison has many benefits. When organisations in the same industry are compared, their processes and measures are similar.

Functional benchmarking looks at organisations that are leaders in their field, even if that field is different from the one your company is in. Bringing best-in-class processes into one's setting is likely to be difficult, which makes the benchmarking process challenging and time-consuming. However, the clear benefits outweigh the problems:

- Good results
- Access to important information
- Building professional networks
- An opportunity to find innovative ways of doing things

In terms of the benefits these different forms of benchmarking offer to an organisation wishing to improve performance, it is worth ranking the types (including subtypes). Those with the most benefits are at the top of the list:

1 Functional best practices – world-class
2 Functional best practices – own country
3 Industry best practices – including non-competitors
4 Competitor best practices
5 Internal best practices

An example of functional benchmarking could be when Gautrain in South Africa used a benchmarking process to make trains cleaner. The result was that a fewer number of people were able to clean a 660-seat train in 8 minutes after the benchmarking exercise. Another example could be when a partner helps Ethiopian Airways find a team that can clean a 250-seat plane in just 9 minutes. Another name for this kind of benchmarking is "non-competitive."

The three types of benchmarking discussed above can all be used in facilities management. Since benchmarking is essential to any organisation's primary goal, three other types of benchmarking may also be helpful. These are strategic, process, and generic benchmarking

Strategic benchmarking is done in a way that makes it possible to compare the strategy direction of different organisations. The process looks at major problems that affect the organisation's strategy, such as people and culture, as well as process issues, like the availability of buildings and people. As before, companies that are best-in-class or world-class will be chosen as benchmarking partners. The process doesn't look at each process but focuses on the main strategic actions and techniques that make it succeed.

Process benchmarking looks at world-class companies' methods, techniques, and business processes, regardless of their core business (they don't have to be competitors). Being good at benchmarking involves identifying comparable metrics and methods.

Generic benchmarking, which is the most commonly used for collecting data, has no set parameters, and is limited to understanding how to change the data obtained and use it. Generic benchmarking tries to find the best-in-class by comparing a company's performance to its competitors and other companies in the same industry that do similar work or have similar problems. In other words, a hotel's accounting department would look at a manufacturing company's accounting department in terms of having the best processes. For instance, BMW, a company that makes cars, can benchmark itself to Toyota, which also makes cars, and Rwanda Airlines can compare itself to South African Airways. This makes it easy to get information because the best companies are more willing to share their stories. However, general benchmarking can take a long time, and companies may need to modify study results significantly before they can set their own standards.

Benchmarking can also be classified based on "what is being compared?" Here we have process benchmarking, performance benchmarking, and strategic benchmarking. It can also be compulsory or voluntary.

Another classification can be strategic, competitive, cooperative, collaborative, and internal benchmarking. Most of the time, cooperative and collaborative benchmarking is used because it is simple to practice. Putting benchmarking into these groups is a better way to find information. When companies use cooperative benchmarking, they ask the best organisations in their field to meet and share their knowledge with their benchmarking team. This is done with little difficulty because these groups don't directly compete with each other. Information only moves in one direction during this process. This is different from collaborative benchmarking because there information can go in many directions. A lot of organisations share information through collaborative benchmarking. It's a meeting where organisations can come up with ideas. It is essential to know that not all collaborative projects are considered benchmarking. This process is sometimes known as "data sharing." Data sharing results do not focus on the process but only the result, while benchmarking focuses on the functions of the organisations (Adewunmi, 2014).

In addition to types of benchmarking, there are also categories of benchmarking (Adewunmi, 2014):

- Metric (performance data comparison) benchmarking compares with quantitative data such as financial results, production figures, and league tables, and concentrates on the results.
- Process (differences in approach) benchmarking compares qualitative data such as management methods and operational techniques and concentrates on the organisational processes.
- Diagnostic benchmarking compares metric and process data with best practice/business excellence models. In this way, organisations' performance differences can be compared with the excellence model to determine whether the score represents a strength or weakness.

Types of benchmarking are internal benchmarking, competitive benchmarking, functional benchmarking, best-in-class/generic benchmarking, external benchmarking, strategic benchmarking, operational benchmarking, business-management benchmarking, consultant study benchmarking, reverse engineering/product benchmarking, process benchmarking, relationship benchmarking, performance benchmarking/result benchmarking, diagnostic benchmarking, hooded benchmarking, open benchmarking, and others[6]. In facilities management, the most common types of benchmarking are internal, competitive, functional, strategic, process, and generic. Table 5.1 shows some main classifications developed to describe or classify benchmarking characteristics.

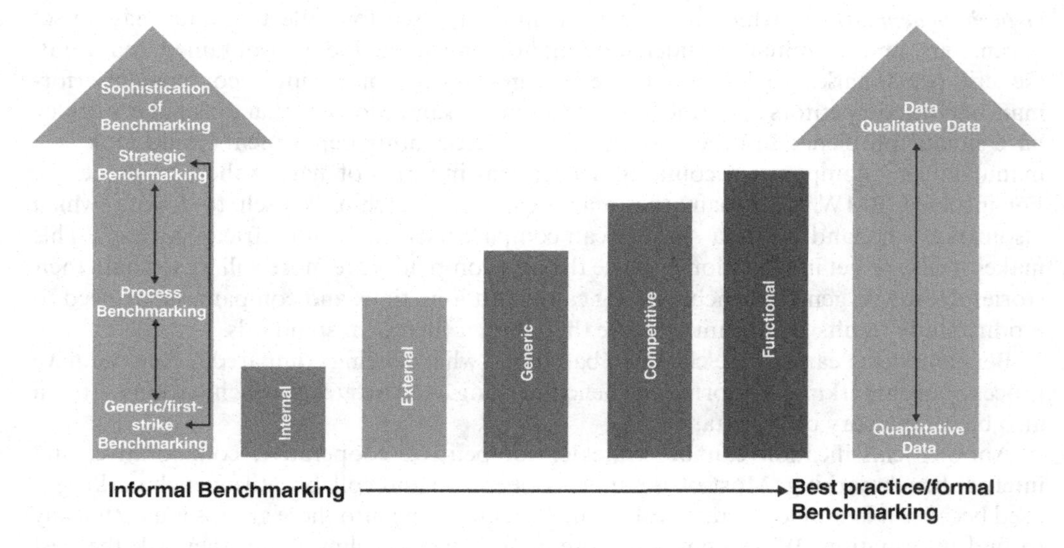

Figure 5.1 Developments in FM benchmarking from internal to functional benchmarking.
Source: Author

Table 5.1 Benchmarking Typologies

Typology	*Definitions*
Internal, competitive, functional, and generic benchmarking	Internal means comparing comparable duties within the same organisation. Competitive means comparing to the best of its closest competition. Functional means comparing methods to those used by external companies that do the same task and use similar methods. Generic process means comparing work processes to innovative and good examples of work processes.
Results benchmarking and process benchmarking	Results benchmarking concerns the comparison data that benchmarking gives you. Process benchmarking affects how results have been achieved so that performance gaps found in the "results" can be filled by investigating and learning from other practices.
Voluntary or compulsory	Considers the differences between public and private sector benchmarking and classifies public sector benchmarking as voluntary or compulsory.
Unilateral and cooperative	The difference is between unilateral (often independent) and cooperative approaches (where peers voluntarily share information). The letter includes "group," "indirect third party," and "database forms."
Implicit benchmarking and explicit benchmarking	The emergence of benchmarking in higher education that gathers data from peers.
Lateral benchmarking	Another word used for generic benchmarking.
International or global benchmarking	Involves comparison with a standard performance by overseas organisations and can be competitive, functional, or generic.

Source: Francis, G., & Holloway, J. (2007). What have we learned? Themes from the literature on best-practice benchmarking. *International Journal of Management Reviews*, *9*(3), 171–189

5.4 Examples of Types of Benchmarking in Africa

Internal benchmarking arises when businesses operating in Nigeria, such as Dangote and Globalcom, try to copy the high-performance levels in distribution, customer support, and service or product quality. Consumers have regularly accepted them as a result of a quality procedure. Table water providers like Edomat Water, Woodland Water, and Water First in Benue State, Nigeria, have benefited from competitive benchmarking. These businesses frequently review their procedures, operations, and quality performance to keep up with competitors like Swan Water and Coca-Cola Eva Water. Agro-based organisations like Olam collaborate to create and exchange industry-specific knowledge and technology and employ comparable procurement and marketing channels with best-in-class businesses in various sectors, and this is an example of functional benchmarking in action. The Nigerian government has used generic benchmarking to provide and implement helpful policies for different economic sectors by copying policies from other countries. For example, Nigerian hospitals compare health insurance programmes to those used in the USA to create an effective healthcare system (Adewunmi, 2014).

Another example is a benchmarking study on medical tourism in Egypt to create a medical tourism strategy to highlight the importance of such benchmarking concerns. The project compared the performance of Egypt's medical tourism industry to global best practices and identified gaps in the industry. It looked at the serious issues needed to develop the "service value chain" for the industry. The results of the benchmarking process have shown how important it is to address several areas for the growth of medical tourism in Egypt. These regions have professionals as a part of their "service value chain" for the medical tourism industry. In this study, another benchmarked location was India, a top medical tourism destination that shared many of the same issues as Egypt, such as urban overcrowding, poverty, and environmental degradation. Also, Jordan has some of the same challenges as Egypt in terms of the use of technology, the development of human resources (especially nursing and support service workers), and marketing issues, but it has still managed to establish itself as a leading location for health tourism in the Middle East. Such developing countries now have a positive status in the market for international medical tourism, supported by a well-organised industry, due to applying the competitive advantage method, which involves making the best use of the points of strength while reducing the negative effects of the time of weakness.[7]

A new World Bank global database with objective and perception-based measures of infrastructure performance from South Africa, Botswana, Lesotho, Namibia, and Swaziland was used to do the first systematic benchmarking of infrastructure performance in the Southern African Customs Union (SACU) countries against a group of comparative countries in the three important sectors of electricity, water and sanitation, information and communication technology. The study showed significant comparative differences in all critical infrastructure areas despite SACU performance changes. Performance gaps were more obvious in rural regions. In-depth estimates of infrastructure performance are anticipated to use benchmarking as a comparison input, mainly concerning scaling-up activities (such as those in South Africa, Lesotho, and Botswana).[8]

5.5 Formal and Informal Benchmarking

Benchmarking is a "formal process that uses comparison approaches and models; informal approaches to benchmarking exist from the experiences of organisations".[9] The explanations of formal and informal benchmarking are as follows:

Informal benchmarking – This benchmarking does not follow a set of procedures or a process. It refers to the benchmarking that everyone does at work, even if they are unaware of it. It involves comparing and learning from how other people behave and practice. Learning from informal benchmarking typically comes from:

- Talking to work colleagues and learning from their experiences
- Consulting with experts with experience using a particular process in different business environments
- Networking with people from other organisations through attending conferences, seminars, and internet forums
- Online databases, websites, and publications that have benchmarking information provide quick and easy ways to learn best practices and benchmarks

Informal benchmarking can be defined as "actively encouraging employees to learn from the experience and expertise of other colleagues and organisations through comparing practices and processes. For example, through best practice tours, conferences, best practice websites and networking".

Formal benchmarking performance and best practice benchmarking are two sources of formal benchmarking. These benchmarking methods may include a subset of functional, competitive, or internal organisation comparisons.

Performance benchmarking compares performance data gathered from looking into similar tasks or processes. When performance comparisons are made, one might compare financial measures like expenditure, the cost of labour, and the cost of buildings and equipment, or non-financial measures like staff turnover, complaints, and call centre performance.

It has been described by Anand and Kodali (2008) as "Comparing performance levels of a process/activity with other organisations – therefore comparing against benchmarks."

5.6 Best Practice Benchmarking

The steps listed in Table 5.2 are best practices in benchmarking and must be included in the existing benchmarking process.

Table 5.2 Best Practice Benchmarking

No.	Steps
1	Determine the data collection method
2	Project future performance
3	Communicate benchmark findings to both management and employees
4	Identify the information sources for collecting pre-benchmarking information by searching different technical and business journals, external databases, and public libraries
5	Narrow the list to a few benchmarking partners by comparing the candidates
6	Prepare a proposal for benchmarking and submit it to management to get their commitment, with a clear explanation of the benefits, costs involved, resources required, etc.
7	Identifying the customers for the benchmarking information
8	Gain acceptance from management and employees through commitment and participation, respectively
9	Evaluate the importance of each subject area based on priorities
10	Determine the purpose and scope of the benchmarking project
11	Collect lower-level details on that partner before contacting them (e.g., location, when they started, number of employees, product line, key managers, market share, revenue and profit, customer satisfaction, etc.)
12	Establish a protocol for performing the benchmarking study and develop a non-disclosure agreement about the information that will be shared and approval for benchmarking between the participating corporations
13	Present your benchmark findings to your management and get their commitment to implementing recommendations
14	Identify the strategies intent of the business or process which is to be benchmarked
15	Sort the collected information and data
16	Identification of possible causes and the practices that are responsible for the gap
17	Establish contact with selected partner(s) and gain acceptance for participation in the benchmarking study
18	Establish a benchmarking report which provides information on the best practices, how it was implemented in the benchmarked company and how it was adapted in the existing organisation and a comparative analysis of the reported benefits

Source: Anand, G., & Kodali, R. (2008). Benchmarking the benchmarking models. *Benchmarking: An International Journal*, *5*(3), 257–291

5.7 Benefits of Benchmarking in Facilities Management

The following are some benefits an organisation can realise from a benchmarking exercise.

Benefits to the organisation include improvements in efficiency and effectiveness, performance, creativity, finances, management of resources, and leadership. Efficiency and effectiveness come from the extra value that benchmarking brings to the organisation regarding lower operational costs, lower personnel requirements, and shorter work completion times. If correctly implemented, benchmarking will cause efficient value management of the supply of facilities services. It differs from simple budgeting because it calls for comparisons with external peer groups. The goal of senior management is to reduce operational expenses because inflation often causes price increases for the same quality of service. Facilities managers may find it very difficult to present their budget requests to the senior executive to whom they report without evidence to back up the submission of a budget that shows the levels at which the rest of the industry is working. Facilities managers often want to know the costs and service levels of managing similar facilities. Benchmarking is a simple way to do this. To change the customer's profit and performance, there is a need for cost reduction and cost-effectiveness. Benchmarking can bring this change. When similar performance is removed, procedures are improved, internal standards are established, and performance is achieved. There is a link between benchmarking and enhanced operational and corporate performance. Benchmarking lets organisations know where they stand relative to other businesses by identifying and learning from best practices worldwide. This enhances performance. Other organisations can be used to show where there are issues and offer potential fixes for each site. It helps organisations better understand their administrative processes and identifies areas that need to be improved. Benchmarking is a way to help enhance performance by helping to form goals that have been successfully achieved. A process improvement that satisfies customer expectations is the final goal of benchmarking.

The innovation that comes from fresh ideas within an organisation and when organisations line up with ever-changing markets is another advantage of benchmarking. Financial gain can be achieved when benchmarking is used to gain funding access. Benchmarking methodologies can provide innovation directly for FM performance and make fresh concepts.

Leadership benefits include a change in leadership style, strategic thinking, and change management. By helping client organisations form objectives, benchmarking may also be helpful in goal-setting. Benchmarking can assist in achieving defined goals. The effectiveness of performance benchmarking depends on selecting reasonable goals, which will also influence how employees behave inside the organisation. When benchmarking supports resource management and people management, resource management also benefits. Benchmarking helps in the sharing of funds. Benchmarking may assist with strategic decision-making on operational approaches in a highly competitive environment by providing awareness of the resource allocation needed for a competitive advantage.

Benefits of service delivery include benchmarking in outsourcing, increasing service standards, ensuring smooth operations, and learning about the advantages and disadvantages of providing services. Benchmarking is used for ongoing quality improvement. A lack of benchmarking for outsourcing will limit the ability to show which providers performed better than others. This results in worries over the monetary incentive or punishment system.

Customer advantages include increased client, user, or customer spending. Benefits from practitioners' attitudes and expertise are known as knowledge benefits.

Strategic and commercial advantages arise when benchmarking helps determine how to make the most significant use of facility resources. Improving FM practices and supporting the organisation's position in the marketplace are necessary. Productivity and asset value are greatly influenced by management at the strategy level. This is also an advantage since it helps organisations show the senior executives to whom they report how well their organisation functions compared to the rest of the industry.

For example, in a benchmarking survey conducted in Nigeria, facilities management departments and organisations engaged in formal benchmarking agreed that doing so helped their organisations correctly map out processes, make strategic plans, aim to be the best in the industry and develop improvement measures. It also addressed issues with customer service delivery that now harm the company and its viability. It assisted in obtaining justifications for those upgrades that could not be completed immediately. Those who do formal benchmarking may use it to develop strategic plans to help their organisations prepare for changes in the general business climate, enhance productivity, and strengthen their competitive position. Enabling people to define objectives and compete for excellence will help them become leaders in their respective industries. The FM sector in Nigeria is led by those that use formal benchmarking. Additionally, creating business case presentations using the information gathered during benchmarking assists management in making the best decisions regarding their assets. Many local organisations lack strategic roles, which is linked to an under-appreciation of FM (Adewunmi, 2014).

In the interview with a South African company, benchmarking was believed to help:

- "Because our department has so many facilities, it may take time to determine what is using up all our resources. As a result of benchmarking, which will provide us with a thorough property profile, our business can keep track of all of its expenses."
- The organisation may identify initiatives that waste money and highlight regions that provide higher productivity rates using the statistical data from benchmarking.
- Identify severe problems with our infrastructure.
- Benchmarking can validate that the best solutions are available if the proper outcomes are reached and within realistic prices, and may assist in implementing energy savings to decrease carbon footprint and reduce expenses.
- Benchmarking may be used to evaluate present strategies and help predict future facilities performance. This will reduce unnecessary and wasteful spending on ineffective strategies.

Table 5.3 Benefits Obtained From Maintenance Benchmarking by Eskom

Action Plan	*Benefits/ Successful Use*
• Building operations interlinks with most benchmarking attributes	• Any organisation hopes for a system that is easy to use
• Identify and register all assets so that an asset number can be used when logging calls	• That will reduce operation costs
• Link assets and their pictures to the system and their location to minimise time spent looking for an asset	• Minimise costs spent on reactive maintenance
• Load/ keep history on assets (how many times has it been broken/repaired)	• Has few disruptions

(*Continued*)

Table 5.3 (Continued)

Action Plan	*Benefits/ Successful Use*
• Liaise with HR personnel to have images of all staff so that it is easier to identify the person logging the call	• The organisation will need buy-in from other departments to ensure seamless implementation
• Give access to the Maintenance Team to log/ change/action requests and limit that to Call Centre Agents	• That will require certain individuals to be trained, such as (Facilities Managers, Call Centre Agents, Building Managers, Building Services Supervisors, End Users and System Administrators)
• Allow automatic feedback/ progress to clients on the status of logged calls to minimise escalation	• Test drive the system with regular users
• Scheduled Planned Preventative Maintenance to minimise reactive maintenance	• The aim, after all, is to get a single point of information
• Load headcount and capacity in the workplace/each section or division	• Any organisation hopes for a system that is easy to use
• Up-to-date reports on open, hold, stop, update and complete work orders	• That will reduce operation costs
• Compare costs to see which section or discipline spends the most money	

Source: Eskom

5.8 Problems of Benchmarking in Facilities Management

Problems that companies trying to benchmark can face include the following:

Organisational barriers may cause benchmarking issues, often from situations, cultures, and individuals. People's problems can include resistance to change, employee reluctance to cooperate and participate when change is necessary due to the stress of leaving comfort zones, the difficulty of learning new skills, or the fear of exclusion. When an organisation's culture prevents learning techniques, including careful problem-solving, experimenting, learning from the past, learning from others, and communicating information internally, this is called a non-learning organisation. This might be because staff aren't familiar with searching out and sharing expertise or because they're afraid of showing organisational weaknesses like a lack of training and development.

Lack of opportunities and incentives for employees to communicate inside and between departments and among all levels of the organisational structure, both formally and informally, are added issues that might result in poor communication practices.

The third obstacle is context, which comes from a lack of a thorough quality culture and employees' inadequate understanding of how to deliver a good or service that meets customers' needs.

Benchmarking project management issues may result from poor project planning and execution, issues with project leadership, and business demands. Barriers to project planning and implementation may result from inadequate employee training, a lack of adequate employee skills to implement benchmarking, a lack of sufficient employee skills to understand organisational processes, and a lack of proper employee knowledge of the organisation's products and services and how they relate to the rest of the organisation. Project planning issues could also come from a poorly defined benchmarking subject, unexpected difficulties or changes brought on by considerable unexpected challenges, or unplanned technical and implementation timetable changes. The main project leadership challenges are senior management's inability to help benchmark and the lack of involvement/commitment to organise and involve concerned employees and managers in benchmarking. The management's inability to efficiently plan the implementation activities and manage the uncertainty and changing expectations that arise during the benchmarking process also adds to poor project coordination.

Business pressure problems include the unavailability of the resources, time, money, and expertise needed to achieve benchmarking targets. Company burdens can also come from opposing demands, challenging goals, or unpredictable situations caused by internal or external business environments. As a result, it becomes essential to re-evaluate the benchmarking process to ensure it meets changing company needs.

Benchmarking data issues come from access/comparison issues caused by challenges in getting and using benchmarking data. Confidentiality concerns, incomparable data, or reluctant partners are to blame for this.[10]

In African countries, the problems with benchmarking are sometimes different from the problems in developed countries. In Nigeria, for example, research showed that the most severe benchmarking issues might come from standards, people, or data. Small organisations have the most severe problems with data issues, competitiveness, and a lack of role models in the market. For medium-sized companies, data difficulties were likewise the most severe issues. The biggest issues for large organisations are data, employees' confidence in new efforts, and top management support. The refusal of benchmarking partners to grasp the value of the project and issues arising from the quality of data gathered were challenges specific to those who use formal benchmarking methods. Another category of benchmarking barriers is known as the "market category," which includes problems from not being able to network with industry peers, market issues, lack of role models in the market, client reluctance to pay for the benchmarking project, and client disinterest in benchmarking (Adewunmi, 2014).

In a country like South Africa, the following are some of the problems related to benchmarking:

Organisations could be hesitant to provide information because most of it is the intellectual property of the service provider. Information may move unusually slowly. Also, benchmarking professionals may not accept the results of the benchmarking. It's possible that they do not want to be held accountable for the information given because they fear it will be used against them in court and that they won't want to testify.

Since some companies have financial difficulties, getting top management to invest in benchmarking will take a lot of persuasion and incentivising. Benchmarking will need the development of employee skills and training. Since benchmarking is still a fairly new concept, not many professionals with expertise work on benchmarking projects, which might imply that the professional is more expensive to use. Benchmarking is helpful but becomes a check-the-box activity if senior management or decision-makers are not involved or buy into it. As a result, despite the approach and suggestions for what may be changed to increase the performance of the FM function, the systems still need to be applied.

Benchmarking has been criticised for being similar to spying on competitors. However, benchmarking is about keeping up with what competitors and the rest of the industry are doing; it is not spying on them.

Some businesses do not employ benchmarking because they believe that if something works, why change it? When a company is financially successful, it tends to oppose change and not be concerned about competitors. Many organisations avoid benchmarking because they need to learn what benchmarking is and believe they have nothing to gain.

Benefits of FM Benchmarking	Problems of FM Benchmarking
Organisation	Organisation
Leadership	Lack of opportunities and incentives
Innovation	Lack of top management support
Knowledge	Project planning
Strategic and competitive	Business pressures
Cost effectiveness	Benchmarking awareness
Efficiency	Financial
Improved performance	Access to data
	Access to peer group

Figure 5.2 Benefits and problems of benchmarking.
Source: Author

The belief is that the peer group should be located in the same area as the organisation conducting the benchmarking for the exercise to produce the best results. This has the disadvantage that it might be challenging for specialised companies to identify the particular peer group. This can prompt them to search overseas for companies with comparable qualities. Also, defining the parameter with the necessary accuracy is essential to reduce or eliminate the margin of error. Due to the various sizes and cultures, this is challenging in practice. This tends to skew the results and give the impression that implementing the systems of the benchmarked organisations is simple. Some of the lessons learned are impossible to execute since benchmarking data only considers the result or the product and pays little attention to the inputs or processes.

5.9 Critical Success Factors in Benchmarking

If well handled, these factors, variables, or characteristics will directly affect customer satisfaction and, thus, the organisation's success. Therefore, the ability to benchmark processes that are appropriate to the organisation's goals is a crucial part of successful benchmarking. Goals, in particular, might include maximising profits and improving customer satisfaction, competitiveness, and performance. The goals may be benchmarked using performance measurements, for example, standard business metrics for reducing cost or improving performance. However, it is crucial to identify some exact measurement standard since it would be difficult to locate a performance difference and conduct benchmarking without it.

The importance of each step varies amongst companies; critical issues are company and industry-specific.

The following are major factors (Adewunmi, 2014):

1 Assistance, along with guidance from the top management
2 Levels of satisfaction among consumers are regularly tested and tracked
3 Create a mind-set of constant quality growth

4 Use comparison data to make short- and long-term business plans
5 Being aware of systems and procedures
6 Systems for systematic planning and implementing
7 A commitment from top management to follow benchmarking suggestions
8 Willingness of the organisation to break down current systems and culture
9 There are enough tools available
10 Staff innovation and the ability to come up with new ideas
11 Staff education and training that is relevant is in place
12 Information and research
13 Managing the process
14 Levels of satisfaction among workers are regularly measured and monitored
15 The existence of a good environment and culture at work
16 Human resources methods that are helpful

5.10 Benchmarking Models

An example of an organisation-based model was the earliest model developed by Camp (1989), which was applied at Xerox Corporation in a continuously looping process to ensure continuous improvement (see Figure 5.3). This model is also helpful in facilities management benchmarking. The Xerox method of benchmarking involves ten steps. They are:

1 Identify what is to be benchmarked
2 Identify comparative companies
3 Determine data collection method and collect data
4 Determine the current performance "gap"
5 Project future performance levels
6 Communicate benchmark findings and gain acceptance
7 Establish functional goals
8 Develop action plans
9 Implement specific actions and monitor progress
10 Recalibrate benchmarks

Camp's model contains planning or preparation, analytical, integration, and action phases.

5.11 A Model for Benchmarking Facilities Management

This model is comparable to Camp's Xerox model, which split the benchmarking process into four main parts. Camp's approach uses the planning, analysis, integration, and action processes to benchmark best practices. In contrast to the Xerox model, this framework has an additional function, one known as pre-planning. Benchmarking over other frameworks is implemented with the help of pre-planning, which includes determining the organisation's need for benchmarking and the challenges with benchmarking. It also illustrates how location and best practice benchmarking relate to one another. This area is crucial in developing countries because access to skilled labour depends on the location and might affect benchmarking. Convincing all parties involved in the benchmarking exercise that it will be effective also entails using communication throughout the benchmarking process. Camp's model was created for the manufacturing industry, but this framework applies to the FM discipline (see Figure 5.4).

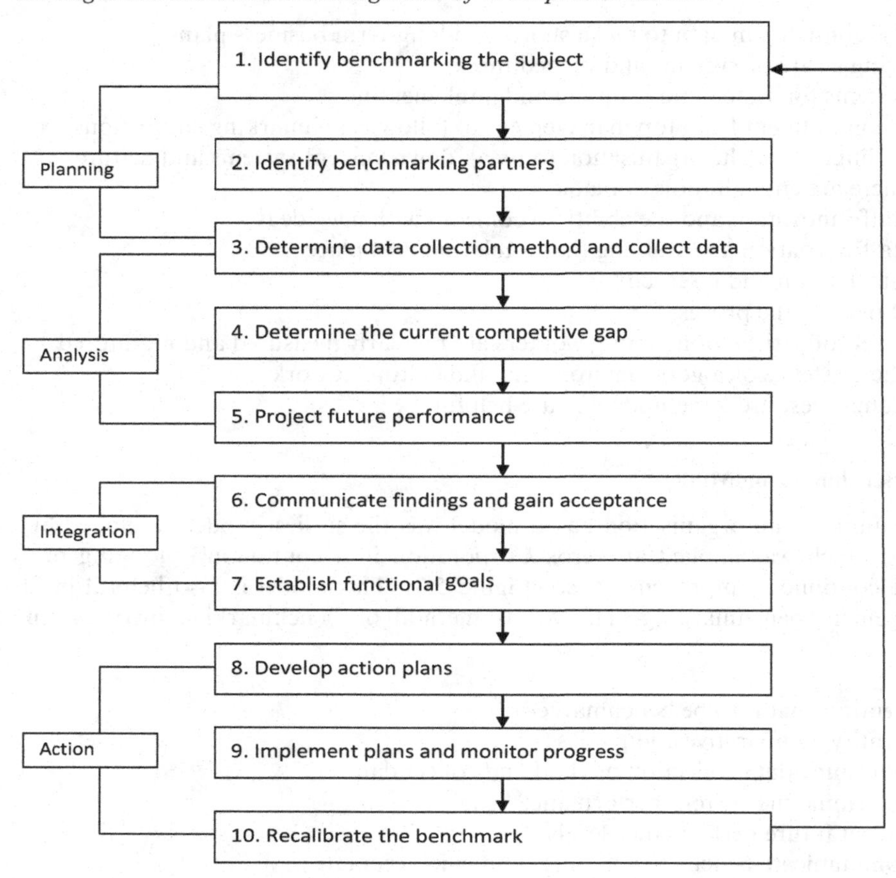

Figure 5.3 Camp's four-step model for benchmarking.

Source: Camp, R. (1989). *Benchmarking: The search for industry best practice that leads to superior performance.* ASQ Quality Press

***Pre-planning*:** The pre-planning process involves assessing an organisation's need for benchmarking and identifying benchmarking's challenges.

Determine if the organisation needs to do benchmarking: This entails assessing the organisation's existing status regarding benchmarking. FM usually has to be connected to an organisation's goals. Here, the facilities manager should decide if benchmarking is something that the organisation needs to do. Finding out what caused the benchmarking exercise is necessary. Was it a market share loss? Do you need to improve? One must decide what to accomplish before the activity begins to get the correct results and perform the training effectively.

Determine the benchmarking issues: The facilities manager must identify these issues. A facilities manager could run into problems such as staff resistance to change, a lack of knowledge of the exercise, trouble obtaining and comparing data from other organisations, and poor planning and execution of the movement. The location of the facilities manager has a considerable impact on the issues encountered during benchmarking. An organisation's resources should be used to fix the problems after the facilities manager analyses and

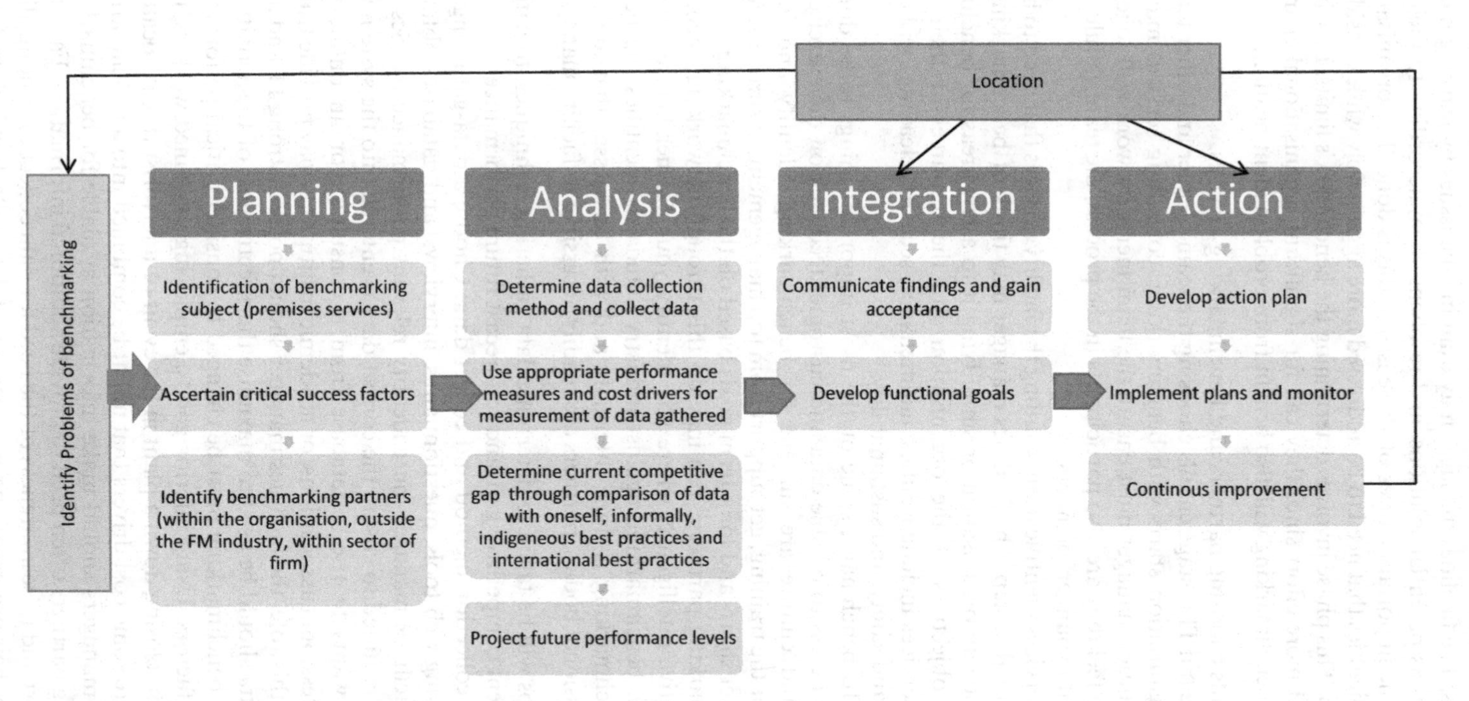

Figure 5.4 FM best practice benchmarking framework.

Source: Adapted from Adewunmi, Y. (2014). *Benchmarking practice in facilities management in selected cities in Nigeria*. PhD thesis, University of Lagos, Nigeria

prioritises the issues. The facilities manager may examine the issues by learning more about them, defining them as potential solutions, and resolving them according to their priority. After determining the important causes of these issues, this is done. The organisation's culture has specific difficulties that need to be addressed more aggressively while considering the expense of doing so. This phase involves determining the benchmark's limits. Benchmarking activities often need more effort since they call for significant organisational changes. The implementation of benchmarking will also benefit from problem identification.

***Planning*:** This entails choosing partners and benchmarking services.
Formation of the team: This stage entails choosing the team's leader and other team members. When the organisation starts with benchmarking, consulting services may also be used. Here, the facilities manager must ensure the team members work well together. For top management to feel they are also participants in the process, the team should include a balance of senior and younger members.

Benchmarking services identification: Finding the facility services that need to be benchmarked is the task of this step. The facilities manager may think of benchmarking all FM facility services inside an organisation or simplifying it to a few areas for benchmarking, depending on the objectives of the organisation and the resources at their disposal. Examples of FM services include utilities, maintenance, security, cleaning, real estate, IT equipment, health and safety, and sustainability.

Identify CSFs: To benchmark, some of the most important things to be done are to ensure that there is access to reliable data and customer feedback on the exercise, provide the environment and culture are suitable for benchmarking, identify and justify the resources needed for the training, get support from top management, ensure procedures are standardised, and set short- and long-term goals based on the benchmarking.

Choose benchmarking partners: You may do this through networking, recommendations, and competitive intelligence, where you identify your partner before approaching them. Additionally, benchmarking against industry standards identifies the best in the field. Further, benchmarking is done internally and with businesses that are not FMs. Through the professional bodies, partners are readily accessible to facility managers.

***Analysis*:** Here, steps will be taken to determine data gathering techniques, to gather data, to identify the present competitive gap, and to forecast future performance.

Determine data collection method and collect data: Choose a data-gathering strategy, then gather data using web tools, questionnaires, interviews, and databases obtainable by consultants. The facilities manager must identify relevant performance metrics and cost factors. Training is required to ensure the correct data is entered into the system when the facilities manager wants to use a database from a consultant or an online resource. Consultancies, professional associations, or academic institutions may provide training. To properly carry out this role, the facilities manager should possess strong IT and analytical abilities. Trusted data should be used to maximise the effectiveness of the activity.

Programmes for benchmarking must be changed to consider critical performance metrics and local cost factors. Using inferior goods, employing untrained workers, installing imported equipment, closing down a plant in the event of a terrorist threat, location costs, and other factors are some cost drivers that might be combined into a benchmarking programme. Facilities managers should make information available for benchmarking since likely benchmarking partners often need to provide additional information. Facilities managers should support and raise awareness of the need for data collection among all relevant stakeholders within their organisations and ensure that data collection is done for ongoing projects. Additionally, there is a requirement to routinely network with other facilities managers and organisations that will help participate in these meetings. To inspire others, facilities managers should publish the findings of their benchmarking exercise online; social media may be used to issue information. Each FM organisation should provide clear information and have business continuity strategies for data. FM professional bodies should help spread up-to-date details on the sector; they could also grant organisations that have made their activities transparent over the relevant period financial incentives to encourage data collection practices.

Determine the existing competitive gap: The facilities manager should determine the reasons behind the performance gap that the benchmarking research revealed. Determine the procedures that must be implemented next to ascertain if the lessons learned may be applied to the corporate culture. The firm's culture should be known to the facilities manager, who should also determine if it is well-defined. The benchmarking exercise helps define the corporate culture if one does not already exist. Since the corporate policy is published and may need to be developed in specific organisations, it is usually best to apply lessons learned there.

Estimate performance levels in the future: After determining the facility's performance gap, this step is taken.

Integration: This entails presenting results and getting support.
Communicate findings and get support: The facilities manager should provide benchmark results to the organisation, win support from it, and then project the financial advantage to win management support. Make sure both quantitative and qualitative benefits are captured. Customer or user satisfaction surveys are often used to measure the qualitative advantages. Value addition, resource efficiency, performance, and process improvement are some significant benefits of benchmarking that may be shared throughout the company. In addition, cost-benefit analysis may be used to measure intangible advantages indirectly through shadow pricing because not all variables are measurable. The facilities managers must thus show their progress towards benefits and get the support of top management for the benchmarking effort.

Establish functional objectives: Based on the findings of the benchmarking process, the facilities manager should establish practical and operational goals. Also, the objectives should be quantifiable to be readily monitored.

Action: This stage includes creating an action plan and putting it into practice.
Create a facility or organisation action plan: The facilities manager should make sure there are protocols and procedures in place so that, in the event of a change, the processes are outlined so that whoever comes in will fulfil their duties. There should be feedback that enables evaluation after milestones. The right individuals should be selected to do their roles and be required to do so correctly. There needs to be a distinct division of labour. This is accomplished by offering suggestions and establishing a deadline for the strategy. Milestones are necessary so that you can assess your deliverables. Because the industry might change, the system can be adaptive.

Implementing the strategy derived from the benchmarking exercise's findings: This is given priority, and the relevant tasks are assigned to various individuals with deadlines for completion. Here, new techniques learned from the benchmarking exercise are taught to task owners, and then the required steps are taken to narrow the performance gap between the controlled facilities. Phased implementation is possible since the necessary finances and resources could become available eventually.

A commitment to ongoing improvement should be made. This is done by keeping track of performance, calibrating or upgrading the benchmarks, including benchmarking into the system, using structural incentives to motivate better performance, and more. The facilities manager should establish clearly defined, structured incentives and punishments (Smart, 1998).

5.12 Protocols of Benchmarking in Facilities Management

A facilities manager must choose what should be benchmarked before beginning the process. Across all recognised FM services, a first-strike exercise can be performed. If a specific operational service component has been identified as problematic, a more thorough benchmarking approach, such as database benchmarking or benchmarking clubs, can be conducted to discover the issue more accurately.

Benchmarking clubs often evaluate their company performance in a private setting. The capacity to thoroughly assess the data, costs, and facility services is the main advantage of a benchmarking club. The club members may choose if using a particular facility's best-of-breed solution inside their organisation makes sense. However, the limitations of the peer group and the impossibility of recognising the best-of-breed within it are the drawbacks of benchmarking clubs. Another issue is obtaining accurate and relevant cost information from club members to ensure a like-for-like comparison. This is mainly because each participating organisation has a separate set of cost standards.

In database benchmarking, the consultant ensures that the client follows the standard process of examining a good database. Benchmarking a database depends on the consultant's ability to choose the appropriate peer group. Consultants must see how well they performed compared to their peer group's costs. The database should reflect the delivery methods and operators diverse organisations use to identify best-of-breeds. The most significant benefit of database benchmarking is how quickly a large peer group can be assembled to follow a standard protocol suited for comparison with the client's company to find best practices and start copying them.[11]

After deciding what will be benchmarked, the following step is to gather data, which depends on the service(s) that will be benchmarked but regularly involves the following elements.

5.12.1 Peer Group

Everyone in the peer group needs to be doing something similar. As an example, if comparing how well services are maintained, that means that all systems have to function at the same level of quality, and the maintenance schedule must be suitable for good performance. Also, everyone in the peer group needs to be motivated, and the activities used as a standard must be done in similar settings. For the peer group to improve performance, one must work with organisations in a different market sector or location (Bogetic & Fedderke, 2006). The age of the building, its location, and the range of engineering services are some of the other criteria that can be used to choose the peer group.

5.12.2 Space Measures

To compare apples with apples and like to like, it is essential to have a common base of space comparison. The standard measurement is typically the amount of square metres.

There are many standards of measurement of space, so it is essential to have measurement converted into a unified form of measurement to get accurate benchmarks. For example, the gross internal area should not be confused with the gross external area, nor should it be confused with the net internal area or net external area. Benchmarking the space use is important as it affects the premises costs, and the floor areas need to be known to compare costs.

5.12.3 Cost Measures

These are costs that have been audited, such as costs excluding VAT. These costs are presented based on the currency used in each country. If benchmarking is done between two countries, then a single unit of cost comparison should be done as exchange rates vary per country. In an organisation, they form the base data, such as the original invoices, which should preferably be available to the benchmarker so that any irregularities about misallocated bills may be found for like–for–like comparison; additionally, having data for a period of time can be useful for cost prediction purpose. Benchmarking costs and space are usually carried out during first-strike benchmarking.

5.12.4 Cost or Resource Drivers

These can influence facilities' costs and cause them to rise or fall above or below feasible levels regarding quality, quantity or unit price. They include:

- Age of the building
- Geographic location
- Extent of mechanical and electrical services
- Performance requirements in terms of delivery of communications, environmental conditions
- Local environmental circumstances
- Management efficiency
- Level of quality control
- High densities of occupation
- Economic resource drivers such as labour costs and local problems in purchasing consumables
- Operating conditions such as access to works, time of the day, or year

5.12.5 Performance Metrics

These relate to the service in question and are used to measure the performance of services. When benchmarking to get like-for-like benchmarking, it is essential to have a uniform basis of comparison using performance measures. The measures can be qualitative or quantitative and show how services perform in areas such as cleaning, maintenance, energy, customer satisfaction, etc. Examples are total maintenance costs per square metre, total number of unexpected outages over the past five data, total cost of cleaning per full-time equivalent, user feedback and complaints. The most commonly used parameters include cost per unit area and per capita (defined as Full Time/Whole Time Equivalents or per workspace). These are high-level measures; other metrics that might be used when the benchmarking is at a more detailed level can include cost per bed (hospitals), student (schools), copy (reprographics), meal (catering), hour (security), and piece (laundry). To benchmark performance, the systems need to be of similar quality regarding how they operate.

5.12.6 Labour Productivity and Labour Rates

For the same task performed in several contexts, labour input and productivity might change significantly. Labour rates related to national indicators like the National Living Wage increasingly determine labour rates. Still, it's essential to recognise that what happens in the economy, such as whether there is employment for specific FM skill sets, can cause differences.

5.12.7 Included and Excluded Items

The protocol must show the items included and excluded from the cost centre. In addition, the protocol should indicate in which cost centre the excluded items can be found. This process of highlighting those areas that often lead to misinterpretation is extremely useful when extracting the information from, for example, the Accounts Department (Bogetic & Fedderke, 2006).

5.12.8 Gathering and Analysing the Data

After collecting it, the data will be collated and processed to ensure it fits the agreed-upon measurement parameters and classification protocol. It is usual for the research to be done twice or three times before the data is finally in a state that can be used for benchmarking. After benchmarking, the results are shown in graphs, maps, and reports that make sense and are simple for readers to understand.

Resource Drivers Example from a Benchmarking Exercise in Eskom

- Performance requirements of terms of delivery of communications, environmental conditions
- Local environmental circumstances
- Management efficiency and level of quality control
- High densities of occupation
- Economic resource drivers such as labour costs and local problems in purchasing consumables
- Operating conditions such as access to works, time of the day or year

FIRST STRIKE BENCHMARKING – ESKOM

This is how the fundamental data acquired during a high-level study may be used to highlight places where something incorrect or missing needs quick investigation. It is helpful when taking into account space consumption and operating expenses. For the month under review, 54 work orders were reported, of which 16 were dealt with and cleared, leaving 18 orders in process and 20 task orders awaiting clearance. The monthly data below shows that most operating expenses in October came from carpentry (see Figure 5.5). Therefore, care must be taken to ensure that the expenditures are kept to a minimum.

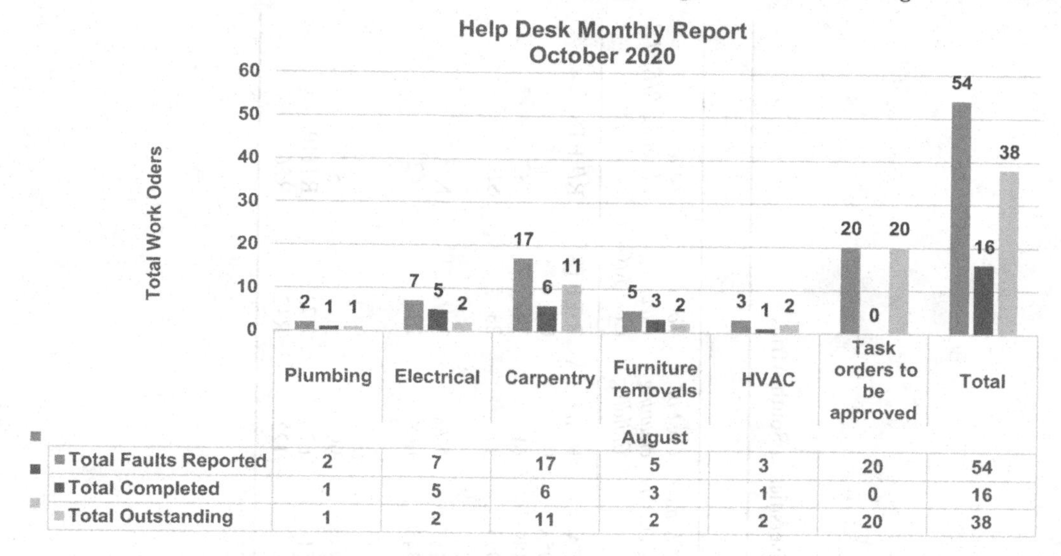

	Plumbing	Electrical	Carpentry	Furniture removals	HVAC	Task orders to be approved	Total
Total Faults Reported	2	7	17	5	3	20	54
Total Completed	1	5	6	3	1	0	16
Total Outstanding	1	2	11	2	2	20	38

Figure 5.5 First strike benchmarking – Eskom.

Source: Eskom

The monthly maintenance backlog and the total backlog up to this point are shown in Fig. 5.5. Unapproved task orders and delays in outsourcing work that cannot be completed internally are to blame for most of the backlog.

5.13 Benchmarking in FM Applications

The benchmarking applications discussed in this section are those used in Africa.

5.13.1 HEFMA

All Higher Education institutions in South Africa submit an annual benchmarking survey. The benchmarking carried out is at the first strike benchmarking level. The services benchmarked include maintenance, cleaning, security and energy. In the data collection exercise, the data collected include: the all-inclusive total floor area of all floors measured, non-assignable floor area (NFA): NFA includes mechanical floor areas, custodial floor areas, the total cost to erect a similar building at today's cost, building cost, professional fees, municipal costs and fixed equipment.[12] Examples of the information on the benchmark costs presented in the report are in Table 5.4 and Figure 5.6 for the participating universities.

5.13.2 IPD

These benchmarks are run through Invest Property Databank Limited Company (IPD) occupiers. The biggest independent real estate database, updated annually with new data, focuses on property performance measurement. It includes the IPD Occupiers Reporting Tool and Global Estate Measurement Standards for benchmarking. Using Global Estate

Table 5.4 Statistical Benchmarking Data on Cleaning and Waste Management for Selected Universities in South Africa

	Cleaning and Waste Management Service								
Name of institution	*Total staff salaries and wages*	*Cleaning Materials*	*General waste removal*	*Contaminated Waste Removal*	*Total Cleaning Expenditure*	*Area Cleaned from Central Fund*	*Cost of cleaning Building*	*Total Cleaning Cost R/GFA*	*Total Cleaning R/EFTS*
	R	R	R	R	R	M^2GFA	R/m^2GFA	R/EFTS	R/EFTS
UNISA	R489696	R0	R0	R0	R0	4896958	0	0	0
UP	R343445	R509800	R1232155	R70909877	R454677	4167489	61	79	982
STELLEBBON UNI	R100568	R208945	R4543443	R45454776	R451209	1165998	17	23	S10
TUT	R500989	R909087	R1098765	N/A	N/A	6345457	N/A	N/A	N/A
UNIVERSITY OF THE WATERSRAND	R200096	R307676	R1009878	R12658870	R547657	1223343	098	232	3624
UJ	R145677	R303212	R5565643	R45768678	R500909	6177899	99	90	1547
MEAN	1175617	6166164	2890988	R37600576	R556750	360151	67	R73	R1110
MEDIAN							122	R12	R265

Source: HEFMA

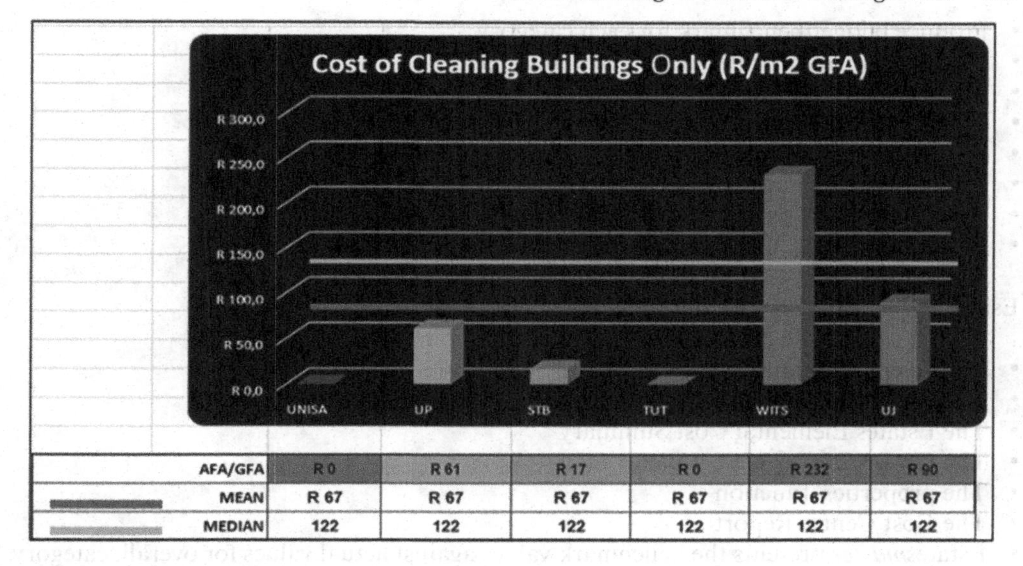

AFA/GFA	R 0	R 61	R 17	R 0	R 232	R 90
MEAN	R 67	R 67	R 67	R 67	R 67	R 67
MEDIAN	122	122	122	122	122	122

Figure 5.6 Cleaning data.

Source: HEFMA

Measurement Standards, customers may produce reliable performance data on their buildings anywhere on the globe. Through a secure online portal, the IPD Occupiers Reporting Tool provides users immediate access to many helpful reports and benchmarking alternatives. IPD Occupiers, which gathers information from Austria, Belgium, Central and Eastern Europe, Denmark, France, Germany, Ireland, Italy, the Netherlands, the Nordic area, Norway, Portugal, South Africa, Spain, Sweden, Switzerland, and the United Kingdom, has one of the biggest databases of properties in the world.[13]

The Green Building Council does IPD benchmarking in South Africa. The IPD South Africa Annual Green Property Index tracks the performance of Green Star Certified Prime & Grade A Office properties to the rest of the IPD Prime & Grade A Office.[14]

5.13.3 Estatesmaster

Estates*master* is a process- and performance-based benchmarking tool that employs normalised data, intelligent decision-making, and custom site-specific benchmarking tools. It is used to correctly assess whole estates utilising types of buildings and isolate single buildings as needed. International Facilities and Property Information Limited, in the UK, produces Estates*master*, a solution with global applicability for cost planning, monitoring, and management. It was developed by Prof. Bernard Williams and his team using information from over two million data entries, records, and more than 40 years of benchmarking consultancy.

Adaptation – The solutions

- Divide buildings into categories
- Answer the Estates*master* questions per category rather than individually if required
- Allocate the benchmark values to each building in each category by GIA

- Produce budget/benchmark for each category
- Produce budget/benchmark for each building
- The building data entry
- The questionnaires
- The Help texts
- Modern building categories
- Answering questions by Building Category
- Answering questions by Individual Buildings

Estates*master* model – The outputs

- The Overall Summary
- The Individual Building Cost Summary
- The Estates Elemental Cost Summary
- The Group Overview Summary
- The properties' function
- The Cost Centre Report
- Estates*master* presents the benchmark values against actual values for overall, category, and individual buildings

Applications of the model
The first-strike benchmarking: Estates*master* allows a glance comparison between budget, actual, and benchmark costs for different sites.

Sensitivity testing of service levels: It has a provision to react to sensitivity tests by using different service level agreements to compute appropriate benchmark and actual cost values immediately.

Global and national cost forecasting: It can show international performance levels.

Current services:

- Maintenance
- Cleaning
- Security
- Energy/CO^2
- Distribution
- Archiving
- Stationery
- Reprographics
- Catering[15]

The application:
In 2014, Prof. Bernard Williams researched the benchmarking model Estates*master* (www.estatesmaster.com) with Ayo Abolarinwa and Associates and a Nigerian institution, Covenant University. The institution had over 200 buildings totalling more than 200,000 m2 of the gross internal area benchmarked using the model to assess maintenance and cleaning/janitorial services.

Maintenance and cleaning/janitorial had similar results, showing that costs were much higher than planned compared to productivity levels at the highest levels achieved

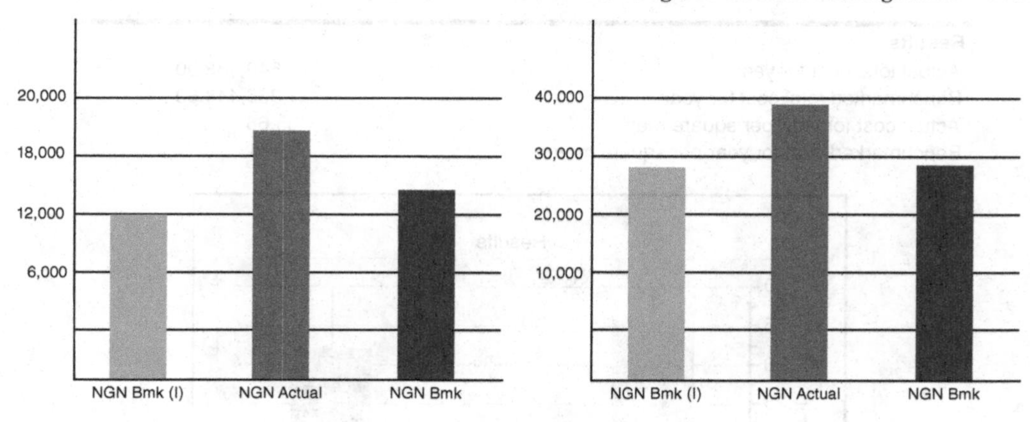

Figure 5.7 Sample cleaning benchmark result from the Nigerian benchmarking study for maintenance and cleaning services showing that costs were higher compared to productivity levels achieved globally.

Source: Ayo Abolarinwa and Associates

internationally. The case study results are shown in Figure 5.7. In both cases, the histograms represent the extreme of the benchmarked best-performance costs, while the centre histogram shows the actual costs.

The institution outsourced all of its labour and is well-known for its openness, contributing to the integrity of the original findings. They had a well-organised internal staff, and both services met "Fairly High" requirements regarding the quality plateau set by the Estates*master* model. Informal comparisons with these services' prices in other organisations were made. Although the Institution's facility productivity was somewhat low compared to the best performance worldwide, these informal comparisons imply that the expenses for comparable services in other organisations were much more significant. This led to the first conclusion that other businesses with internal personnel may be less effective and that external contractors may unnecessarily drive up the cost of local labour by a significant amount more than is reasonable in a usual commercial market. Nigeria's lack of reliable facilities benchmarking techniques is likely to blame for the inability to fulfil best performance cost and productivity objectives. The lack of maturity in Nigerian facilities management, which can cause a lack of control over productivity levels, is the primary result that accounts for the 40–50% productivity gap between a high-performing, insourced Nigerian facilities management system and that in best-performance international systems.

5.13.4 IFMA: Operation and Maintenance Benchmarks for IFMA

IFMA publishes annual benchmarks for facilities and has been doing that since 1987. The report has information on janitorial costs, maintenance costs, utility costs, occupancy and space costs, IT costs and facilities costs. The online survey was first done in 2023 in Africa. The benchmarking covers the operation and maintenance costs of janitorial, maintenance and utility costs. The cleaning, maintenance, and utility costs comprise a significant part of the total FM costs.

Results	
Actual total cost for year	1,546,488.00
Benchmarked total cost for year	1,385,444.69
Actual cost for year per square metre	35.66
Benchmarked cost for year per square metre	31.94

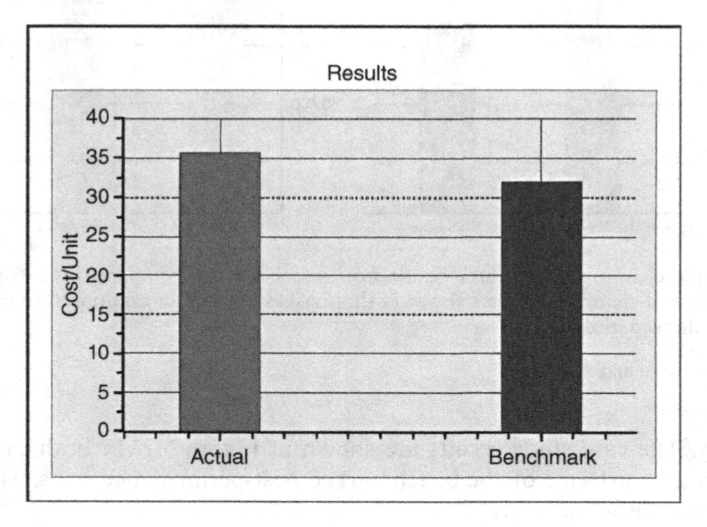

Figure 5.8 Estates*master*.

Source: International Facilities and Property Information Ltd

International Case Study on Estates*master*

The Estates*master* is an intelligent decision support system for benchmarking. The system has been in use for over four decades. Organisations have used it for facilities policy and budgeting, retendering and renegotiating service contracts, and global cost management. This system is only being developed in Africa and used for first-strike benchmarking, while the international version can be used for performance benchmarking.

The case study was to re-tender or renegotiate the services contract

A central Government Department in the UK deliberated whether to re-tender a major service contract covering five buildings or re-negotiate with the incumbent provider. Using the Estates*master* programme, they found they were paying close to the best performance for the provided scope and quality. Therefore, they re-negotiated the contract without unnecessary costs and disruptions on both sides.

5.14 Conclusion

The information obtained from benchmarking can be used to make decisions in strategic management.[16] Some of the other strategic benefits of benchmarking in FM are that it helps determine how to make the most significant use of facility resources, it improves FM practices and supports the organisation's position in the marketplace, and it helps organisations show top management performance relative to the others in the industry. A successful application of benchmarking includes comparison with having experience with the previous implementation, choosing the project to focus on and then finding partner organisations, working well with others from various disciplines, getting support from the top management, and committing to the budget. Internal, competitive, and generic benchmarking are three types of benchmarking in facilities management. This chapter discussed the benefits and problems of benchmarking as well as the models and applications of benchmarking.

5.15 Guidelines

- Assess an organisation's need for benchmarking
- Assessing the organisation's existing status regarding benchmarking
- Determine possible benchmarking issues.
- Formation of the benchmarking team
- Identifying benchmarking partners
- Benchmarking services identification
- Identify CSF to benchmark
- Determine steps that will be taken to determine data gathering techniques, gather data, identify the present competitive gap, and forecast future performance
- Programmes for benchmarking must consider critical performance metrics and local cost factors
- Facilities managers should network to facilitate benchmarking
- Facilities managers should publish the findings of their benchmarking exercise online to encourage others to benchmark
- Each FM organisation should provide information clearly and have business continuity strategies for data
- Facilities managers should have business continuity strategies for data
- FM professional bodies should help spread up-to-date details on the sector
- FM associations could also grant organisations that have made their activities transparent over the relevant period financial incentives to encourage data collection practices
- Determine the existing competitive gap
- Determine the organisation's culture to determine the organisation's benchmarking fit
- Estimate performance levels in the future
- Present results and get support
- Communicate findings and get support from the organisation
- Capture and present both qualitative and quantitative benefits
- Establish functional objectives that should be quantifiable and readily monitored
- Create a facility or organisation action plan by having the right protocols and procedures in place
- There should be feedback that allows evaluation of milestones
- Have milestones to assess your deliverables
- Implementing the strategy derived from the benchmarking exercise's findings
- Train task owners to use new methods learnt from benchmarking
- Implementation can be phased depending on the organisation's resources
- A commitment to ongoing improvement should be made
- Monitor performance and update benchmarks
- Having rewards and punishment structures in place to improve performance

5.16 Questions

1 Why is benchmarking important?
2 What is benchmarking?
3 What is the difference between benchmarking and performance measurement?
4 What are the types of benchmarking?
5 What are critical success factors for benchmarking?
6 What are the benefits of benchmarking in your organisation?
7 What are the problems of benchmarking in your organisation or country?
8 How would you go about the benchmarking process in your company?
9 What platforms are most suited for benchmarking in your organisation?

Notes

1 Loosemore, M., & Hsin, Y.Y. (2001). Customer-focused benchmarking for facilities management. *Facilities*, 19 (13/14), 464-–475.
2 Enoma, N. (2008). *Developing key performance indicators for airport safety and security: A study of three Scottish airports*. PhD thesis, School of Built Environment, Herriot-Watt University, Edinburgh, UK.
3 Smart, D.A.J. (1998). *Examination of benchmarking practices in South Africa firms*. Masters' thesis, University of Witwatersrand, South Africa.
4 Prašnikar, J., Debeljak, Ž., & Ahčan, A. (2005). Benchmarking as a tool of strategic management. *Total Quality Management and Business Excellence*, *16*(2), 257–275.
5 Adewunmi, Y. (2014). *Benchmarking Practice in facilities management in selected cities in Nigeria*. PhD thesis, University of Lagos, Nigeria.
6 Adebanjo, D., Abbas, A., & Mann, R. (2010). An investigation of the adoption and implementation of benchmarking. *International Journal of Operations & Production Management*, *30*(11), 1140–1169.
7 Helmy, E. (2011). Benchmarking the Egyptian medical tourism sector against international best practices: An exploratory study. *Tourismos*, *6*(2), 293–311.
8 Bogetic, Z., & Fedderke, J.W. (2006). International benchmarking of infrastructure performance in the Southern African customs union countries. *World Bank Policy Research Working Paper*, 3987.
9 Anand, G., & Kodali, R. (2008). Benchmarking the benchmarking models. *Benchmarking: An International Journal*, 5(3), 257–291.
10 Adewunmi, Y., Omirin, M., & Koleoso, H. (2015). Benchmarking challenges in facilities management in Nigeria. *Journal of Facilities Management*, *13*(2), 156–184.
11 Wauters, B. (2006). The added value of facilities management: Benchmarking work processes. *Facilities*, *23*(3/4), 142–151.
12 HEFMA. (2016). *Benchmarking report*. HEFMA.
13 Zhou, F. (2014). *International benchmarking in facility management - Comparison of different national benchmarking pools*. PhD thesis, Der Technischen Universität Graz, Austria.
14 IPD South Africa Green Property Index 2017.
15 Estates*master*. Benchmarking and Cost Prediction Programme.
16 Prašnikar, J., Debeljak, Ž., & Ahčan, A. (2005). Benchmarking as a tool of strategic management. *Total Quality Management and Business Excellence*, *16*(2), 257–275.

Bibliography

Adebanjo, D., Abbas, A., & Mann, R. (2010). An investigation of the adoption and implementation of benchmarking. *International Journal of Operations & Production Management*, *30*(11), 1140–1169.

Adewunmi, Y. (2014). *Benchmarking practice in facilities management in selected cities in Nigeria*. PhD thesis, University of Lagos, Nigeria.

Adewunmi, Y., Omirin, M., & Koleoso, H. (2015). Benchmarking challenges in facilities management in Nigeria. *Journal of Facilities Management, 13*(2), 156–184.

Anand, G., & Kodali, R. (2008). Benchmarking the benchmarking models. *Benchmarking: An International Journal, 5*(3), 257–291.

Bogetic, Z., & Fedderke, J.W. (2006). International benchmarking of infrastructure performance in the Southern African customs union countries. *World Bank Policy Research Working Paper*, 3987.

Camp, R. (1989). *Benchmarking: The search for industry best practice that leads to superior performance*. ASQ Quality Press.

Enoma, N. (2008). *Developing key performance indicators for airport safety and security: A study of three Scottish airports*. PhD thesis, School of Built Environment, Herriot-Watt University, Edinburgh, UK.

Estates*master*. Benchmarking and Cost Prediction Programme.

Francis, G., & Holloway, J. (2007). What have we learned? Themes from the literature on best-practice benchmarking. *International Journal of Management Reviews, 9*(3), 171–189.

Helmy, E. (2011). Benchmarking the Egyptian medical tourism sector against international best practices: An exploratory study. *Tourismos, 6*(2), 293–311.

HEFMA. (2016). *Benchmarking report*. HEFMA.

IPD South Africa Green Property Index 2017.

Loosemore, M., & Hsin, Y.Y. (2001). Customer-focused benchmarking for facilities management. *Facilities, 19*(13/14), 464–475.

Prašnikar, J., Debeljak, Ž., & Ahčan, A. (2005). Benchmarking as a tool of strategic management. *Total Quality Management and Business Excellence, 16*(2), 257–275.

Smart, D.A.J. (1998). *Examination of benchmarking practices in South Africa firms*. Masters' thesis, University of Witwatersrand, South Africa.

Wauters, B. (2006). The added value of facilities management: Benchmarking work processes. *Facilities, 23*(3/4), 142–151.

Williams, B. (2000). *An introduction to benchmarking facilities & justifying the investment in facilities*. International Property and Facilities Information Ltd.

Zhou, F. (2014). *International benchmarking in facility management - Comparison of different national benchmarking pools*. PhD thesis, Der Technischen Universität Graz, Austria.

6 Emergency Preparedness and Business Continuity

6.0 Introduction

Emergency preparedness and business continuity are concerned with crises and how to respond to such incidents when they happen. They involve preparing the right measures to deal with such incidents and getting critical functions back into operation as soon as possible. Since strategic planning involves predicting the future, having the right strategy entails having preventive measures for dealing with risk and disasters in organisations, which will preserve value in the organisation. Business continuity helps with strategy since it helps with planning and skills, is proactive, and helps identify vulnerable areas. Risk can prevent an organisation from operating, and the ability of operations to quickly return to normal when there is disruption will improve customer satisfaction. Integrating business continuity into strategic planning helps with organisations' long-term survival since they can respond to crises when they happen. Business continuity can be added through training, testing, maintenance, and updating continuity plans. Business continuity can be integrated into strategic plans since both focus on the long-term survival of organisations, minimise risk, reduce supply chain vulnerability, involve top management, involve the ability to recover quickly, comply with regulations, are embedded in the different functions of the organisations, and focus on the business environment and customers. The integration can be limited by cost, belief that organisations are not susceptible to risks, lack of trained personnel, and fear of change.[1,2] This chapter focuses on emergency preparedness and business continuity.

6.1 History of Business Continuity

As a concept that emerged in the 1970s, business continuity management focuses on technical systems, decision-making orientation, and prevention. It was developed due to the IBM 360 and Model 370, which provided organisations with integrated management information systems. The US Foreign Corrupt Practices Act (FCP) in 1977 introduced the idea that soft systems could cause crises, and the 1988 Illinois Bell Hinsdale central office fire prompted a more strategic approach to crisis management planning. The emergence of business continuity planning in the 1990s led to the development of business continuity law, Presidential Decision Directives, and standards for various sectors. The 9/11 attacks in 2001 in the US led to significant changes in business continuity management practices, including introducing laws, rules, and recommendations requiring businesses to have business continuity planning expertise. The internationalisation phase of business continuity management saw the emergence of competing norms and regulations covering regional or

DOI: 10.1201/9781032656663-7

national boundaries, as well as the founding of the US Disaster Recovery Institute (DRI) and the UK-based Business Continuity Institute (BCI).[3]

6.2 The Difference Between Emergency Preparedness and Business Continuity

Emergency preparedness is the facilities department's reaction to certain types of incidents. Since most of them are temporary, the responses are tactical. An organisation may need to respond to a more severe situation, which may take two or three days and involve other agencies like the local fire department and police. The emergency response strategy accounts for potential crises. It takes care of the event and the time immediately after the occurrence to restore important organisational activities to a minimal level.

On the other hand, business continuity plans are strategic and quickly restore the company to regular operations after an event. This strategy handles the results of a major incident and ensures that the organisation can go on and maintain long-term recovery. While regular services and activities are being restored, the plan must manage any physical damage and loss of productivity from an event. Emergency preparedness responds to emergency physical events, while business continuity handles business disruptions.

6.3 Essential Elements in Emergency Preparedness

The six essential elements included in any emergency response plan are:

- Assignments for emergency routes and emergency escape plans
- Procedures for workers to follow while performing (or stopping) essential plant activities before evacuation
- Procedures for employees to follow once an emergency evacuation has been completed to account for all personnel
- Rescue and medical responsibilities for staff who must carry them out
- The preferred means for reporting emergencies, including fires
- The names of the individuals or departments to be consulted about further details or an explanation of duties under the plan, as well as their specific work titles

Employers must also set up and install alarm systems that alert people to the need to take action by the emergency action plan or give people enough time to react to guarantee that everyone can safely leave the workplace or the immediate work area. Signals must be utilised for each purpose if the alarm is used to notify firefighters or for any other cause. The business must also hire and educate many individuals to help evacuate workers and inhabitants before implementing the emergency action plan.[4]

6.4 Emergency Preparedness in FM

Most companies face six main areas of risk regarding security: Workers, tools, data, information and knowledge, information systems, public relations, and grounds.[5] Facilities management creates, supplies, and takes care of a workplace so that a company can reach its primary business goals. It also includes handling the risks that a company faces on its premises. However, a building may be at risk because of its position, form, construction, or how it is used. For example, going through the front door of a property could be dangerous. Many sites are in places with many people, which makes them vulnerable to security problems. A chemical attack could also make a building dangerous in minutes. It might be hard to keep things safe around construction materials and equipment in a building prone to fire, and explosions can also be a risk. Data records in information systems can get damaged. There are risks when a business has facilities in more than one building or location, and those facilities may include different areas for non-core services like parking, drinks, dining, entertainment, leisure, and relaxing. There may also be dangerous chemicals in the buildings, and anyone who needs to get into them could be in danger. There could be trouble if attackers make threats. The facilities manager's job is to ensure safety measures are put in place, checked, and regularly updated.[6]

Facilities can't know when events will happen, but they can get ready for them. Disasters can occur in a building, like when the building systems don't work, or someone does something illegal. Ideally, a facilities manager would have a detailed, step-by-step plan for all situations. Being ready for disasters is an essential skill for facilities managers. According to the Royal Institution of Chartered Surveyors' (RICS) description of the facilities manager's job in health and safety, they are responsible for safety and security, acts of terrorism, environmental disasters, violence in the workplace, chemical or biological events, pandemic emergencies, and data protection. They must also ensure the building's construction and tools are up to code.[7]

As part of disaster planning, including running emergency operations centres, evacuation plans are made and updated. These strategies are needed to ensure an organisation's processes go smoothly. Before putting the building through a disaster assessment, facilities managers should take a tour and review any design or infrastructure faults again. In this way, the facilities manager can plan for evacuating the building. There must be references to business continuity in all Business Continuity Plans.[8] These plans must be firm and offer specific services and support to internal users. A disaster recovery plan is vital for a business to stay open after a disaster. To save lives and keep a company running, planning how to handle emergencies is essential. Without these critical chances for employees to learn first-hand, saving lives and improving recovery would be more complicated or impossible. It would be helpful for FM staff to have more information, best practices, and disaster plan forms. Top management's commitment to the strategy, the ideas, the answers, and the funds needed to carry out the project are the most essential parts of putting emergency plans into action. The FM team needs to communicate these concerns to top management and their boards until they are sure they are ready for a disastrous building loss (Hardy et al., 2009).

Facilities experts and executives need to make business continuity models to meet business and disaster preparations. Facilities managers today work in various complex and unique settings and are responsible for managing virtual offices, offices with multiple sites, and sometimes many different groups. A business continuity model lays out how a company will handle interruptions and develop good ways to get back to normal after they happen.[9]

6.5 Benefits of Emergency Preparedness in FM

- Makes planning, responding to, and recovering from disasters easier. Information is used to implement emergency recovery plans, keep track of lost property, and make claims
- Keeps correct records and is kept up to date with everyday activities
- Gives the information needed to make prompt choices, reducing downtime as much as possible
- Gives information on how to get better insurance coverage[10]

6.6 Problems of Emergency Preparedness in FM

- These are barriers to interactions with people due to their language, ability to read and understand communication materials such as newsletters.
- Outdated plans may affect preparedness standards
- Assessing effects and requirements may be difficult due to the scale of the disaster, which can cause wide disruption in services
- Another problem is accessibility, as managers in specific locations may struggle to access resources and transportation. Also, evacuation can be a problem caused by the scale of the event. Some employees may also not cooperate in the relocation of people from the site in which the event took place
- Health and safety problems caused by poor sanitation during the event
- Rehabilitation and recovery may be complicated and costly
- Problems from improper documentation of roles and responsibilities
- Inadequate coordination in disaster management may cause substandard and variable levels of preparedness since the facilities manager has to work with other departments and organisations that may have different preparedness criteria and priorities
- Insufficient resources, such as equipment that can be used in the process
- Lack of proper training for disaster management will cause low preparedness standards

6.7 Business Continuity

Over a decade ago 5% of business continuity management (BCM) professionals in Africa worked in the facilities management field. Business continuity was mainly used in large organisations and the financial services industry (KPMG, 2013). During and post-pandemic the relevance of BCM has increased.

6.7.1 Problems With Business Continuity Planning in Africa Before COVID-19

- In Africa, two-thirds of organisations had a BCM programme, but only about half evaluated or implemented most BCM components. This suggests that organisations needed more time to react to an interruption in operations in a coordinated manner or have suitable recovery options.
- IT disaster recovery was prioritised, just like in other parts of the world. African business leaders focused on IT and data disruptions. IT disaster recovery was the most widely used BCM component, showing how reliant organisations were on complex information systems that they could not afford to have fail.
- Business continuity planning was often only partially implemented. Many African organisations didn't conduct business impact analyses to identify crucial business procedures and suitable recovery plans. Many of these organisations lacked Business Recovery plans or Crisis Management strategies. Additionally, BCM education, awareness, and performance assessment must be improved.
- The adoption of BCM standards was low. Fewer than 2% of African organisations had their BCM programme accredited, and less than 6% had plans to approve their BCM programme. Furthermore, 60% of these organisations needed to follow a BCM standard to promote the execution of the BCM programme. Implementing standards helps African organisations follow global business continuity best practices and reduce the weaknesses in their BCM systems.[11]

6.8 Disasters

Disaster management at the regional, social, and in-use levels can all be used to determine how bad a disaster was. If disasters can't be handled and controlled, they will continue to significantly affect the world. Since environmental and natural disasters are happening worldwide, disaster prevention is integral to solving problems. Disasters can be dangerous for a country because they can destroy its current assets, take resources away from growth, and make it take longer to recover. This shows that countries need a complete plan for dealing with disasters that includes ways to stop them, reduce their effects, get prepared, respond, recover, and develop new things that can help in case of a disaster. Another thing they should do is go over every part of the disaster management cycle. It would help with handling disasters if we looked at more than just what happened after they occurred.

Earthquakes, storms, volcanic events, tsunamis, wildfires, floods, landslides, and drought are just a few of the natural disasters people worry about. Additional dangers, like terrible accidents, are also a possibility. These events keep killing people, destroying communities and the economy, and harming the earth. Also, new disaster risks that have grown because of more crime and violence in society, like carjacking, terrorism, unrest, and war with regular weapons can be to blame. These have sometimes put too much pressure on society and governments, putting more strain on foreign aid funds and reducing the world's ability to respond to disasters. New threats could come from dangerous drugs or materials, and nuclear and atomic sources could be causes for concern. The disaster site itself could also be hazardous. Disaster risks can come from the modern loss factor, which is the link between current disaster threats and the losses they can cause.

Heavy floods in the Democratic Republic of Congo and Rwanda in 2022, which killed at least 574 people, are an example of a disaster. In South Africa, 544 people died in flooding, and 87 people died in flooding in West Africa, also in 2022, with 63 of those deaths happening in Nigeria. In 2023, an earthquake with a magnitude of 6.8 hit Marrakech, Morocco. It killed over 3,000 people and destroyed many houses and infrastructure. A terrorist group called al-Shabab struck the Westgate Mall in Kenya in 2013 (World Meteorological Organisation, 2024; Institute for Security Studies, 2023; Voice of America, 2023). More than 150 people were hurt, and 67 were killed. In South Africa, strikes happen over things like trade union problems, unemployment and injustice, the use of foreign workers, the effects of apartheid, and the fear of job cuts. Nigeria often experiences power cuts because the national grid has been broken down by waste and insufficient investment in the power sector. The Famine and Early Warning Systems Network (FEWS NET) says that the Horn of Africa has been through some of the worst droughts in history for almost three years. Since the end of 2020, Ethiopia and Somalia have had five failed rainy seasons, which has forced many people to move and has killed animals. Many people around the world still remember the 9/11 attacks, when, four planned Islamist suicide attacks were carried out against the US by al-Qaeda extremists. Another event was the COVID-19 pandemic, caused by the severe acute respiratory syndrome coronavirus 2 (SARS-CoV-2). This event is also sometimes called the coronavirus pandemic. It was during an outbreak in Wuhan, China, in December 2019 that the new virus was found. There were efforts to stop it, but they failed. This let the virus spread to other parts of Asia and then to the world by 2020. It was called a PHEIC (public health emergency of international concern) by the World Health Organisation (WHO) on January 30, 2020. Figure 3.1 shows the types of disasters, threats of disasters, and disaster countermeasures.

Types of Disasters	Threats of Disasters	Disasters Countermeasures
Fires. Floods. Wind/Rain/Snowstorms. Earthquakes/Hurricanes/Tornadoes. Structural/Roof collapse. Power failures. Hazardous/Toxic chemical and vapour release. Elevator breakdown HVAC failure. Telephone/Telecommunications failure. Crime/Bomb threats/Terrorism. Transportation accidents, including air, rail and road. Data and information storage and retrieval collapse. Medical/Health emergencies. Loss of life.	Injury, Damage to and destruction of property, Damage to and destruction of the environment, Disruption of production, Disruption of lifestyle, Loss of livelihood, Disruption to essential services, Damage to infrastructure and disruption to systems, Economic loss, and Sociological and psychological after-effects.	Warning measures Education and awareness Building regulations and standards Planning Risk assessment Facility survey Relocation and evacuation Monitoring systems Response measures to emergencies Providing resources Health and safety Effective organisational emergency services Training in handling the impact of speficic hazards

Figure 6.1 Types and threats of disasters and disaster countermeasures.

Source: Author

6.9 Emergency Preparedness Legislation

For disaster reduction and development planning integration, national legislation is both required and needs to be revised. African parliaments approved many new disaster management legislations in the 1990s and 2000s. The rules and regulations considered essential for Disaster Regulatory Management for each country in Southern Africa are included in Table 6.1. However, certain countries, like Botswana, Eswatini, and Lesotho, lack national disaster management laws. These countries have made considerable progress in addressing disaster risk concerns through national coordinating mechanisms.[12]

In Nigeria, the National Emergency Management Decree of 1999 is the relevant legislation, while the National Emergency Management Agency (NEMA) is the national agency. The National Disaster Management Organisation Act 517 of 1996 is one of the laws enacted in Ghana, and the federal agency is known as the National Disaster Management Organisation (NDMO). The National Disaster Management Bill (2002), the National Policy on Disaster Management (October 2001), and the National Disaster Management Agency (NDMA) are all now under consideration in Kenya (International Federation of Red Cross and Red Crescent Societies, 2021).

6.9.1 Additional Legislation

According to a 2016 Business Continuity Institute (BCI) study, these issues come under the remit of occupational health. The six professional practices outlined in the BCM life cycle are the foundation of the 2013 BCI good practice standards, representing an international best practice.

Table 6.1 The Guiding Legislations for Disaster Regulatory Management (DRM) in Southern Africa

Country	*Legislation Guiding DRM*	*Guiding Policy Documents*
Angola	Basic Civil Protection Law of 2003, as amended by Law 14/20 of 22 May 2020	Presidential Decree No.29/16 of 1 February 2016 approving the national plan for preparation, resilience, response and recovery from natural disasters for 2015-2017 and Presidential Decree No.30/16 of 3 February 2016 approving the strategic plan for Disaster Prevention and Risk Reduction
Botswana	None	National Policy of Disaster Management (1996) National Disaster Risk Management Plan (2009) National Disaster Risk Reduction Strategy 2013-2018 of 2013
Eswatini	Disaster Management Act 1 of 2006	National Emergency Response, Mitigation and Adaption Plan 2016-2022
Lesotho	Disaster Management Act 2 of 1997	Multi-Hazard Contingency Plan 2015-2018
Malawi	Disaster Preparedness and Relief Act of 1991 (Chapter 33: 05 of the laws of Malawi)	Disaster Risk Management Policy (2015)
Mozambique	Law on Disaster Risk Reduction and Management (Law 10/2020) and Regulation approving the Law on Disaster Risk Reduction and Management (Decree 76/2020)	National Policy on Disaster Management 1999 National Disaster Risk Reduction Master Plan 2017- 2030 of 2016
Namibia	Disaster Risk Management Act 10 of 2012 and Disaster Management Regulations of 2013	National Disaster Risk Management Policy 2009 National Disaster Risk Management Plan 2011
South Africa	Disaster Management Act 57 of 2002	National Disaster Framework 2005
Zambia	Disaster Management Act 13 of 2010	National Disaster Management Policy 2015 Disaster Management Operations Manual 2015
Zimbabwe	Civil Protection Act (Chapter 10:06) of 1989	

Source: International Federation of Red Cross and Red Crescent Societies (2021)

Acts similar to this may be found in Nigeria, such as the Nigeria Security and Civil Defence Corps Act (2003). They help with crowd and traffic management, provide emergency medical services, including first aid, and distribute emergency supplies to victims during crises. Their duties include protecting and rescuing the civilian population during an emergency, executing rescue missions, and controlling egregious incidents. They also constantly monitor all federal, state, and local government sites, infrastructure, and initiatives. This includes for any individuals engaging in a riot, civil disturbance, insurrection, or social unrest, who should be investigated and turned to the Nigerian police for further examination. They give the required warnings when there is a risk and remove the civilian population from danger. They help restore and maintain order in troubled places during national emergencies.

The provisions of the Red Cross Society of Nigeria Act (2004) include:

- In times of war, the society provides volunteer assistance to non-combatants who are sick or injured in the armed forces, as well as to prisoners of war and civilians who are suffering from the effects of war, by the spirit and terms of the conventions outlined in the Schedule to the Act.
- The country that agrees to the agreements above shall perform all the obligations of a national society. In war or peace, efforts should continue to enhance well-being, protect against illness, and reduce suffering.

Under the authority of the Minister and the Act that established it, members of the Fire Service Act (2004) perform any other humanitarian or other tasks that may be required of them in addition to their primary duties of putting out, controlling, and preventing fires, keeping people safe, and protecting property. They are to reduce the number of fires in the country and encourage interagency cooperation in distributing and maintaining fire prevention equipment.

In Ghana, there is the Ghana National Fire Service (Act 537, 1997) that the Ghana Red Cross established to manage and prevent fires, and Act 10 (1958) for disaster relief.

6.10 Building a Business Case, Evaluating Risk, Selecting a Strategy, and Developing a Plan

6.10.1 The Stages in Business Continuity Planning

There are five stages in business continuity planning.[13]

Stage 1: A business continuity strategy is required, which defines the scope of the task and what resources are needed to finish it.

Stage 2: An impact analysis is required, which looks at how vulnerable the business might be to unforeseen interruptions, how long the consequences might last, and how much revenue the business could lose.

Stage 3: A risk assessment is required to assess the likelihood of an event on the business.

Stage 4: A plan is made to map out the steps that might need to be taken to protect the business from the threat that has been discovered.

Stage 5: The plan must be communicated to ensure that everyone who needs to know about it does and that it is tested and kept up to date.

6.10.2 Recovery Strategy

The recovery strategy shows how to go about planning recovery. It tells those making recovery plans "what" they need to know. The "how" comes from each person's ideas. An approved method doesn't do anything that goes against the company's plans for recovery. A recovery plan is used to get back to a level of service that is at least acceptable for the business. This basic level of service lets the company plan for a long-term recovery while still giving its customers goods and services. Individual recovery plans will be used for other parts of the company.

Disaster recovery planning, a part of business continuity planning, can make it less likely that a major disaster will destroy an organisation. Business continuity is another way to run a company to handle "in-process disasters" and keep working. Disasters that destroy buildings don't happen very often. Instead, process-related local disasters happen more often. Business continuity adds value to a company by making backup plans for when a critical business function needs to be stopped. It also forces the company to look at its most essential processes and reorganise them so they can get back to normal quickly. Using more straightforward methods is cheaper, more reliable, and more effective. The emergency recovery plan is better because of the business continuity plan.

6.10.2.1 Selecting a Recovery Strategy

The goal of a recovery strategy is to get what was lost in the event of a disaster. It determines the future costs and capability of the overall programme. It involves considering the organisation's people, facilities, systems, and equipment. All plans for recovery need to consider the resources that would be used to recover what was lost so that the organisation can carry out its essential functions and how quickly these resources can be made available for critical functions to work again. These plans should be written to fulfil the recovery time required and the solution selected. A poorly chosen strategy will need all procedures to be rewritten when replaced. A recovery strategy must consider time and expenses. The recovery time objective (RTO) and its Business Impact Analysis (BIA) determine how the resources can respond to time demand. Rapid recoveries are often preferred until the initial and ongoing costs are complete. The RTO happens from the instant the incident occurs. Time spent on declaring a disaster is time lost toward the goal of recovery. Organisations must develop a different recovery strategy for each essential area and situation:

- *Information Technology*. Getting back up and running with a data hub, internal and external network links, and phone service.
- *Work Area Recovery*. Getting back a place for workers to work safely connected to the returned data centre and having a computer, phone, printer, and other tools.
- *Pandemic*. During a public health situation, it could take 18 months or longer to keep the business going
- *Business Continuity*. Maintains the flow of goods and services to customers even when essential parts of the organisation's processes go wrong.
- *Manufacturing*. Getting the flow of goods back on track after a crisis.
- *Call Centres*. Keeping in touch with the customer during the trouble.
- During the recovery project, the recovery strategy must be communicated, and all team members must know how to achieve the organisation's recovery timeline most cost-effectively. It's the start of every recovery plan.

6.10.2.2 Work Area Recovery Strategy

A work area recovery strategy is essential to planning for business continuity and disaster recovery. In a disaster, a group can move to this place. It's critical to think about the basic needs of the alternative work location and set selection goals based on what's vital to the company.

The manager should ensure the new site has desks, office furniture, and services that are easy to set up. Employees who do essential work for the company must return to work immediately. Moving to various locations is easier when special transportation is used. Hotels should be used for other things besides large-scale recovery from work. Hotels want to save money like any other company and don't want to pay extra monthly for internet capabilities.

Creating a work recovery site needs:

- A budget for the move
- A setup that would not take time
- A location far enough away not to be influenced by the disaster.
- Office furniture
- Provision of comfortable facilities for the amenities
- Location of complementary office functions in the same place for better interaction
- Employees' backup data should be provided
- Alternative communications for systems
- Documents for the business

6.10.2.2.1 RECOVERY OPTIONS

A Cold Site

A cold site is a space an organisation moves to during a disaster and which is empty with a short-term lease. This would need equipment, furniture, and an internet connection to set it up. Cold sites are often the least expensive recovery alternative but can be expensive due to the setup and equipment costs.

A Warm Site

It is a space that the organisation moves to which also is part of the organisation. It allows for maximum control of the recovery effort, testing of the plan, and employees. Some organisations split their operations so the facility can cover the essential functions in a crisis. A warm site is a work site with some of the required equipment but also requires some preparation before a disaster. That preparation takes time. Such spaces have security and phone and internet connections available for immediate use.

A Contracted Hot Site

A contracted hot site essentially has the full functionality of the organisation. Facilities include complete computer setups, system access, and recent systems backups. While a contracted hot site is convenient and can be ready in hours, it is likely the most expensive alternative. Organisations that need immediate availability, such as utility companies, financial institutions, and government agencies, use hot sites. These organisations work with agents to contract for a set number of available seats and frequency of data syncing. In a large-scale disaster, this space may be far away since other subscribers may have already taken over the closest recovery sites.

Table 6.2 Example of a Work Area Recovery Strategy for an International Organisation in Kenya (Hypothetical)

Recovery options	*Option 1 Head Office*	*Option 2 Outside Head Office*	*Option 3 Outside Nairobi, Kenya*
Use site when	There is a local incident - A new office relocation site	This happens when people cannot temporarily access the Head Office. Location 1 Location 2 Location 3	An incident that needs relocation outside Kenya Location in Dubai Location in Germany

Source: Author

6.10.2.3 Pandemic Strategy

A pandemic emergency plan should say that the organisation's behaviour should keep up to a level that lets it stay in business. Something must be done to stop the spread of illness inside and outside the company. The company will have to pay more because of steps taken to prevent the spread of diseases. A pandemic plan lasts for 18 to 24 months.

Due to a severe acute respiratory syndrome (SARS) outbreak in Toronto in 2003, the WHO asked that anyone who didn't need to go there to cancel their trip. The city lost many conferences, so hotel usage rates dropped to half what they usually were. Even though there weren't many confirmed cases of SARS, it cost many funds. In 2020, COVID-19 broke out in Wuhan, China, and affected every part of the economies of many countries.

Pandemic emergency steps require different strategies for essential stakeholders:

Employees

- The organisation's sick policy must be relaxed so that sick people are not forced to enter the workplace
- Anyone who is sick is encouraged to stay home, and they should also stay home if they have a sick family member
- Areas used must be cleaned thoroughly often to reduce infection from outside
- Employees who travel to locations with a high rate of pandemic infection should work from home for the first week of their return
- Access to medical aid for those who need treatment
- Providing support groups for those who need psychological and emotional support
- Report to immediate supervisor if they feel they are infected

Managers

- Regular screening and checks for those who may have been infected
- To reduce infection rate from outside, managers should ensure that areas, where users access the facility, are thoroughly cleaned
- Provision of isolation facilities for infected staff
- Risk assessment to control the spread of disease
- Training of staff on the suitable protocols and use of PPEs
- Provision of required budgets for PPE and essential equipment
- Managers should show appreciation to those who comply with the protocols
- Communicate with the employees regarding their responsibilities
- Have and enforce that there are standards for ventilation to reduce the spread of disease
- Managers should provide complimentary hand sanitation for users at all entrances
- It may be essential to bring individual sanitation supplies for an extended time
- Managers should consider reducing the number of staff that come to work at once
- Managers should ensure that customers sanitise all returned products

Vendors

- Videoconferencing and other electronic tools are used to meet vendors
- Carefully choose meeting places with a low incidence of infections

An example of a pandemic strategy from a public sector organisation in Windhoek, Namibia

The Pandemic Emergency Plan is designed to control the potential spread of disease within the organisation.

- It starts when the state public health authorities where headquarters are located declare a pandemic emergency.
- Limitations on the number of sick days offered to each staff in the organisation's sick leave policy are suspended.
- Employees are encouraged to stay at home with ill family members.
- Managers must thoroughly sanitise all areas where employees are in close physical contact with customers or vendors daily. Each employee is to be provided with personal sanitation gloves and face masks.
- Any employee returning from a business trip to an infected area can work from home through a VPN for several days before entering the office.

6.10.3 Business Continuity Strategy

A business continuity strategy helps a company keep running so people don't notice when essential functions stop working. It also gives the organisation the tools to quickly and easily return to normal. There might be a break in service, but keep it short. Depending on how long it takes to get the organisation back up and running, fix and replace machines and systems, and get any buildings harmed by the loss event back to work, this business interruption could last a while. FM is an integral part of business continuity planning for business survival. It can begin before or after a disaster or event. In this plan, it's essential to consider the danger the business faces during an incident. The company's assets should also be protected during an unplanned incident. Using backup tools and skilled workers cuts down on each risk point.

There are five initial steps in the business continuity planning process:

- Initiation of the business continuity planning process and the assessment of risks
- Conduct a business impact analysis
- Analysis of critical processes linked to the identified key business impacts
- Creation of an initial business continuity plan
- Testing and revision of the business continuity plan and management system.
- Updating of business continuity plan and management arrangements

Requirements for developing a business continuity plan:

- When making a business continuity plan, consider who, what, where, when, why, how, and how much. Be aware of the area that needs the most attention.
- Finishing up core and business support services, focusing on the differences between in-house and external tasks as well as tasks that would usually be in different groups.
- Planning or ways to communicate.
- Be careful not to cover too much of what is doable at the expense of what is essential or unstable; focus on what is inconsistent.
- Protect real-time communication lines with a backup plan.
- Write down goals for business continuity, such as how long it will take to return and how well it will work.
- Take care of your health first.
- Make a detailed (in order) imagined situation to help think of ideas and plan things.

- Focus on options that make sense and are specific enough to support the planning.
- After making the initial business continuity plan, the next step is to create a plan for how it will be implemented and managed.
- The ability to use it on the day of the event and during a crisis so people can get around and deal with the shock and other problems that came up because of the event all at the same time.
- Who makes the choices, how does the business continuity plan work, and when does it start?
- Who is going to back up? Does a single person make the choice? If so, what is the instant chain of communication? How and by whom is an open plan selected?
- How and by whom can this be read when the real world changes significantly from the imagined situations used to make the plan?
- What can be planned for after the event, and what needs to be done right now (this depends on how big the event is)?
- Do you believe your team and know what the most critical business problems are?
- Set aside the resources and determine what the business continuity plan will cover.
- The mental "space" to do this well under pressure is needed.
- Figure out how a situation and the BCP might affect the staff, current projects, and their ability to access tools, supplies, paperwork, and storage.[14]

Example of business continuity strategy for power failure in an office building in Lagos, Nigeria
The following structure was implemented to ensure the recovery time objective (RTO) set in the organisation's business continuity plan are achieved and to minimise interruption to business activities.

Power structure in the office building
Total Consumption Load = 72.017kW
Real power (P) =72.017kW
Power factor (PF) = 0.8
Apparent power(S) = Real power (watt)/ (1000xPower factor)

$$S_{(kVA)} = P_{(W)} / (1000 \times PF)$$

S= 72017/ (1000 x 0.8) = 90kVA
Apparent Power(S) = 90kVA

Total Consumption Load per day =1,667.938kWh/day

An emergency backup generator was sized to serve the facility's power needs. It has the following components (see Figure 6.2).

An automatic transfer switch (ATS) is a self-acting, intelligent power-switching device (panel) governed by dedicated control logic. It ensures electric power is delivered continuously from one of two power sources to a connected load circuit.

Cut-out fuse – Range 400 amps. This fuse unit is installed for various domestic and commercial purposes for the smooth passage of electricity and protects against various hazardous accidents.

Maximum Demand meter (MD) that monitors the consumption load in Power Distribution systems.

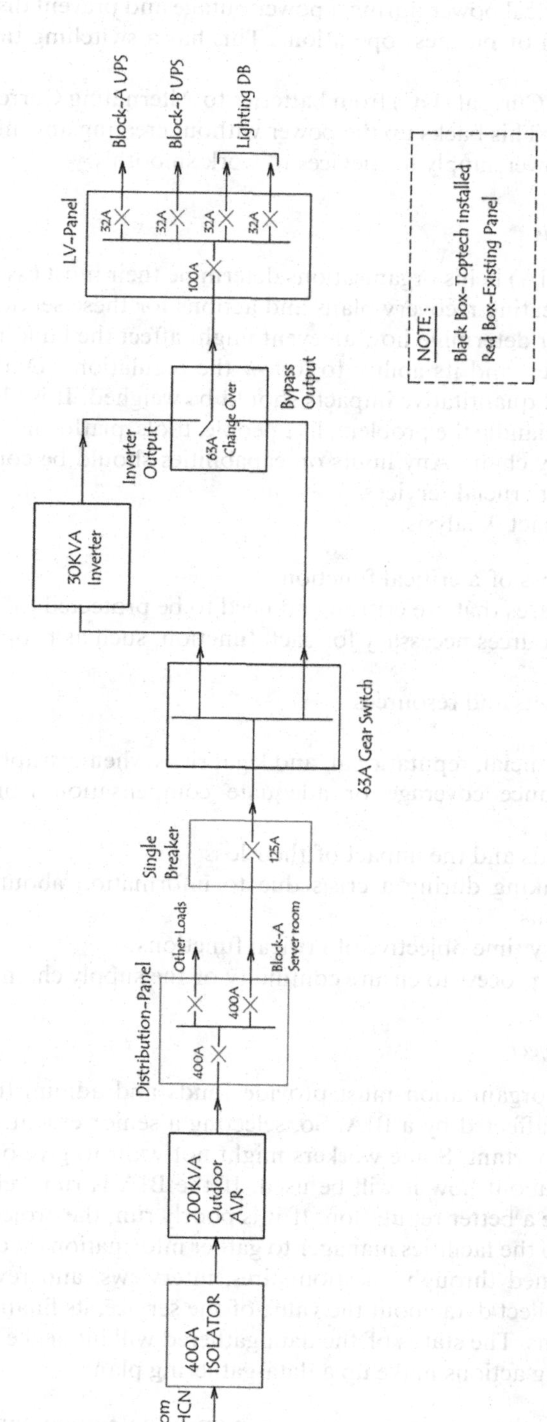

Figure 6.2 Power structure in an office building.

Source: Anonymous Source

Generators: Supply electrical power during a power outage and prevent discontinuity of daily activities or disruption of business operations. This has a switching time of around 2 minutes.

Inverter: Converts Direct Current (DC) from batteries to Alternating Current (AC) while storing power in the battery. This backs up the power without creating any interruption.

UPS: Offers a backup power supply for devices to work smoothly.

6.10.4 Business Impact Analysis

Business impact analysis (BIA) helps organisations determine their most essential services. It is also the first step in creating recovery plans and actions for these services. This study can help a facilities manager determine how an event might affect the building's processes, its ability to provide services, and its ability to follow the regulations. During the whole review, both qualitative and quantitative impacts should be weighed. It is also essential to list the resources needed to handle the problem, like people, tools, platforms, infrastructure, information, and the supply chain. Any limits on capabilities should be considered when determining how they affect crucial services.

Benefits of Business Impact Analysis:

- Quantifies costs of the loss of a critical function
- Identifies business processes that are critical and need to be protected
- Identifies the critical resources necessary for each function, such as people, equipment, and software
- Protection of critical assets and resources
- Determination of risks
- Prevents exposure to financial, reputational, and legal risks when disruption occurs
- Helps to obtain insurance coverage or adequate compensation from insurers in case of loss
- Identifying crucial records and the impact of their loss
- Helps with decision-making during a crisis due to information about the potential impacts of various actions
- Determining the recovery time objective of critical functions
- Involves suppliers in the process to ensure continuity of the supply chain

6.10.4.1 Managing a BIA Project

The highest levels of the organisation must provide funds and administrative support. Every part of FM will be affected by a BIA. So, selecting a senior executive to move the project forward is very important. Some workers might not want to give out information because they are worried about how it will be used. If the BIA is run well, the business continuity project will have a better reputation. If it is poorly run, the project will fail.

Data collection will help the facilities manager to gather information on critical services. Information can be obtained through questionnaires, interviews, and reviews of documents. The essence is to collect data about the value of the service, its financial status, and compliance with regulations. The status of the data gathered will influence how successful the project is. The following actions make up a data-gathering plan:

- Identify who will receive the questionnaire using an up-to-date organogram
- Develop the questionnaire for data collection
- Provide training on how to respond to the questionnaire

- Follow up with each staff member to ensure timely completion of the questionnaire
- Review responses with respondents if the responses are not clear or are incomplete
- Conduct review meetings with employees and service providers to discuss responses
- Compile the information gathered for recommendations

6.10.4.2 Identify Respondents

The organogram will help to identify respondents. The supervisors and Heads of Departments should be encouraged to complete questionnaires. Top management support will help with the success of the process.

6.10.4.3 Develop the Questionnaire

The BIA questionnaire should first be pre-tested on one of the organisation's departments. With this department's help, the questionnaire's directions, questions, and any missing questions may all be made better. It's best for the questions to be easy for the reader to understand, and to keep things consistent, the questionnaire might include a list of terms.

Questions can be answered in two different ways: qualitatively or quantitatively. Qualitative data comprises variables that can't be measured, like gender or colour. Quantitative data contains values, like a time or a sum of money. It is easy to look at quantitative data because it can be combined. The questionnaire will come with written instructions as soon as everyone agrees on all the questions. The steps and answers should be explained to fill out every response on the questionnaire. A phone number should also be captured so that the manager can answer any questions.

Content of a BIA questionnaire:

- Purpose of the BIA exercise
- Contact respondent information, such as name, location, work position, phone number, and email
- Department information, such as duties and size
- Information on service providers and the services they provide
- Exposure or vulnerability of facilities to risk
- Recovery complexity, which is information on how difficult it would be to recover the FM processes to an adequate level of service following a prolonged disruption or outage
- The process information section is to identify FM critical processes and their frequency
- Processing periods determine if performing this process is especially critical during particular months or if the impact would be higher during specific months
- Process Unavailability Impact asks how many hours each critical process can be unavailable before an impact occurs to the organisation and users
- Processes Deferrable are the FM processes that do not have to be performed until a specific timeframe
- Manual workaround procedures are questions on the length of time a manual workaround can be followed before an impact happens
- The alternative facilities section seeks to find if alternative sites can be used for a task
- Backlogged work shows how long it will take to handle the backlogged work for each process once facilities are available
- Dependencies are internal (same company) or external (outside) resources that either supply the organisation with information, support, and/or data or depend on information, support and/or data from the organisation

- Resources that FM processes depend on to function should be identified
- The reports section should identify the report's name, the report's criticality, a brief description, the report's type, the report's source, and the timeframe in which the report must be produced for the organisation to continue operations when there is a disruption
- Potential impacts on an organisation if a process cannot be performed; the impact can be financial, operational and regulatory

6.10.4.4 Unavailability Impact Table

Table 6.3 shows the impact of business activities in prioritising resources to restore critical activities.

6.10.4.5 Collect the Data

Once the questionnaire has been written, it should be sent to respondents, and the facilities manager should respond to questions that need clarification from the respondents. The questionnaires should be returned within a stated time frame. Following up on the respondents' surveys is essential to increase the response rate.

6.10.4.6 Reporting the Results

Once all questionnaires have been returned, reports will be written and organised into an order of reports with each function.

The example in Table 6.4 shows a work group report for a facilities department. Each column shows the impact of a function being unavailable at that time.

N/B: The denomination of the currency will depend on the country.

6.10.5 Recovery Time Objectives (RTOs)

There is also the setting of recovery time objectives (RTOs). These determine when the effect on the facility's operation is unbearable, such as when crucial services are lost. It is used to develop continuity and recovery solutions. An example can be found in Table 6.5.

Example of a strategy selection for a bank in South Africa

The business interruption scenarios listed in this item are based on the outcomes of a business impact analysis and risk assessment study conducted by the FM business unit on water loss (see Table 6.6).

Under normal circumstances, should the municipal supply fail:

- Implement Stage 1 Water Savings plan
- Turn off irrigation, fountains, and ponds

Where storage capacity is reduced to 75%:

- Implement Stage 2 Water Savings plan
- Extend borehole runtime by 20%
- Restrict Kitchens to water-saving offerings
- Override HVAC

Table 6.3 Unavailability Impact Table

Criticality scales	*Impact categories*				
	Health and safety consequences	*Possible occupant reaction*	*Enforcing compliance with regulations*	*Finance wasted*	*Building occupant concerns*
Very low	Possible minor injuries	Dissatisfaction in one area			
Low	Possible minor injuries	Dissatisfaction in certain areas			
Medium	More than minor injuries	Impacts on performance that are significant			
High	Major injuries to a small group	General refusal to cooperate			
Very high	Likely to put lives at risk and cause major accidents	Trade union action			

Source: Author

Table 6.4 Work Group Report

WORK GROUP REPORT						
Work group: FM services						
FM Function	1 hour	4 hours	1 day	2 days	1 week	2 weeks
Electricity	$0	$50,000	$100,000	$200,000	$500,000	$550,000
Water	$0	$0	$15,000	$25,000	$100,000	$300,000
Lift operations	$0	$0	$0	$0	$20,000	$50,000

Source: Author

Table 6.5 Example of an RTO for the Provision of an Uninterrupted Supply of Utilities

Mission Critical Activities	*Dependencies*	*Overall Rating*	*Recovery Time Objective*	*Strategy Selection*
provision of an uninterrupted supply of utilities	Electricity	3	36 hrs	Utilise standby generators
	Water	4	3 hrs	Use storage tanks
	Air-conditioning	4	3 hrs	Adequate arrangements
	Elevator operations	4	1 day	Depends on the source and period of outage, either relocate to DR site
	Diesel supply	3	36 hrs	Storage capacity for 36 hrs, and SLA in place with Dell Carriers, source diesel from local suppliers, SLA Diesel Electric to fill up
	HOD	2	1 week	Succession plan in place
	Staff	1	1 month	Supplement staff shortage by sourcing temporary staff
	Building Management System	1	1 week	Use contractors and implement manual workaround
	Contractor	1	1 week	Source temporary skills to provide necessary service

Source: Anonymous

During total water supply shutdown, where storage capacity is reduced to 50%:

- Implement Stage 3 Water Savings plan
- Invoke DR for smaller buildings
- Close all kitchens

Where storage capacity is reduced to 25%:

- Implement Stage 4 Water Savings plan
- Invoke DR for smaller buildings

Table 6.6 Business Continuity Strategy Selection for a Bank

Business Continuity Strategy Selection	Business unit:		Facilities Management		Sub Business Area:	N/A
Interruption/Loss scenario:	Loss of Water (for period > 2 days)		Recovery Time Objective (RTO)			*3 hours*
Continuity Strategy:	Supplement water shortage utilising water stored in domestic & JoJo tanks (storage capacity ~ 2 days operation @ 250KL per day) thereafter once the new 1000KL tank is installed we will have an additional 4 days.					
Mission Critical Activity	Immediate Response (0 – 15 minutes)	Action by:	Continuity Response (15 – 60 minutes)	Action by:	Recovery Response (1 -3 hrs)	Action by:
Provision of an un-interrupted supply of water for all operations	Notify FM	FM	Notify staff of the situation at hand (via e-mail) and request the sparing usage of water.	FM	If relocation is required, convene a central FM Team & relevant others at the primary control room. Provide an incident brief.	Operational Risk Manager
	Contact Johannesburg Water & enquire the cause of fault and estimated outage period – provide BU Head with feedback result	FM	Additional water can be brought to Campus via water tank trucks. FM has an adhoc agreement with Rescue Rod Services.	FM	Relocation: Discuss, agree upon & list the business units/ functions to be relocated. Prioritise list. List BU's/personnel capable of working from home	Central FM Team with, COO's and IT Heads
			Advise Marsh Vikela (Insurance Co.) that we will be drawing water from the fire tankers.	FM	Notify BU Heads regarding decision(s) made, actions to be taken & timelines	FM Team

(*Continued*)

Table 6.6 (Continued)

	Dependent on estimated outage period (>5 hrs), notify Operational Risk Manager Increase Borehole pumping time.	FM	In conjunction with Operational Risk Manager, based on estimated longer than 3 hrs & alternate arrangements implemented, make a decision regarding relocation of critical staff and functions to DR Site	FM	Notify staff and coordinate actions as directed by FM Team / Coordinators.	BU Heads

Approved by ________________ Position: ________________ Date: ________________

Source: Anonymous

6.10.6 Risk Assessment

A risk assessment carefully finds and manages the risks that affect FM by putting the proper controls in place. Risk management reduces loss and uncertainty, which is especially important since disasters can happen anytime. It is essential to look at the risks and plan what to do in case of a catastrophic event or uncertainty. To keep organisations running smoothly, there should be plans for what to do if something unexpected happens. It's hard to know when these risks will happen; when they do, they can be expensive. If the risk comes true, it could have a significant effect, so the facilities manager may need help from other experts. Plans for things like business continuity and emergency recovery are documented and used in risk assessment.

6.10.6.1 Benefits of Risk Assessment

There are many benefits to facilities managers, contractors, and the organisation, including the following:

- Improves communication in organisations
- Improves management reporting
- Improves understanding of the risks and making business case
- Improves stakeholder relationships
- Helps make clever purchasing decisions
- Encourages a learning culture
- Increases the acceptance of opportunities
- Improves productivity at performing FM tasks
- Encourages better planning for facilities
- Reduces cost
- Reduces compliance costs
- Improves resource provision and use
- Achieves organisation's objectives
- Improves accountability
- Increases stakeholder confidence in facilities
- Reduces the potential for litigations
- Achieves a more organised method of comparing and accepting risk

6.10.6.2 The Skills of the Facilities Manager and Risk Management in Business Continuity Plans

When there are disasters, the scope of facilities management, which reflects the life cycles of facilities, means that risk assessment in certain aspects of FM will be useful. Business continuity planning is an aspect of strategic planning and risk management. It happens through having business interruption strategies by identifying potential disasters and business interruptions that could happen to the business based on its location. Also, facilities managers should provide input when designing the facility on how the building can minimise interruptions in an uncertain event, provide safe measures, and ensure that buildings are managed subject to standards and legislations that ensure fewer disruptions.

In operations and maintenance, facilities managers maintain buildings with fewer disruptions from critical services such as electricity and water. IT should continue functioning in uncertain events to maintain business continuity when employees move to a temporary space or work from home. The outsourced service contracts should be checked for capability to ensure minimal disruptions to services provided. The performance of services should also be monitored for business continuity and performance levels can be checked using service level agreements. Furthermore, there should be compliance with regulations and standards that promote business continuity.

Aspects of the operations useful in a business continuity plan include security, since it ensures facilities are safe. Many emergency procedures are found in health and safety regulations.

6.10.6.3 Risk Management Process

6.10.6.3.1 RISK IDENTIFICATION

This is the process of discovering, identifying, and recording risks. Risk identification determines which risks might affect the operations of facilities. Facilities managers must know about events that disrupt the organisation by identifying risks. Many methods can be used to identify risks. After identifying risks, they should be inspected, and there should be a discussion with other team members and professionals connected with safety.

Potential risks might include:

- Fall of a person from height
- Mechanical lifting operation
- Fall of object/material from a height
- Noise
- Fall of a person on the same level
- Biological agents
- Epidemic
- Strikes
- Terrorism
- Changes in legislation
- Changes in market condition
- Disruption in communication
- Loss of data
- Corruption
- Human error
- Contract risks
- User commitment
- Energy prices
- Manual handling
- Ionising radiation
- Uses of machine
- Vibration
- Operation of vehicles
- Hand tools
- Fire
- Adverse weather
- Electricity
- Chemical substances

- Drowning
- Stacking
- Excavation work
- Housekeeping
- Stored energy
- Lighting
- Explosion (chemical dust)
- Confined spaces
- Contact with cold/hot surfaces
- Cleaning
- Compressed air

6.10.6.3.2 RISK ANALYSIS

This is multiplying the chance of a risk by its effect to determine the risk associated with an identified risk. Chance of risk occurring, objectively or subjectively, can be stated quantitatively or qualitatively and is the definition of likelihood. Therefore, the probability might be determined using past data and the subjective opinions of experts. In addition, the meaning of a consequence is results of a risk that will impact the organisation's goals. It is possible to refer to a risk's effect and outcome interchangeably. The resulting value may be quantitative, qualitative, or something in between (Torabi et al., 2016).

Determining the degree of each risk is a crucial task that requires accuracy. When no risk data is available, the qualitative technique may subjectively assess a risk's probability and consequences via verbal descriptions of the risk's effect. Some techniques include word clouds, focus groups, surveys, consultants, interviews, and brainstorming sessions. In semi-quantitative approaches, a risk level or value is calculated by adding subjective risk weights that are provided. The acceptability of risk criteria may be established using this number. Using data from potential outcomes, risk probability, and variables and sub-factors for outcomes and likelihood are all part of quantitative risk approaches. It includes both the possibility and the consequence of a certain degree of risk. The risk assessment will be completed correctly if each risk level is accurately evaluated. Lifecycle cost analysis, influence diagrams, decision trees, fault trees, event trees, consequence analysis, and probability analysis are a few examples.

When sub-factors are combined, they give a more accurate measure of the impact factor than when used separately to determine the effect of risk. Here are some of them:

- The risk's ability to cause more risks
- Its rate of growth
- The financial and human losses it causes

They also include the time and resources needed to get the organisation back up and running after the risk event. One part of the risk effect is the cost of a risk coming true. This is because any risk can use up an organisation's resources and cause losses in both funds and persons. All risks can make it hard for the team to do its work. When determining the risk impact, you should also consider how much time and money it will take to get the organisation's primary functions back to normal. Another part of the risk is the path that makes the event happen. A risk may lead to additional risks that go against the way it was supposed to go. So, the chance that risk will lead to more problems should be considered a risk effect. If a risk can't be seen coming, it might have a more significant impact than one that is easy to see. How easy it is for a company to identify each risk helps them respond better. So, an extra risk factor, risk increase, should be considered part of the effect. The growth factor shows how a particular risk might change in the future. A risk with a low growth factor might go away very quickly. On the other hand, a risk that doesn't have much of an effect but has a high growth factor would become more critical over time. Therefore, the risk assessment process ranks the hazards by how likely they are to happen and the effect they could have if they do.[15]

Next, we use the risk assessment phase to determine how likely it is that the consequences will happen. This can be done by rating the probability of them occurring as "frequently," "probably," "occasionally," "remotely," or "improbably," or by doing it quantitatively like "once in 10," "once in 100," or "once in 1,000," as shown in Table 6.8. This list of probabilities is the list of risk probabilities.

Giving an event high importance makes sense if it has adverse consequences and a significant likelihood of happening. On the other hand, it makes sense to provide a low priority if the risk will have little effect and improbable probability. To do this, we can develop specific criteria for what is acceptable and then rank all the risks found in order of importance to safety and reliability goals. This can be done using the proper risk indexes based on risk severity and probability, as shown in Table 6.9.

	CONSEQUENCES				
	INSIGNIFICANT	MINOR	MODERATE	MAJOR	CATASTROPHIC
LIKELIHOOD	1	2	3	4	5
A (ALMOST CERTAIN)	H	H	E	E	E
B (LIKELY)	M	H	H	E	E
C (POSSIBLE)	L	M	H	E	E
D (UNLIKELY)	L	L	M	H	E
E (RARE)	L	L	M	H	H

E	EXTREME RISK, IMMEDIATE ACTION
H	HIGH RISK, ACTION SHOULD BE TAKEN TO COMPENSATE
M	MODERATE RISK, ACTION SHOULD BE TAKEN TO MONITOR
L	LOW RISK, ROUTINE ACCEPTANCE OF RISK

Figure 6.3 Risk assessment matrix.

Source: Author

If a risk has been assigned with a risk index of 1A, 1B, 1C, 2A, 2B, or 3A, it must be corrected immediately. If the risk index is 1D, 2C, 2D, 3B, or 3C, something might need to be done to fix it. Similarly, a risk with an index of 1E, 2E, 3D, 3E, 4A, or 4B would be tracked for a low-priority action to be taken to correct it, or it might not let any action be taken. On the other hand, a risk with a rating of 4C, 4D, or 4E might not even need a review for action.[16]

6.10.6.3.3 RISK EVALUATION

Risk evaluation uses factors to compare an estimated risk and measure its importance. The next step after risk analysis is determining the best way to deal with the risks. Facilities managers need to reduce risks. Facilities managers need to know which risks impact operations most because they will probably find that more resources are required. Based on the risk and building analysis outcomes, the organisation can now determine which risks could significantly affect its goals beyond what it can handle. Finding the link between the known risks and the critical FM tasks can help find the risks that have the most significant effect on the objectives of an organisation and the steps that need to be taken by management. A known risk may change how resources are used. Risks may be more or less likely to affect specific resources depending on how the company is set up, its location, and its

Table 6.7 Assessment of Risk Severity and Categories

Risk consequences	*Risk severity*	*Category*
Death, system loss, or great environmental damage, etc.	Catastrophic	1
Severe injury, great occupational illness, major system or environmental damage, etc.	Critical	2
Minor injury, occupational illness, system or environmental damage, etc.	Marginal	2
Minimal injury, occupational illness, insignificant system or environmental damage, etc.	Negligible	4

Source: Rao Tummala, V.M., & Leung, Y.H. (1996). A risk management model to assess safety and reliability risks. *International Journal of Quality & Reliability Management*, 13(8), 53–62

Table 6.8 Assessment of Risk Probabilities and Levels

	Risk probability		
Risk categories	*Qualitative*	*Quantitative*	*Level*
Frequent	Frequently likelihood of occurrence	The probability is greater than 10^{-1}	A
Probable	This happens many times in the life of an item	The probability is between 10^{-2} and 10^{-1}	B
Occasional	Sometimes happens in the lifetime of an item	The probability is between 10^{-3} and 10^{-2}	C
Remote	Unlikely but can happen in the life of an item	The probability is between 10^{-6} and 10^{-3}	D
Improbable	Very unlikely to occur	The probability is less than 10^{-6}	E

Source: Rao Tummala, V.M., & Leung, Y.H. (1996). A risk management model to assess safety and reliability risks. *International Journal of Quality & Reliability Management*, *13*(8), 53–62

Table 6.9 Priority Matrix Based on Risk Severity and Risk Probability

Risk probability	*Catastrophic*	*Critical*	*Marginal*	*Negligible*
(A) Frequent ($x > 10^{-1}$)	1A	2A	3A	4A
(B) Probable ($10^{-1} > x > 10^{-2}$)	1B	2B	3A	4B
(C) Occasional ($10^{-2} > x > 10^{-3}$)	1C	2C	3C	4C
(D) Remote ($10^{-3} > x > 10^{-6}$)	1D	2D	3D	4D
(E) Improbable ($x < 10^{-6}$)	1E	2E	3E	4E

Source: Rao Tummala, V.M., & Leung, Y.H. (1996). A risk management model to assess safety and reliability risks. *International Journal of Quality & Reliability Management*, *13*(8), 53–62

infrastructure. When figuring out the risks, it's essential to consider how the resources are exposed, which shows how the risks affect them.

Once the facilities manager knows about the risk, they should try to lower it by delaying the work that causes it, especially if there aren't any safety steps that can eliminate it or make it doable. Facilities managers may also try to lower the chance of the event by removing lawsuits that might be linked to using the facility. Changing the effects of a risk might be possible by giving people alternate choices, like how a lack of water during interruptions would affect people. When an organisation hires another organisation to work for it, the risk is usually outsourced to other organisations.

Table 6.10 Example of a Risk Evaluation and Review for COVID-19

Business continuity plan components	*Description*	*Minimum update frequency*
Purpose and objectives	The purpose is to give the organisation a strategy and instructions for the coordination, preparedness, planning, mitigation and readiness to respond to COVID-19 and other infectious diseases. Also, ensure operational functionality and business continuity.	Evaluation takes place every 2-3 weeks but 3 months maximum
Immediate response	This component entails: • The process flow • Threat levels • Initial response procedures to COVID-19 • Declaration of disaster procedure • Damage assessment team information • Damage assessment checklist • Contact list • Crises communications	Every 2-3 weeks but 3 months maximum

(*Continued*)

Table 6.10 (Continued)

Business continuity plan components	*Description*	*Minimum update frequency*
Recovery procedures	Business risk identification • Work from home and Covid-19 • Self-isolation/quarantine procedures • Pandemic plan • Disaster recovery	Every 2-3 weeks but 3 months maximum
Self-isolation quarantine procedures	Do and Don'ts for self-isolation • Screening and testing	Every 2-3 weeks but 3 months maximum

Source: Anonymous

Example of COVID-19 response plan for IBM

IBM had a goal to provide employees, partners, and customers useful information about IBM's first steps towards getting back to work after the COVID-19 outbreak. The company implemented data-driven, evidence-based practices and policies to protect IBM employees as they return to work in stages. IBM has created a set of guidelines and standards that are the same everywhere. The policies and standards come from government and public health organisations like the World Health Organisation and the Centres for Disease Control. The guidelines are about going back to work, which are called Wave 1. The following assumptions were made that: people will continue to be easily infected and new outbreaks will happen; there aren't many fast tests available to find infections or antibodies to show immunity; and strong contact tracing isn't widely used yet. The next waves will be decided by an in-depth assessment of the health and policy situation in each market.[17]

6.10.6.3.4 RISK CONTROL

Risk control is done to put risk assessment decisions into action. Facilities managers should have strategies for dealing with risks after other measures have been taken and should write these plans down. When looking at cost-effective solutions, the most economical option should be carefully thought through when several can handle the same level of risk. You might need to do a cost-benefit study to find the best choice. When deciding which option is the most cost-effective, managers should consider how much work it takes to handle risk, as well as cost, time, resources, and legal requirements. It is essential to look at how vital risk control is. In some situations, the right mix of control methods may be helpful. Engineering controls, warning signs, and teaching on personal safety equipment are often used when there is no way to eliminate the risk. The hierarchy of control in an organisation setting can be used. The facilities manager should make and write down a risk containment plan before implementing any chosen risk management measures. This plan should include a timeline and details on how resources will be shared so that everyone knows what they are responsible for and what they need to do. The paperwork should have a budget, reasonable objectives, and time frames for meeting those goals.

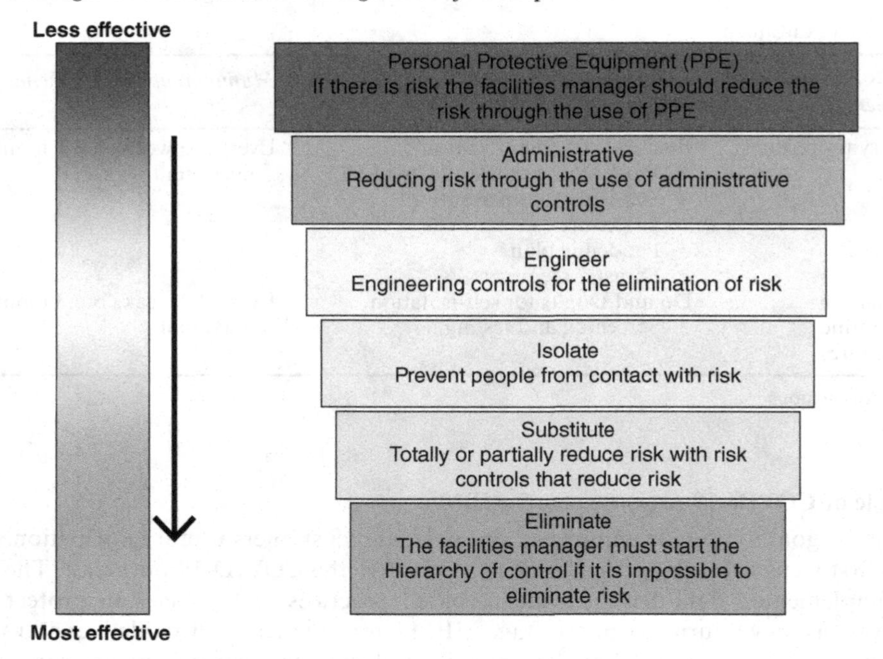

Figure 6.4 Hierarchy of controls, includes moving from minimising risk options to eliminating risk.

Source: Author

When considering controls, we must first consider if the hazard can be eliminated. If it cannot be eliminated, consult the Hierarchy of Controls (see Figure 6.4) and work through the minimising of risk options:

- Eliminate – Remove the risk
- Substitute – Replace all the risk or part of the risk with something with a lower risk
- Isolate – Preventing a person from coming into contact with the risk
- Engineer – Providing controls to remove the risk
- Administrative – These controls could include procedures, training, additional workers, or changes in how the job is performed

Personal Protective Equipment (PPE) is only considered a last resort to eliminate any remaining risk after considering all these other options. Hard hats, safety glasses, gloves, breathing gear, shoes, clothes, helmets, and more could all be PPE.

6.10.6.3.5 RISK MONITORING AND REVIEW

Risk monitoring means monitoring risks to ensure that changing situations don't change their importance. Risk management must be done often to keep up with new risks because some risks don't change. Monitoring risks is essential to keep them up to date, find out how much risk there is still, understand the risk, look at how it is treated and works, and keep an eye on the quality of decisions made about risks. It also helps keep the risk record current and creates information about significant events and project progress while projects are being evaluated. Some of the most critical steps are finding risks, making general risk scores, keeping them up to date, making action plans, and choosing what to do in response. It also makes it easier to share this kind of knowledge with all levels of management, even the top ones (BIFM, 2017). Committees, commissions of inquiry, risk records, physical inspections, policy reviews, and the evaluation of programmes and strategies can be used to assess the risks.

6.11 Conclusion

Emergency preparedness is the facilities department's reaction to certain types of incidents. On the other hand, business continuity plans are strategic and quickly restore the company to regular operations after an event. Facilities experts and executives must make business continuity models to meet business and disaster preparations. Business plans are partly or hardly implemented in Africa; there are few business continuity standards and organisations that conduct business impact analysis. Many of these organisations lacked Business Recovery Plans or Crisis Management strategies. Additionally, BCM education, awareness, and performance assessment must be improved. Disaster management at the regional, social, and in-use levels can all be used to determine how bad a disaster was. There are five steps in business continuity planning. Selecting a recovery strategy involves considering the organisation's people, facilities, systems, and equipment. All plans for recovery need to consider the resources that would be used to recover what was lost so that the organisation can carry out its essential functions and how quickly these resources can be made available for critical functions to function again. These plans should be written to fulfil the recovery time required and the solution selected. A business continuity strategy helps a company keep running so people don't notice when essential functions stop working. It also gives the organisation the tools to quickly and easily return to normal. Business impact analysis (BIA) helps organisations determine the most essential services. It is also the first step in creating recovery plans and actions for these services. This study can help a facilities manager determine how an event might affect the building's processes, its ability to provide services, and its ability to follow the regulations. A risk assessment carefully finds and manages the risks that affect FM by putting the proper controls in place. It also means putting the likelihood and impact in order of importance and finding out what caused the risks.

6.12 Guidelines

- Need to have business continuity plans in place for emergencies
- Get top management support for BCP projects
- Managers should be involved in change management behaviours of staff in the organisation regarding business continuity and hazards
- Ensure that there is a long term view about the process
- Departments should have an understanding of how hazards affect operations
- Managers should ensure that BCPs are embedded from the top layers to the bottom and that the process gets support of everyone to guarantee success
- Use formal and informal communication to send messages about the importance of BCP
- Appoint champions to promote the concept in the organisation
- Get collaborations on BCP beyond the organisation
- Managers and organisations must put aside resources for business continuity plans
- Managers should conduct a business impact analysis of all facilities-related functions

- Managers should have knowledge about regulations and standards that guide emergencies and disasters and manage facilities to comply with such regulations
- Managers must have knowledge, processes, priorities and documentation of possible threats to business operations before they happen
- Managers can use consultants if there is a gap in skills to implement the project
- Managers should be versatile in risk assessments
- Managers need skills in management, analytical, communication, leadership, and innovation to embed business continuity in the strategic management of facilities
- Managers must ensure the personnel in their department and suppliers are trained for the success of BCP.

6.13 Questions

1. What is the difference between emergency preparedness and business continuity?
2. What are the elements of emergency preparedness?
3. Describe what a pandemic risk assessment should cover
4. Explain the criteria for a pandemic risk assessment
5. What is a work area recovery strategy?
6. Develop a business impact analysis questionnaire for your organisation
7. Describe the probabilistic, quantified, and qualitative methods used to evaluate risk
8. Identify a risk scenario and determine its risk assessment severity and probabilities
9. In a relatively simple way, if a grading system is created, consider, for likelihood and impact, which grades would be awarded to each of these brief situations:
   a. A terrorist attack on a business (assuming the business is more likely a terrorist target than a relatively general one)
   b. Building sinking due to nearby use of heavy machinery
10. Provide an example of situations when a hierarchy of controls has been used in your organisation
11. What are risk evaluation, risk monitoring, and review?
12. What overarching legislation implies employers require risk assessment in your country?
13. Which regulations specifically stipulate the use of risk assessments?

Notes

1 Sawalha, I.H. (2011). *Business continuity management and strategic planning: The case of Jordan*. PhD thesis, University of Huddersfield, UK.
2 Herbane, B., Elliott, D., & Swartz, E.M. (2004). Business continuity management: Time for a strategic role? *Long Range Planning*, *37*(5), 435–457.
3 Herbane, B. (2010). The evolution of business continuity management: A historical review of practices and drivers. *Business History*, *52*(6), 978–1002.
4 Gustin, J.F. (2010). *Disaster & recovery planning: A guide for facility managers*. River Publishers.
5 BWA. (1994). *Facilities economies*. Bernard Williams Associates, Building Economics Bureau Limited.
6 Loosemore, M., & Then, S.K. (2006). Terrorism preparedness of building facilities managers. *Australian Journal of Emergency Management*, *21*(3), 22–29.
7 Dissanayake, D.M.P.P., & Fernando, N.G. (2012). Role of facilities manager in pre-disaster risk reduction phase: A literature synthesis.. Accessed February 2023 at CIB 2005 Full Paper Model (uom.lk).
8 Molloy, S. (2021). Occupational Health and Safety. University of Witwatersrand Lecture Notes.

9 Castillo, C. (2004). Disaster preparedness and business continuity planning at Boeing: An integrated model. *Journal of Facilities Management*, *3*(1), 8–26.
10 Moeid, M.A. (2004). *Assessment of automating facilities management current practices; Potential framework models for Saudi Arabian universities.* Masters' thesis, King Fadh University of Petroleum and Minerals, Saudi Arabia.
11 KPMG. (2013). Business continuity in Africa: Building resilience in a volatile environment. KPMG.
12 International Federation of the Red Cross and Red Crescent Societies. (2021). Legal Preparedness for International Disaster Assistance in Southern Africa; https://disasterlaw.ifrc.org/sites/default/files/media/disaster_law/2021-06/20210531_IDRL_SouthernAfrica_SUMMARY_ONLINE.pdf.
13 BIFM. (2017). *Risk assessment*. BIFM.
14 Holmwood, B, Fowler, I., Puybaraud, M. & Hardman, C. (n.d). FM and business continuity planning: Are managers ready for it?
15 Torabi, S.A., Giahi, R., & Sahebjamnia, N. (2016). An enhanced risk assessment framework for business continuity management systems. *Safety Science*, *89*, 201–218.
16 Rao Tummala, V.M., & Leung, Y.H. (1996). A risk management model to assess safety and reliability risks. *International Journal of Quality & Reliability Management*, *13*(8), 53–62.
17 IFMA. (2024). IFMA Knowledge Bank. IFMA.

Bibliography

BIFM. (2017). *Risk assessment*. BIFM.

BWA. (1994). *Facilities economies*. Bernard Williams Associates, Building Economics Bureau Limited.

Castillo, C. (2004). Disaster preparedness and business continuity planning at Boeing: An integrated model. *Journal of Facilities Management*, *3*(1), 8–26.

Dissanayake, D.M.P.P., & Fernando, N.G. (2012). Role of facilities manager in pre-disaster risk reduction phase: A literature synthesis. Accessed February 2023 at CIB 2005 Full Paper Model (uom.lk)

Gustin, J.F. (2010). *Disaster & recovery planning: A guide for facility managers*. River Publishers.

Hardy, V., Roper, K.O., & Kennedy, S. (2009). Emergency preparedness and disaster recovery in the US post 9/11. *Journal of Facilities Management*, *7*(3), 212–223.

Herbane, B., Elliott, D., & Swartz, E.M. (2004). Business continuity management: Time for a strategic role? *Long Range Planning*, *37*(5), 435–457.

Herbane, B. (2010). The evolution of business continuity management: A historical review of practices and drivers. *Business History*, *52*(6), 978–1002.

Holmwood, B, Fowler, I., Puybaraud, M. & Hardman, C. (n.d.). FM and business continuity planning: Are managers ready for it?

IFMA. (2024). IFMA Knowledge Bank. IFMA.

Institute for Security Studies. (2023). Loss and damage funding vital after DRC and Rwanda floods. Accessed January 2024 at https://issafrica.org/iss-today/loss-and-damage-funding-vital-after-drc-and-rwanda-floods

International Federation of the Red Cross and Red Crescent Societies. (2021). Legal Preparedness for International Disaster Assistance in Southern Africa; https://disasterlaw.ifrc.org/sites/default/files/media/disaster_law/2021-06/20210531_IDRL_SouthernAfrica_SUMMARY_ONLINE.pdf

KPMG. (2013). *Business continuity in Africa: Building resilience in a volatile environment*. KPMG.

Loosemore, M., & Then, S.K. (2006). Terrorism preparedness of building facilities managers. *Australian Journal of Emergency Management*, *21*(3), 22–29.

Moeid, M.A. (2004). *Assessment of automating facilities management current practices; Potential framework models for Saudi Arabian universities*. Masters' thesis, King Fahd University of Petroleum and Minerals, Saudi Arabia.

Molloy, S. (2021). Occupational Health and Safety. University of Witwatersrand Lecture Notes.

Rao Tummala, V.M., & Leung, Y.H. (1996). A risk management model to assess safety and reliability risks. *International Journal of Quality & Reliability Management*, *13*(8), 53–62.

Sawalha, I.H. (2011). *Business continuity management and strategic planning: The case of Jordan*. PhD thesis, University of Huddersfield, UK.

Torabi, S.A., Giahi, R., & Sahebjamnia, N. (2016). An enhanced risk assessment framework for business continuity management systems. *Safety Science*, *89*, 201–218.

Yusuf. (2023). Kenya Marks 10 Years Since Westgate Mall Attack, Reflecting on Security Progress; Voice of America, January 2024 at https://www.voanews.com/a/kenya-marks-10-years-since-westgate-mall-attack-reflecting-on-security-progress/7278239.html#:~:text=One%20Saturday%20in%202013%2C%20four,more%20than%20150%20were%20injured

World Meteorological Organisation. (2023). Africa faces disproportionate burden from climate change and adaptation costs. Accessed January 2024 at https://wmo.int/news/media-centre/africa-faces-disproportionate-burden-from-climate-change-and-adaptation-costs

Part II

Sustainable Management of the Workplace

7 Sustainable Management of Workplace Facilities

7.0 Introduction

Sustainability in facilities management involves preserving facilities with a long-term view to enhance cost savings and bring about energy and water conservation, the efficient use of materials and resources, increased comfort levels, reduced pollution, and responses to climate change demands and legislation compliance. It uses environmental management systems, water conservation, energy efficiency, waste recycling, energy performance ratings, and green purchasing practices. In developing countries there is often a struggle with environmental issues, including imminent desertification, urban physical quality deterioration, land degradation, deforestation, soil erosion, and flooding. Most challenges are typically encountered in urban areas, which frequently function as centres for organisations' economic and administrative operations. These organisations employ a considerable number of facilities managers. Additionally, inadequate waste management, pollution resulting from hazardous pollutants, a lack of security, poverty, and unemployment affect urban areas. To support the development of sustainable communities, facilities managers can promote the adoption of eco-friendly technologies, implement effective waste management practices to minimise waste, and actively encourage recycling among institutions.[1] Sustainable facilities management contributes to achieving the UN's 17 sustainable development goals. The facilities manager does this by ensuring responsible consumption through sustainable procurement, helping to reduce carbon footprint and emissions, maintaining peace and safety, job creation, gender balance, sustainable maintenance, water conservation, and the use of innovation and technology.[2] This chapter focuses on sustainability.

7.1 Sustainable Development

Sustainable development (SD) has three pillars, emphasising its social, environmental, and economic dimensions. It has to do with meeting the needs of future generations when we are still determining what those needs will be. The global community agrees with the idea that SD is important. In FM, the sustainable parts of practice must consider how it will affect the building throughout its life, from design to construction to operation. Facilities managers must consider the life cycle of the services they provide to reduce environmental impact. This occurs during the planning, procurement, operations, transportation, and end-of-life of the services and materials used to provide the service.

Considering what happens at each stage in the life cycle of a building, firstly, the design stage of sustainable FM includes understanding clients' needs, project goals, and objectives. The facilities managers work with other professionals from the design stage and set

DOI: 10.1201/9781032656663-9

goals. The facilities manager offers an advisory role on sustainability, culture, maintenance, innovation, flexibility, and health and safety in the construction of buildings right from the design stage. This stage involves planning development to meet sustainable building specifications and criteria. It also involves the optimisation of construction materials for sustainability and cost-effectiveness. The project's carbon footprint of C02 emissions is planned during this phase. This stage involves conducting a life cycle analysis to understand the environmental impact of various activities in the project and for cost-effectiveness. It also includes training professionals on sustainability principles for the project. Secondly during the construction stage, facilities managers are also involved in considering green policies and standards, green procurement, and supervising the project to ensure that it aligns with sustainability criteria and green performance requirements. Thirdly, at the operations phase, sustainability practices include environmental performance, building energy management systems, water use management, and recycling (see Figure 7.1).

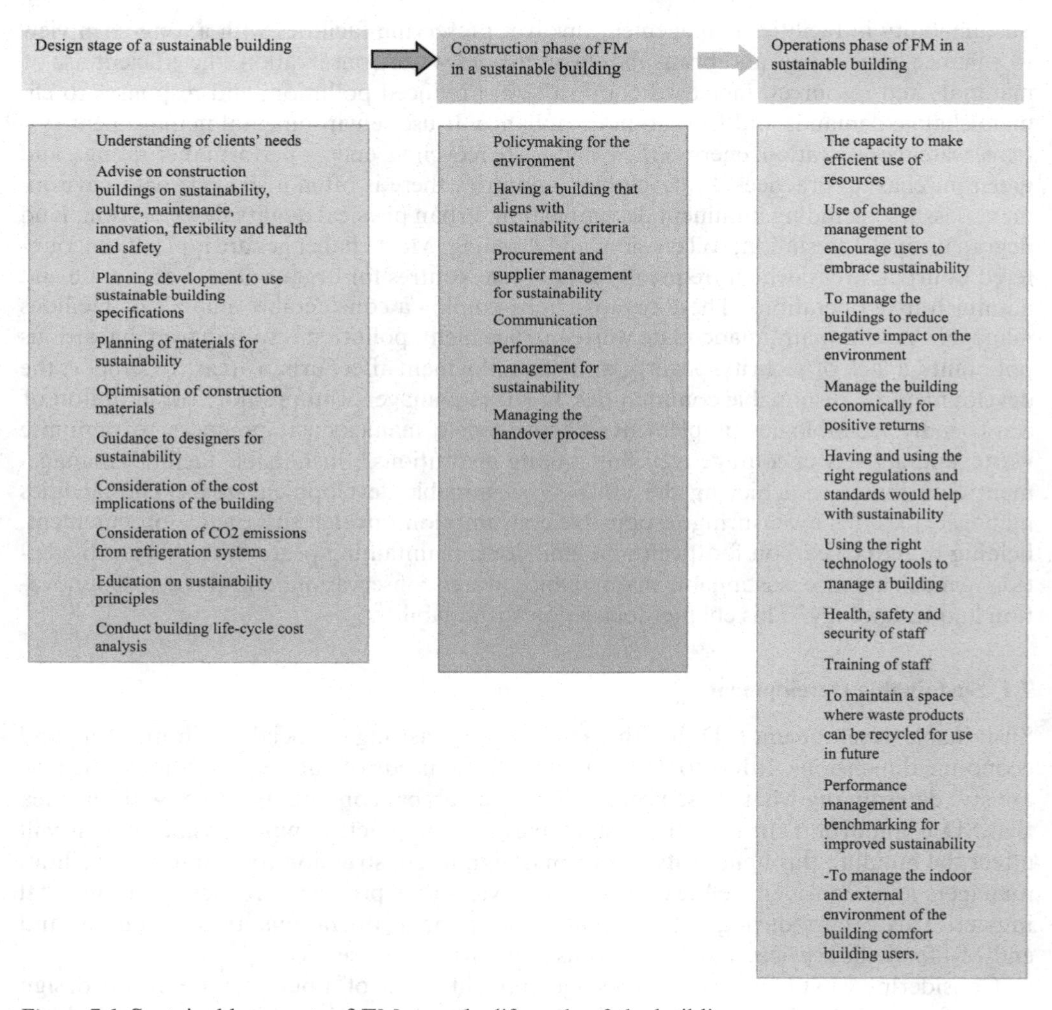

Figure 7.1 Sustainable aspects of FM over the life cycle of the building.

Source: Author

This chapter looks at environmental management. Environmental management from the perspective of the life cycle of a building involves all environmental aspects of each stage. Environmental management systems also have a life cycle approach through their frameworks; for example, the clauses in ISO14001 follow a life cycle perspective in the management of the environment. According to ISO 14001:2015, the life cycle perspective is in environmental management in that:

> a systematic approach to environmental management can provide top management with information to build success over the long term and create options for contributing to sustainable development by controlling or influencing the way the organisation's products and services are designed, manufactured, distributed, consumed and disposed of by using a life cycle perspective that can prevent environmental impacts from being unintentionally shifted elsewhere within the life cycle.

Green purchasing and the net zero carbon concept are also sustainable activities using the life cycle approach. Green purchasing is essential since many environmental effects happen early in a product's life cycle, and it is useful to reduce concerns, save costs, and conserve resources more efficiently. Taking into account the complete life cycle of a product is useful before making a purchase. The net-zero carbon concept ensures that buildings are managed for energy efficiency and use environmental practices such as water use and recycling and are concerned with occupant awareness, environmental performance, improved HVAC systems, and reduced carbon emissions.

7.2 Environment Management

Environmental management guides growth to make the most of opportunities, avoid dangers, reduce issues, and prepare people for problems they cannot prevent by making them more adaptable and resilient. Environmental management deals with how human beings affect the world. It seeks to determine what is good for the environment, what cannot be done because of physical, economic, social, or technological issues, and what alternatives are available.

Environmental management must do three things:

1 Set goals
2 Determine if these goals can be met
3 Develop and use the tools to do what it believes is possible

Environmental management shows the following characteristics:

- It embraces sustainable development
- It means a world changed by people
- it requires a multidisciplinary approach
- It has to bring together and balance different views on development
- It aims to bring together science, social science, policymaking, and planning
- It is a proactive process
- It understands how important it is to meet and, if possible, exceed basic human needs
- It takes a long time and looks at global concerns
- It puts more emphasis on protecting than taking advantage of factors

Reasons for environmental management include the following:

- *Cost savings* – it might be better to stop or solve problems than to deal with the issues that arise, like pollution, species going extinct, human deaths, and expensive lawsuits
- *Practical reasons* – people or administrators try to avoid problems out of fear or political reasoning
- *Compliance with legislation* – regulations and agreements at the national or international level may be needed to protect the environment
- *A shift in ethics* – the media or campaigners working alone or in groups may bring new laws, agreements, or perceptions.
- *Macroeconomic reasons* – environmental management may help the economy grow by creating a market for pollution control tools, recycled materials, safer and more reliable sources of energy and raw materials, or by "affecting externalities" positively

Environmental management aims to better care for the environment by combining science, policymaking, planning, and social growth. Its goals are:

- To protect and, if possible, improve existing resources
- To stop and solve environmental problems
- To set limits and to create and support institutions that work in ecological research, monitoring, and management
- To warn of threats and find opportunities
- To improve "quality of life" as much as possible
- To see new technologies or policies that are useful

Environmental management can be divided into many fields, including:

- Sustainable development issues
- Environmental assessment, modelling, forecasting and "hindcasting" (using history for future scenario prediction), and impact studies
- Corporate environmental management activities
- Pollution recognition and control
- Environmental economics
- Environmental enforcement and legislation
- Environment and development institutions (including NGOs) and ethics
- Environmental management systems and quality issues
- Environmental planning and management
- Assessment of stakeholders involved in environmental management
- Environmental perceptions and education
- Community participation in environmental management/sustainable development
- Institution building for environmental management/sustainable development
- Biodiversity conservation
- Natural resources management
- Environmental rehabilitation/restoration
- Environmental politics
- Environmental aid and institution building[3]

7.3 Environmental Management Systems

Although organisations in most developed countries have included environmental protection policies made essential by government agencies since the early 1970s, these restrictions generally focus on controlling water and air pollutants and waste disposal. Many companies started using voluntary pollution prevention practices in the 1970s and 1980s when they wanted to take into account complicated, expensive, and quickly-changing environmental regulations. These practices reduced or removed the sources of pollutants from manufacturing processes rather than controlling them after emission. Several organisations started including their environmental management practices into more comprehensive systems as more companies accepted pollution-escaping strategies and realised links existed. Governments, industry associations, and international organisations know the benefits of creating standards for companies to follow as guides.[4]

An example of an environmental management system (EMS) is the ISO 14001 standard, which describes implementing an efficient EMS in a company. It aims to assist companies in maintaining their financial success while paying attention to their environmental obligations and effects. It can also help with sustainable growth and reduce the adverse impact of that growth on the environment. The incentives and barriers for ISO 14001 certification are in Tables 7.1 and Table 7.2.

The ISO 14001 standard and the EMAS standard in Europe are the two most popular sets of guidelines for EMS creation and approval. As the sustainability plan usually includes looking at the effects and control of resource use, setting up these processes is part of a facilities manager's role.[5] Organisations can use the guidelines in ISO 14001 to create and implement an EMS that includes the company's environmental policy, the environmental aspects of its operations, legal and other requirements, a clear set of goals and targets for environmental improvement, and a set of environmental management programmes.

To follow ISO 14001, a system must have a clear structure for assigning responsibility for environmental management, training, knowledge, and competency courses for all facility workers and ways for the EMS to be communicated inside and outside the facility. It also needs to have a way to keep track of environmental management paperwork, a way to ensure that paperwork is handled correctly, and steps for making sure that operating

Table 7.1 Incentives for ISO 14001 Certification

Environmental performance:	**Regulatory compliances**:	**Competitive advantage**:
Improved management of environmental impacts Reduced environmental risk Reduced pollution	Easier to prepare for regulatory inspections Improved compliance with government regulations	Increased international trade opportunities Access to a new market Increased competitive advantage Greater market share
Financial impact:	**Stakeholder management**:	
Decreased insurance costs Decreased permit costs Greater access to capital	Public demonstration of environmental stewardship Communication with the community Marketing/Advertising opportunity Increased shareholder value Customer requirement	

Source: Marjanovic, L., Mclennan, P., & Hamid, A.J.B.S. (2014). An investigation of the value proposition for environmental management systems implementation in facility management operations in Singapore. *International Journal of Facility Management*, *5*(2)

Table 7.2 Barriers Towards Seeking ISO 14001 Certification

Management: Lack of top management support Lack of time to implement/manage EMS Lack of understanding of ISO 14001 requirements	**Financial**: The design cost of ISO 14001 EMS The annual cost of maintaining an ISO14001 EMS
Regulatory: Potential legal penalties from voluntary disclosure Uncertainty with regulatory agencies' use of audit information It might not assure better environmental performance	**Stakeholder**: Greater scrutiny from the public

Source: Marjanovic, L., Mclennan, P., & Hamid, A.J.B.S. (2014). An investigation of the value proposition for environmental management systems implementation in facility management operations in Singapore. *International Journal of Facility Management*, *5*(2)

controls of environmental effects and emergency preparedness and response are followed. Regarding environmental management, ISO 14001 calls for a checking and corrective action system that includes monitoring, measurement, reporting non-conformance, and taking corrective and preventive action. For continuous improvement, it calls for EMS reports and a management review process where top management regularly checks the environmental management system's usefulness, efficiency, and sufficiency. Based on Annex SL, Figure 7.2 below shows the Plan-Do-Check-Act (PDCA) cycle of the entire ISO environmental management system.

The facilities management sector influences the environment greatly, as many activities that influence the environment occur in buildings managed by facilities managers. Hence, applying ISO 14001 principles to any facilities management project or company can provide many benefits. The following are some of the clauses in ISO 14001 and how each clause relates to FM.

Clause 1: Scope
This section talks about the standard's purpose of helping FM groups reach the goals of an EMS. A "life cycle perspective" is part of its environmental management.

Clause 4: Context of the Organisation
"The organisation's context is the clause that other standards are built on." The facilities manager can find and understand the people and things that influence the EMS for parts of the building environment, such as services, or by looking at possible pollution. The manager must discover internal and external problems important to the goal. The facilities manager must also find "interested parties" that are good for EMS. These could be users, service providers, or non-government organisations. It is to establish, implement, maintain, and keep improving the EMS for FM processes to meet the standards' requirements.

Clause 6: Planning
This section discusses how the facilities manager will deal with the risks and likelihoods mentioned in Clause 4. It focuses on creating and using a planning process instead of a method. It does this by looking at certain factors and the risks they pose. When thinking about risk, there is a need to match up thoughts about possible effects and opportunities. Setting measurable environmental goals is another important part of this section. "Planning

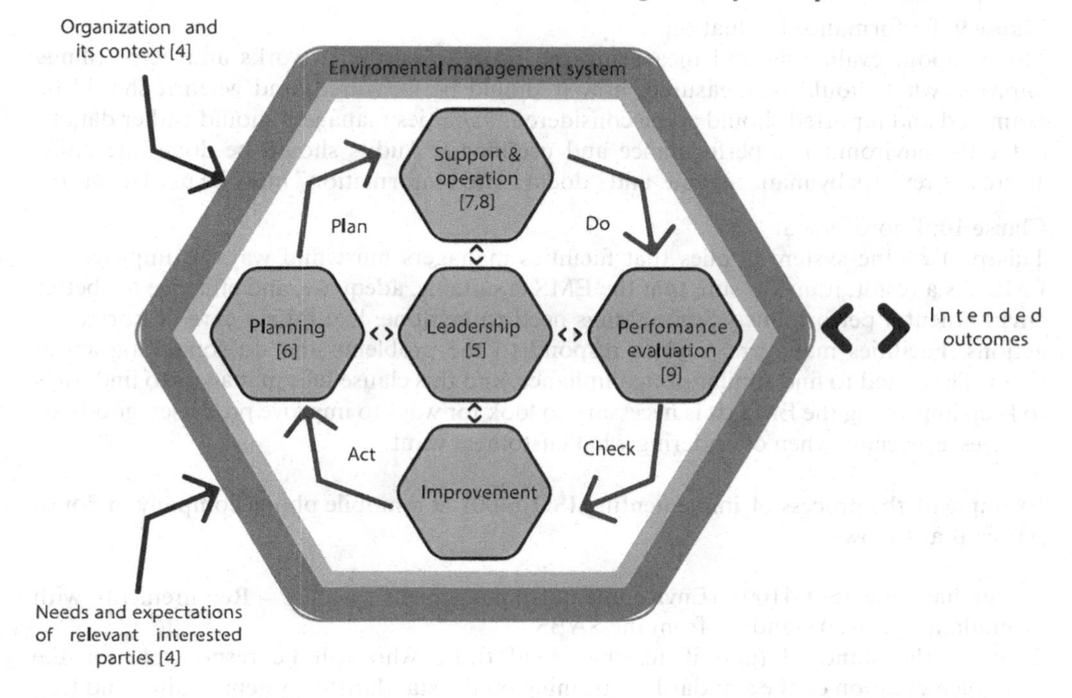

Figure 7.2 The PDCA cycle.

Source: BSI. (2015). ISO 14001:2015 Your Implementation Guide. Accessed August 2023 at ISO-14001-implementation-guide-2016.pdf (bsigroup.com)

of changes should be done systematically". The facilities manager should identify who makes changes and what might happen.

Clause 7: Support
This section puts the plans and steps into action that help the FM section meet its EMS aims. The facilities manager will have to ensure that the people doing work that affects their environmental performance are qualified, can meet their compliance responsibilities, and get the proper training. The facilities manager also needs to ensure that everyone knows about the environmental policy, how their work might affect it, and what will happen if they do not follow the EMS. There are rules about "documented information related to creating, updating, and managing certain data." This clause concerns how people working in FM should obtain training, awareness, and skills in environment, performance, policy, and systems.

Clause 8: Operation
This section discusses implementing plans and procedures to help the FM section reach its environmental goals. From a life cycle point of view, specific needs have to do with having control or influence over processes that are handled and some practical parts. Any team that does work on-site should think about the actual or possible effects on the environment. Because of this, protecting the organisation's funds invested in buildings is essential. In addition, it includes purchasing goods and services and ensuring that environmental concerns are considered at the right time when designing, delivering, using, and getting rid of an organisation's goods and services.

Clause 9: Performance Evaluation
This is about evaluating and measuring the EMS to ensure it works and helps things improve; what should be measured, how it should be measured, and when it should be examined and reported should all be considered. Facilities managers should gather data to judge the environment's performance and usefulness. Audits should be done internally. There are reviews by management, and "documented information" must be used as proof.

Clause 10: Improvement
This part of the system implies that facilities managers must find ways to improve the EMS. As a result, it makes sure that the EMS is suitable, adequate, and effective for better environmental performance. Some things need to be done, like taking care of corrective actions. Facilities managers need to respond to the problems and do something about them. They need to find similar noncompliance, and this clause tells managers to find ways to keep improving the EMS. It is necessary to look for ways to improve processes, goods, or services, especially when considering what customers want.

Example of the process of implementing ISO 14001 at a mobile phone company in South Africa is as follows:

1 Purchase the ISO 41001 (Environmental management systems – Requirements with guidance for use) standard from the SABS
2 Study the standard (and if feasible, send those who will be responsible for the implementation of the standard for training on the standard) – System auditor and lead auditor courses are available in South Africa
3 Develop an audit checklist based on the standard
4 Conduct an internal audit to determine the gap between the policies, procedures, processes, and systems in place at the company and the requirements of the standard

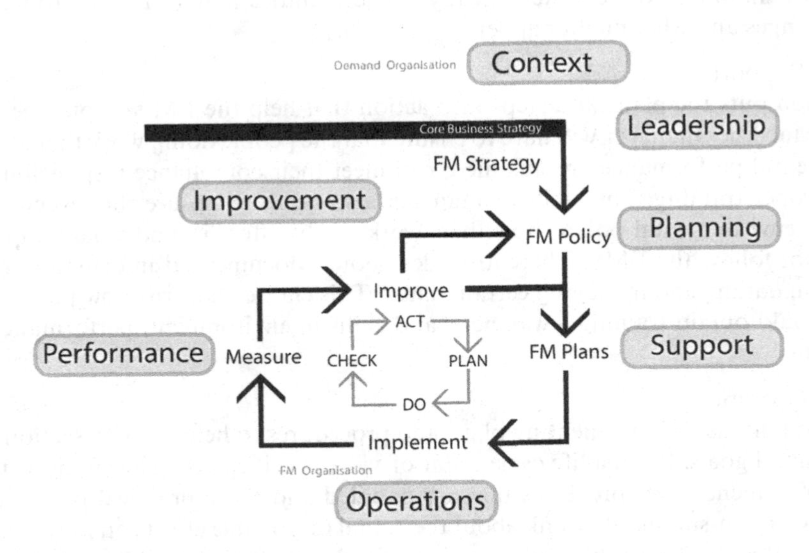

Figure 7.3 ISO 14001 Implementation process.

Source: BSI. (2015). ISO 14001:2015, Your Implementation Guide, Accessed August 2023 at ISO-14001-implementation-guide-2016.pdf (bsigroup.com)

5 Develop a plan which:
   - Identifies the different gaps (if any)
   - Indicates how each gap will be closed
   - Indicates who will be responsible
   - Indicates by when each gap will be closed
6 Execute the plan to completion
7 If time and resources permit, conduct another internal audit to confirm that all identified gaps have been closed
8 Procure the services of an accredited company to certify your company against ISO 14001

The advantages of implementing ISO 14001 in the company can be summarised as follows:

- Many Requests for Proposals (RFPs) issued to the Facilities Management industry now state certification for ISO 14001 as a prerequisite. This means that ISO 14001 certification keeps the company market relevant
- ISO 14001 certification ensures compliance with NEMA (National Environmental Management Act), a legal requirement
- ISO 14001 certification also demonstrates that the company is a responsible corporate citizen

7.4 Green Purchasing in Facilities Management

Sustainable procurement improves efficiency and effectiveness and enhances the partnership between employees and suppliers. It can help with risk management, protect a company's image, save cost, create value, make it more future-proof and transparent, and set a brand apart. The supply of products may cause greenhouse gas emissions. Organisations interested in sustainability should focus on product life cycles and procurement, including material extraction, manufacturing, eventually distribution, and end-of-life management. Since many effects happen early in a product's life cycle, we may reduce concerns, save costs, and conserve resources more efficiently by taking into account the complete life cycle of a product before making a purchase. A life cycle strategy enables reducing or even doing away with expensive end-of-life management procedures. By balancing the environmental, social, and economic aspects of procurement, a purchaser can take care of the specific needs of each organisation efficiently and effectively. Before allowing for specific product and service opportunities in FM, it is important to consider a planned sustainable purchasing programme. Green purchasing can be done for services such as cleaning through a green cleaning policy, lighting through LEDs, and installing motion sensors to reduce energy usage. It can affect buying green building materials such as carpet and flooring, furniture, paint, and hand dryers.[6]

Technology also affects many parts of sustainable procurement, like energy consumption and communication. Making procurement choices to ensure energy use is sustainable can have obvious effects on facilities management, like moving to motion sensor lights to lower the lighting costs. However, in sustainable procurement, technology is used to understand better supply and demand to avoid overproduction or joining packages to ensure that containers are fully utilised using prediction analytics. Additionally, when procuring things for a building, finding out where the parts come from and making sure they are made in a way consistent with the company's sustainability standards shows how sustainability affects procurement as a whole.

7.5 Environmental Performance in Africa

The decisions made by the owners, planners, and builders should be used to rate the environmental performance of a building. In many countries, legislation determines how energy-efficient a building is. The BRE Environmental Assessment Method (BREEAM) in the UK, the Leadership in Energy and Environmental Plan (LEED) in the US, and the Green Star in Australia are all used to determine how well a design will protect the environment.

7.5.1 Green Building Rating System

Green buildings are those that have a lower environmental effect than traditional buildings. The consumption of energy and resources, waste production, pollution, and indoor air quality are the environmental factors considered when calculating the effect. The green building approach considers the above mentioned environmental factors and does not consequently compare to building energy efficiency, as is occasionally wrongly believed. In addition, green building is a method that applies to every stage of a building's existence, not simply the design and construction stage[7].

The Global Green Building Council, a worldwide action network that seeks to improve the building industry, has reportedly led the charge to make Africa's commercial real estate market sustainable, according to the Alpin Report (2021). Having begun in South Africa, it promotes green building methods and offers tools, training, and expertise to improve construction fundamentally. Ghana, Kenya, Mauritius, Namibia, Rwanda, South Africa, Tanzania, and Zambia are among the African countries that are members of the Green Building Councils (GBCs).

The GBC South Africa (GBCSA) Green Star SA rating is one of the tools used. It is based on the Australian version. It looks at the South African context in particular. This grading tool sets standards for making developments less harmful to the environment by giving project points for meeting criteria like reducing waste in landfills and making designs that use resources more efficiently. The tool has been updated for use in Nigeria, Kenya, Ghana, Rwanda, Namibia, Mauritius, and Uganda. The GBCSA works with the Green Building Councils in these countries to certify projects.

Another example is LEED certification, which is a rating system that is widely used worldwide. Some remarkable projects that have been certified by LEED are Knowledge City and Cairo Business Park in Egypt, Procter & Gamble in Nigeria, Residencia Dorthea in Uganda, and Medina Tower in Libya. The Alpin Report (2021) says South Africa is the top in green construction on the continent. Green Star South Africa and EDGE are the most famous green building systems in South Africa. They work with world rating systems like WELL and LEED. Because of the South African National Standard 10400 Part XA, which says that buildings should use as little energy as possible, Green Star and EDGE were changed to fit the needs of the South African context. The government is still working towards its goals for 2030.

There are already two local rating systems in Egypt. They are TARSHEED and the Green Pyramid Rating System. Some projects in Nigeria already have the LEED green building certification. The Nigerian National Building Code, made by the National Council on Housing and Urban Development, and the Building Energy Efficiency Guideline (BEEG), released in 2016 through the Nigerian Energy Support Programme, both support the country's plan for sustainable development. They do this by setting guidelines for energy efficiency, using local, safe building materials, and ensuring people's health and safety are considered. In Nigeria, it is essential to consider both social and economic factors when making a performance evaluation tool. Social factors include health and safety, participation, education, knowledge, social cohesion, fairness, and comfort. Some parts of the economy are maintenance, management, market value, affordability, and durability. Some projects in Morocco are certified as LEED green building certification.[8]

7.5.2 Green Star Rating

7.5.2.1 Green Building Council Rating of South Africa

The Green Star SA grading systems recognise and reward environmental excellence in the real estate market while objectively assessing green buildings in South Africa and throughout Africa. They are created for each form of building and aid architects, designers, and builders in making better-built environments. The tools are in nine different areas, each with credits that address environmental and sustainability issues related to building design, construction, and use. Different weightings can show the environmental problems in the various building classes. The areas examined include land use and ecology, management, energy, transportation, indoor environmental quality, water, materials, emissions, innovation, and socioeconomics.

Green Star SA environmental impact categories:

Management
Indoor Environmental Quality
Energy
Transport
Water
Materials
Land Use and Ecology
Emissions
Innovation

The categories have credits as subdivisions related to environmental performance. Points are awarded for each credit based on the project's ability to meet the tool's objectives. The rating scale used is as follows:

Overall score	Rating	Outcome
10–19	One star	Not eligible for formal certification
20–29	Two star	Not eligible for formal certification
30–44	Three star	Not eligible for formal certification
45–59	Four star	Eligible for four star/rewards best practices
60–74	Five star	Eligible for five star/South Africa excellence
75+	Six star	Eligible for six star/rewards world leadership

7.5.2.2 LEED Certification

The US Green Building Council-established LEED, or Leadership in Energy and Environmental Design, is also a certification method for creating green buildings. LEED certification is globally recognised as a mark of leadership and success in sustainability. The latest LEED rating system consists of five different areas addressing multiple projects:

- Building Design and Construction
- Interior Design and Construction
- Building Operations and Maintenance
- Neighbourhood Development
- Homes

The four levels of certification are:

- Certified (40–49 points)
- Silver (50–59 points)
- Gold (60–79 points)
- Platinum (80+ points)

The LEED certification process needs to meet, at a minimum, these requirements:

- Comply with environmental regulations and standards
- Meet the threshold of floor area requirements
- Meet a minimum of building occupancy in terms of the number of users
- Maintain a reasonable site boundary
- Be a permanent building
- Share energy and water usage data
- Must have a minimum building-to-site area ratio

LEED Certification Credit Categories: To earn credits to achieve one of the above categories, the project must meet specific criteria and goals within the following categories, showing maximum number of points achievable:

- Location and transportation (0/16)
- Materials and resources (0/13)
- Water efficiency (0/11)
- Energy and atmosphere (0/33)
- Sustainable sites (0/10)
- Indoor environmental quality (0/16)
- Integrative process (0/1)
- Innovation (0/6)
- Regional priority credits (0/4)

LEED for Neighbourhood Development provides additional credit categories, such as smart location and linkage, neighbourhood pattern and design, and green infrastructure and buildings.

7.5.2.3 Green Pyramid Rating System

Egypt's Green Pyramid rating system (GPRS) was created by the "Housing and Building National Research Centre" in 2009. The main goal of the GPRS is to make buildings in Egypt last longer by making more people aware of how critical green buildings are for Egyptian cities and the environment. The most crucial part of GPRS is evaluating new facilities after they have been built and during the planning phase. In the GPRS scoring system, seven parts make up the point-weighting method. These are shown in Table 7.3. For green buildings in GPRS, there are three stages of certification: Silver Pyramid, Golden Pyramid, and Green Pyramid.

Table 7.3 shows that the Egyptian Green Pyramid rating method considers seven factors. Each part includes prizes and codes that can be earned when used in the design. Aside from that, each part counts for a certain amount towards the building's certification. The 2017 (second) version of the GPRS shows that these weights have changed. Rethinking the importance of each part and what it brings to a "green building" made the change possible. Daoud et al. (2018) stated that the GPRS needs to change the weights of the different groups to reflect the current problems in Egypt, such as energy efficiency and water, and to use materials and resources better.[9]

Examples of LEED-Certified Buildings in Selected Countries in Africa

NEC Addis Ababa, Ethiopia
Size of the building: 155,200 sq ft
LEED-NC 2.2
Rating system: LEED BD+C
Scorecard:
Sustainable cities 7/14
Water efficiency 2/5
Energy and Atmosphere 4/17
Materials and Resources 6/13
Indoor Environmental Quality 7/15
Innovation 2/5
Certified 28/69[10]

Table 7.3 Green Pyramid-weighting Aspects

Aspect	*Green pyramid assessment % in V1*	*Green pyramid assessment % in V2*
Water efficiency	**30%**	**20%**
Energy efficiency	**25%**	**32%**
Sustainable sites	**15%**	**10%**
Materials	**10%**	**12%**
Management protocols	**10%**	**10%**
Indoor environmental quality	**10%**	**16%**
Innovation and value-added	**Bonus**	**Bonus**

Source: Daoud, A.O., Othman, A., Robinson, H., & Bayati, A. (2018). Towards a green materials procurement: Investigating the Egyptian green pyramid rating system. In *3rd International Green Heritage Conference*

Vodafone SV-C2 Building, Egypt
Size of the building: 13,200 sq m
Rating system: LEED v4.1 O+M: Existing Buildings
Scorecard:
Location and Transportation 10/14
Sustainable cities 0/4
Water efficiency 8/15
Energy and Atmosphere 28/35
Materials and Resources 7/9
Indoor Environmental Quality 18/22
Innovation 0/1
Certified Gold 71/100[11]

Ridge Hospital, Accra, Ghana
Size of the building: 248, 862 sq ft
Rating system: LEED 2009 Healthcare
Scorecard:
Sustainable cities 10/18
Water efficiency 6/9
Energy and Atmosphere 12/39
Materials and Resources 9/16
Indoor Environmental Quality 8/18
Innovation 4/6
Regional Priority Credits 4/4
Certified Silver 53/100[12]

Example of Green Building Council-Rated Building in Namibia
FNB Building, Independence Avenue, Windhoek, Namibia
Type of certification: 5-Star Green Star office building
Size of the building: 15,943 sq m
Total score: 64 points

The scorecard includes management, indoor environmental quality, energy, transport, water, materials, land and ecology, emissions, and innovations. The sustainable building features of the building include:

- The Parkside building is estimated to be at least 50% more energy efficient than a comparable office building built and operating according to building regulation minimum.
- Every major energy consumption component is metered and monitored; luminaires used in the office space work, each with an embedded sensor. The lighting system can be reconfigured without the need to rewire the system.
- The water system installed in the building exceeds the "most efficient" benchmark set by the GBCSA. This is due to the installed wide rainwater and graywater capture and filtration system and low flow fittings and fixtures.
- The occupants experience high thermal comfort, a link to the outside world due to external views, manually controllable blinds on external facades and high fresh air rates.[13]

Example of Green Building Council-Rated Building in Kenya

Garden City Village Phase 1, Garden City Thika Road, Nairobi, Kenya
Type of certification: 4-Star Green star Multi-Unit Residential Design v1
Size of building: 15,113 sq m
Total score: 46 points

The scorecard includes management, indoor environmental quality, energy, transport, water, materials, land and ecology, emissions, and innovations. Sustainable building features include maximum electrical demand reduction.

The following design initiatives were adopted to reduce the maximum demand on the electrical supply infrastructure:

- A solar hot water system was installed with hot water storage vessels insulated following SANS-204: 2011. The development has installed non-electric cooking hobs and ovens. Gas appliances were installed with a higher energy rating than electrical appliances.
- Occupant User Guide: A comprehensive building user guide given to the building tenants to enable the occupants to optimise the dwelling's environmental performance. Occupant Amenity Water: Portable water consumption has been reduced by 28% through rainwater harvesting and filtration and the installation of water-efficient fittings.
- Volatile Organic Compounds: Interior finishes such as paints, adhesives, sealants, and bamboo flooring with low or no VOC were used.
- Local Connectivity: The development is integrated with the surrounding built environment amenities to encourage effective car-based trip reduction.
- GHG Emissions Heating and Cooling: Energy modelling ensures minimised greenhouse gas emissions through increased passive thermal performance, passive thermal performance, and lighting efficiency (GBCSA, 2023).

7.6 Concept of Zero Carbon

The move to low or no carbon emissions is becoming a big issue. Four South African cities – Thekwini Municipality, the City of Cape Town, the City of Johannesburg, and the City of Tshwane – have publicly stated their intention to achieve low to zero carbon building profiles for all new and existing structures by 2030 and for all city-owned structures by 2050, through the C40 platform.

The Green Building Council of South Africa says that a low to zero carbon building, also called a net zero carbon building, is highly energy efficient and is powered by renewable energy sources on- and off-site. It could include offsets that would lower the effect of a part of the energy consumption of a building.

Energy efficiency is the most essential part of building with net zero carbon emissions. A building with net zero carbon pollution uses more green energy sources to help it reach its goal. Before showing off green energy systems or offsets, lowering the energy a building needs is essential. This will keep these systems' prices and installation to a minimum. However, planning a building with energy efficiency in mind can cut its energy use by a significant amount. For instance, it lets natural light in so that less lighting is needed. It plans good airflow to cut down on the need for cooling. Depending on the type of building and sub-sector, the building industry uses a wide range of tools that use much energy. Net Zero Certification can be obtained through the GBCSA. The GBCSA's Net Zero Certification process has

Carbon, Water, Waste, and Ecology categories. These categories credit projects that cancel entirely out (net zero) or positively restore (net positive) their environmental impacts. In addition to any Green Star certification a project may already have, net zero approval.

Different things happen in other African countries when it comes to zero carbon. Namibia has promised to cut its pollution by 91% over the next five years as part of COP26. One of these programmes is the Southern Corridor Development Initiative (SCDI) in the Karas area of Namibia. The SCDI includes a group of Karas-area building projects that may influence Namibia's economy if well planned and carried out. The portfolio has farms, green hydrogen facilities, ammonia assets, rail concessions, port concessions, renewable energy plants, and mining centres nearby. Namibia and Rotterdam have signed a Memorandum of Understanding, and Namibia and Germany have signed a Joint Communiqué of Intent. It went beyond policy talks and created a great sale for a 40-year concession that included more than 5,700 km2 of land that could be used for green energy.

As of June and July 2022, Egypt still needed to have a net zero objective and submit a long-term strategy (LTS) to the UNFCCC. Egypt's 2050 National Climate Change Plan was released in May 2022, but it lacks an emissions reduction goal.

In the case of Nigeria, it should establish yearly carbon budgets and plan to reduce emissions to zero. The president signed a climate bill into law, pledging to develop a strategy to decrease emissions, adapt to climate change, and set yearly and five-year carbon budgets. It declared a 2060 net zero aim during the COP26 climate conference in Glasgow. The Nigerian government appears hesitant to take the necessary steps to get on the emission reduction pathway despite its goal to meet its net zero target by 2060. The failure to perform the Climate Change Act is one of the causes. The Act, among other things, gives the Federal Ministry of the environment the power to establish a five-year carbon budget or the maximum amount of greenhouse gas emissions permitted over a given period. Nigeria's revised nationally determined contributions (NDCs), filed in July 2021, needed to correctly satisfy the requirements, according to the Climate Action Tracker (CAT), whose net zero target information needs to be included.

Nigeria needs to make a clear effort to hold low-carbon technology or management choices, according to a 2013 World Bank assessment by Cervigni (2013). The study predicted that emissions would double by 2035 as a result of changes in carbon emissions due to the following factors: a slower rate of conversion of forests to cropland due to the clearing of a large portion of the forested area in recent decades, a rapid expansion of electricity generation (mainly from thermal power technologies); and increased demand for passenger and freight transportation due to the need to support planned GDP growth. If all the low-carbon solutions considered were gradually implemented over time, emissions would become stable at about 300 million metric tonnes of CO2 equivalent (Mt CO2e) annually, preventing acquiring a total of 3.8 billion Mt CO2e during 25 years. According to estimates, the power sector can reduce carbon emissions by around 50%, the oil and gas industry by 20%, and the transportation and agricultural sectors make up the remaining 30%. Agroforestry, avoiding deforestation, and conservation agriculture are some of the leading low-carbon solutions in agriculture and land use. Reduced gas flaring, improved oil storage management, and increased energy efficiency in oil and gas facilities are all improvements in the oil and gas business.

For instance, Shell admitted in the Farther Africa report from 2021 that its spill-prone activities in Nigeria, where it has been extracting oil for 50 years, challenge its green plan. For over a decade, the company has been increasingly selling onshore assets in West Africa as it tried to move beyond stubborn issues, including pollution brought on by broken pipelines and the resulting legal disputes with local populations. Since Shell announced its target to become a global leader in renewable energy and finally close down its oil and gas operations to reach net zero carbon emissions by the year 2050, the problem has grown more urgent. Ghana's vice-president stated in 2022 that the country was devoted to achieving net zero carbon emissions by 2070.

As part of the Paris Agreement, Kenya promised to fight climate change by reducing carbon emissions by 30% by 2030. The Kenya Climate Change Action Plan of 2013 is a five-year plan based on the National Climate Change Response Strategy (NCCRS) 2010. Kenya has signed the UNFCCC and is in the group of countries called "Non-Annex I," which means they do not have legally binding goals to reduce their greenhouse gas emissions. However, Kenya has taken steps to mitigate the effects of climate change, doing what it feels is right to do to deal with it and take responsibility for lowering carbon pollution. Kenya is trying harder to reduce its carbon footprint. In 2010, the country showed off its National Climate Change Response Strategy. The following year, it had its National Climate Change Action Plan. The action plan from 2013 to 2017 was on how to take action and reduce carbon emissions so that a developing country can be more climate-resilient.

Strathmore University in Kenya is the first education institution in Sub-Saharan Africa to have a carbon footprint that is equal to zero. With 2,400 panels on the roofs of six buildings, it is the biggest rooftop solar project in the area. Kenya is on the equator, so it gets sunlight all year long. To meet its energy needs, Strathmore began a project to build a 600 kW rooftop solar photovoltaic (PV) power system linked to the power grid. Strathmore did this by using a "green line" of financial aid set up by the French government. Due to its ability to make more energy than it needs, the system sells the extra energy to the utility company under a power purchase agreement. With this investment in clean energy, schools can show the way regarding energy and environmental problems.[14]

In Rwanda, for example, the government said it would have no CO2 emissions by 2050. But it hasn't yet said what kind of decarbonisation plan is allowed under the Paris Agreement. Still, the long-term is considered when making its climate strategy. To meet its 38% reduction goal for emissions by 2030, the government is working with partners in and outside the country to bring in green investments that will last. In May 2020, Rwanda was the first country in Africa to turn in a new climate action plan. The government will probably spend $11 billion over the next ten years on steps to reduce and respond to climate change, split into $5.7 billion for prevention and $5.3 billion for adaptation. The Rwanda Green Fund (FONERWA) was created to ensure that the country's funding of the goals stays ordered. Since it began, the Fund has raised $217 million for green projects. Rwanda was among the first countries to eliminate plastic bags and other single-use items between 2008 and 2019. Rwanda works to keep its forests healthy and to plant new trees in parts of the country, so 30.4% of its land is covered in forests. To meet the goals of the Kigali Amendment to the Montreal Protocol, the government put in place a National Cooling Plan that would, over time, eliminate or cut down on the use of HFCs, potent greenhouse gases used in cooling systems. Rwanda invests much of its earnings into renewable energy, climate-smart agriculture, sustainable urbanisation, and e-mobility.

7.7 Zero Carbon Stages for Facilities Managers

The following are ways whereby facilities managers can ensure buildings have net zero carbon emissions:

Involve the facilities manager from the design stage

Including the facilities manager in the design process will help ensure that ideas about energy use and occupants' comfort are considered. The facilities manager should work with architects and builders to find answers to building problems.

Creating an energy use and emissions baseline

Before developing a carbon neutrality strategy, the baseline carbon emissions, energy efficiency, and utility costs, such as water use, waste, and all fuel types, must be determined. This will allow one to compare all decreases, leading to repeatable data-gathering methods.

Start with simple goals

Facilities managers should look for energy-efficient sources. Changing pre-set temperatures on things like HVAC systems, lowering the temperature in an air conditioner by 1° Celsius, or turning it "off" when the room temperature becomes a certain level could be part of it. An intelligent IoT system in the cloud can give real-time information about energy consumption and allow smart energy management.

Sustainable facilities management

To be more environmentally friendly, facilities put better services under their buildings, reduce energy waste, track all their equipment, and use the Internet of Things (IoT) to lower their carbon footprints and emissions. To reduce carbon emissions, facilities managers also inform occupants how to lower their carbon footprints and add green features that will last. Following regional, government, and international emission requirements makes the net zero emissions goal much more likely to come true.

Employing techniques for ongoing development

Facilities managers make many complex expenditures to help reduce building emissions. The following are included: Sensor-based lights should be switched off when no one uses a room or office – use Internet of Things (IoT) sensors to determine whether office spaces are occupied and switch off lights or HVAC systems when unnecessary. Facilities managers can establish and apply rules and procedures throughout their portfolios and analyse spaces and offices.

Occupant awareness

The facilities manager should look into energy-related habits and ways of life that can be used in performance models to make building occupants more aware. They should teach the occupants about energy efficiency through human resources, communication campaigns, staff or team events, or easy-to-read screens showing energy use data.

Preventive maintenance

Facilities managers should ensure that buildings and services are properly maintained to reduce a building's carbon footprint. Poor performance can cause equipment failure, increased maintenance requirements, energy waste, and downtime. Preventive maintenance will reduce the energy consumption of equipment like HVAC systems. Facilities managers can track and get information on planned maintenance and budgets, reports, and project issues with alerts to meet monitoring requirements[15].

Understand embodied carbon

Embodied carbon emissions must be assessed in buildings. Embodied carbon refers to the carbon emissions linked to manufacturing and transporting the materials used in infrastructure and building construction. This includes the emissions produced when concrete, steel, and wood are mined, manufactured, and transported. Reducing embodied carbon is a main part of an overall carbon neutrality strategy. Therefore, a main part of any decarbonisation programme is reducing the energy usage of the building portfolio. An efficient approach is needed to conduct facility evaluations and energy audits and put the energy conservation measures (ECMs) suggested in the energy audit reports into action. Getting extra funds from energy cost avoidance transferred to the sustainability programme needs prioritising when facilities go through repairs.

Construct an ongoing energy monitoring system to ensure that systems retain increased energy efficiency

Buildings must have passed an energy audit process to sustain performance, including finishing repairs and ECMs. Ongoing energy monitoring is required for systems to continue enhancing their energy efficiency. Implementing a monitoring-based commissioning system to regularly analyse fault detection data analytics to maintain and enhance the performance of systems is helpful. When the electricity bill comes, instead of determining a problem and then spending time looking for the source, by determining the cause, system drift can be avoided by determining what caused it. A monitoring-based commissioning system may also automatically determine carbon counts and check for continued energy efficiency when used correctly.

Convert systems that burn fuel to electric ones

The electrification of building systems is the next stage. This involves changing systems that consume fuel to electric ones. For instance, using natural gas has cheaper utility costs than electricity. There are improvements in on-site renewable energy systems if facilities are more energy-efficient and all fuel-burning systems replace electric ones. This usually involves photovoltaic installations and adding on-site renewables since the power provided to a site would likely be produced using renewable energy. However, electric distribution systems cannot provide the power needed to supply energy to 100% electric buildings. This is a big issue since many buildings will follow their sustainability objectives using a mixed-fuel strategy. In addition, facilities may have features that improve combustion equipment's efficiency and air quality. For example, high-efficiency condensing boilers would include combustion systems and control procedures. Switching to all electricity may not be possible through improvements from using fuel-based facilities of a carbon neutrality strategy.

Use certificates for renewable energy to compensate for emissions

Net zero emissions can be achieved by obtaining renewable energy certificates, which can be used to balance emissions from mixed-fuel buildings and consider a building with embedded carbon. When a renewable energy source produces and transmits one MWh of electricity to the grid, Renewable energy certificates (RECs) are issued. According to the REC, the power is produced and used with little to zero emissions. It also comes from a clean source. The building can become carbon-free by getting RECs since it uses certified sustainable energy[16].

Examples of net zero buildings in Africa

As part of its response to international efforts to address climate change, the American University, Cairo (AUC) joined the Race to Zero campaign in 2022. This is a UN-backed global effort to make the world zero-carbon. Egypt wants to cut emissions by half by 2035 and achieve net zero emissions by 2050. AUC did a study focusing on its New Cairo Campus, where most of its activities occur. The Clean Air-Cool Planet Carbon Calculator (CA-CP) method was used to determine AUC's emissions. This method uses an Excel file to determine an annual total carbon footprint. The data that was used was reported from 2012 to 2022. The carbon emissions for the period of the study were provided. Its carbon footprint in 2022 was mainly transportation, electricity for lights and other equipment (non-HVAC), water supply, paper, refrigerants, solid waste disposal, and heating, ventilation, and air conditioning (HVAC). Standardised methods show that between 2012 and 2022, AUC's carbon footprint decreased by 4,103 MT CO2e, or about 10%, going from 41,031 MT CO2e to 36,928 MT CO2e.

In 2022, HVAC made up about 44% of the carbon footprint. This is probably because that institution is in the desert and needs air conditioning for over half the year. Most of these CO2 emissions come from the energy used for cooling. Since 2012, HVAC emissions have dropped by 20% due to using building management systems to make the buildings warmer or cooler, setting the temperature so that it changes based on the weather outside, installing more accurate fan coil units (FCUs) and variable air volume (VAV) thermostats for each room, and making the control algorithms better.

In 2022, about 30% of the carbon footprint came from transportation. Most of these emissions came from people driving their cars and taking buses to get to work. Because the community travels every day, it has a big effect on the total carbon footprint. From 2012 to 2016, CO2 emissions from travel grew mainly because more people drove their cars to work. But from 2016 to 2019, emissions from private cars driving went down by 21%. This was probably because petrol prices increased, more people used ride-sharing services, and commuters drove less. Because more people taught and worked from home online and because of COVID-19, private car emissions went down a lot in 2020 and 2021. People began driving to and from campus again in 2022 when the university started running properly again, causing more emissions.

In 2022, electricity was about 20% of the carbon footprint. Between 2012 and 2022, non-HVAC emissions dropped by 24%. This was from consistent attempts to save electricity, like reactivating and reprogramming motion sensors in all buildings' long hallways and the COVID-19 pandemic. A little over 1.25% of the carbon footprint in 2022 came from refrigerants. Because of increased maintenance and more stand-alone air-cooling units, the amount of refrigerant used and the CO2 emissions that went along with it increased from 2012 to 2019. However, overall, these emissions went down by 19% between 2012 and 2022 because of better performance.

Natural gas use constituted about 0.18% of the carbon footprint in 2022. The study found that natural gas used for home and lab purposes was over-reported from 2015 to 2020. This was possible because the methods had been better since 2016. The institution's water use resulted in up to 1.83% of the carbon footprint in 2022. There has been a reduction in water after 2012 since cleaned water was preferred to domestic water in the irrigation of the university's greenery. In addition, other forms of water conservation were explored.[17]

In South Africa, 2 Pybus Road, Sandton (GBSCA Net Zero Carbon – Level 2; Level 2: Building & Occupant Emissions (measured)) is an office building meant to look like chambers for lawyers. The offices are more cell-like, and each person has their own space. An L-shaped building with a glass front is on the north side. A punch window is used in the office rooms that face away from Pybus Road. The building has eight office levels, six parking levels, and parks on the podium level. Pedestrians can get into the building at Level 4 and take the lift to the top level. Some sustainable features of the buildings include checking the indoor air quality to see how well indoor pollutants are being monitored and controlled for the comfort and well-being of the building users. The Building Users' Guide, the Preventative Maintenance Management Plan, the Landscaping Management Plan, the Hardscape Management Plan, and the Pest Management Plan are used. Also used is a Green Cleaning Policy that meets the standards of the Green Star that sets green operating rules for tenants (GBCSA, 2023).

Net zero energy structures projects in Nigeria, such as the African Development Bank regional office in Abuja, are being developed. The building will include solar panels, energy-efficient lighting, and a rainwater collection system. Some private developers are also embracing the net zero energy trend. For instance, Constructability Limited, a real estate developer in Abuja, is creating a net zero-energy commercial building with roof insulation, solar panels, and rainwater harvesting. The building will have environmentally-friendly building materials, such as Camhirst's 3D-printed concrete, and eventually reduce construction waste by implementing circular economy concepts.

Case Study: Arup

A recent study by Arup used six case studies from Arup projects to look at whole-life carbon assessment of buildings based on the World Business Council for Sustainable Development (WBCSD) Framework. The case studies included four office buildings in London, the UK, a mixed-use building in Copenhagen, Denmark, and a residential timber tower in Amsterdam, the Netherlands. The work shows that getting all the needed data from all the stages of a building project's life is still hard. Even so, information must be used at the beginning of the decision-making process when there is the most opportunity to lower carbon emissions over the product's life.

The case studies, all about some form of low-carbon design, show that clear goals could be set and that the world's building-related emissions could be cut in half within the next ten years. The case studies also help us learn more about the main factors leading to the decarbonisation of the built environment. For example, more than half of the emissions from new buildings may come from the carbon already in the building, with 70% coming from just six materials. Around 20% of all the emissions a building produces over its lifetime come from the maintenance and repair of its installations. Because of this, lowering these emissions and keeping the focus on doing so is very important.

Outcomes

Upfront embodied A1-A5:

The six case studies show that the upfront embedded carbon averages between 500 and 600 kgCO2e/m2. A global goal in this range could be set immediately, showing that global progress can be made.

In-use and end-of-life embodied B-C:

The case studies show that we need to fully understand the in-use embodied carbon, which is more than 300 kgCO2e/m2 (carbon dioxide equivalent emissions) on average when using the standard accounting methods. Adopting circular economy principles instead of replacing all essential parts, which is currently thought to happen, needs more attention to plan out this effect. The company requires a more precise and accurate picture of the decarbonisation of materials over time so that informed decisions have the most negligible impact over the whole life. The case studies also show that end-of-life embodied carbon doesn't significantly affect the overall numbers. The only time it does is when organic materials like wood are considered, and more information is still needed about the different end-of-life incidents that could happen.

Operational:

From about 75 to 220 kWh/m2/year, the practical energy use changes significantly between the case studies. Because the carbon content of the grid varies from place to place, these units are in terms of total energy use instead of greenhouse gas emissions. Most of the case studies use estimates to estimate energy consumption. Getting better data on actual energy use is necessary to confirm these supposed numbers. To see if buildings are on track to meet the general emissions reduction goals, there is a need for a clearer understanding of how the supply grid will become less decarbonised throughout the building's life.

Key challenges and opportunities

One important thing is how hard and time-consuming it was to make the case studies. Collecting the same whole life cycle assessment (WLCA) data for all tasks was necessary. There was a need to speed up the process of making and sharing open WLCA data immediately. It is especially worrying that the carbon intensity data for building materials and parts worldwide is not always available or consistent. Based on the case studies, only six materials relate to about 70% of all upfront embodied emissions. It seems likely to have lower embodied carbon pollution through research, development, and knowledge sharing through industry focus and collaboration.

Key points

Commit to WLCA on all projects:

- Measure everything, at all stages, on all projects
- Consistent methodology and approach
- Process of open source sharing of data

Develop consistent and transparent carbon intensity and benchmark data:

- All components, systems, and materials must have a carbon intensity certification
- Collect and share in-use energy consumption data
- Better understanding of supply chain and national energy grid decarbonisation trajectories

Define explicit targets:

- Clear, simple global targets adopted across the buildings industry
- A valid approach to residual carbon emissions
- Supportive international and country-specific policy and legislation

Define net zero buildings:

- Clear and precise definition of net zero buildings aligned with overall global decarbonisation, emerging net zero definition and the Paris Agreement
- Establish wider collaboration
- Individual organisations taking action is not enough
- Rapid industry-wide systems change is required
- All stakeholders across the value chain must play their part[18]

7.8 Energy Efficiency

Not only is energy efficiency a concept, but it is also being considered a policy by commercial, industrial, and other organisations. Energy efficiency is a general concept without a single quantitative measure. Energy management includes teaching tenants about energy conservation, developing an energy strategy built on leadership and communication, and reducing energy usage. The general agreement is that, although energy cannot be quantified, there should be some energy efficiency performance since excessive energy usage does occur. Apart from the energy efficiency situation, there is also the understanding that a single, really bad carbon footprint in one location has an effect on the whole planet. Since there is a local and global energy problem, more energy-efficient policies and practices must be implemented. In addition, some elements, including national legislation, corporate policies, and training and skill sets, may support efficient processes. Energy-efficient methods are necessary to mitigate the negative effects and prevent further harm in light of the ongoing global energy consumption issue. Regarding energy efficiency in facilities management, improved building maintenance techniques must be used.

According to research by Adewunmi et al. (2019), organisations mostly utilise energy-efficient lighting and equipment. Such outcomes might result from increased worries about energy use and its environmental effects. Promoting energy- and cost-effective techniques is necessary, including de-lamping, changing out lamps and luminaires, installing occupancy sensors, and harvesting daylighting when illumination in buildings exceeds industry-recommended light levels. Energy conservation is becoming a concern for owners and designers, making using energy-efficient lighting more and more crucial.

One is the labelling of appliances practice. Many organisations also employ improved natural ventilation by installing curtain walls, which, when opened, allow for natural lighting and ventilation. Managers may sometimes use load reduction to decrease operational expenses, including peak demand and energy expenditures. Data is a major factor in the application of this kind of activity. Despite being essential to achieving energy efficiency, occupant awareness was seldom used. There is a favourable correlation between energy efficiency awareness campaigns and lower levels of energy use. Leadership, motivation, and communication skills are required to advance energy efficiency.[19]

7.9 ESG and Facilities Management

First, FM companies are pressured to use environmentally friendly methods to help fix the problems caused by people's actions that damage society and the environment. When choosing service providers, clients and other stakeholders are putting more emphasis on environmental, social, and governance (ESG) factors. FM companies that show they care about the environment and people may get more contracts and keep clients while also making the world a better place.

Second, an FM company's ESG success can greatly affect its bottom line. If a company doesn't deal with ESG risks, it could lose money and damage its image. On the other hand, those who take the initiative to handle ESG issues may see lower running costs, a better brand image, and access to new markets and investment possibilities.

Third, FM companies have a special duty to encourage eco-friendly practices and lessen the damage that buildings and facilities do to the environment. By using ESG practices and spreading sustainability to clients and other important people, they can help make the built world more sustainable and responsible.

E – Environmental relates to the environment and includes waste reduction, energy efficiency, water conservation, emissions reduction, green building design and operations, climate risk mitigation, and biodiversity protection.

S – Social looks at staff issues, community management, occupational health and safety, good occupant health, employee welfare, community engagement and impact, and diversity and inclusion.

G – Governance relates to the decision-making processes, reporting, and ethics, and requires ethics and transparency, risk management, and stakeholder engagement.

7.9.1 Integrating ESG into FM

The following is a framework by the International Facilities Management Association (IFMA) for integrating ESG to FM:

- Learn about ESG factors and how important they are
- Learn more about the ESG goals and aspirations of your organisation
- Figure out what tools, such as technology or people, to add ESG to processes

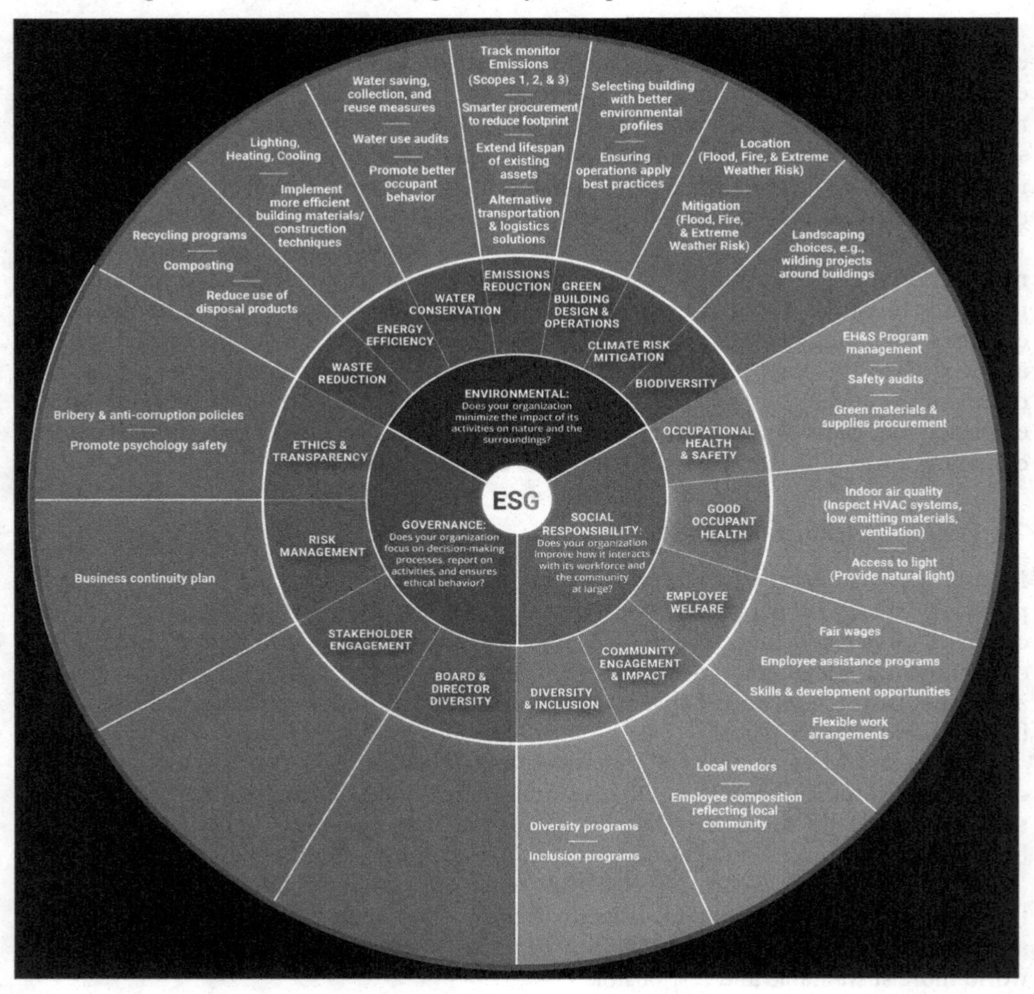

Figure 7.4 ESG.

Source: IFMA. (2024). *ESG and facility management*. Accessed August 2024 at ESG and Facility Management (ifma.org)

- Start working on an ESG approach for real estate and FM plans
- Ensure access to data, and encourage openness and transparency
- Implement new tools, train employees, or change business processes to ensure ESG plan work on a large scale
- Check in on and review ESG success regularly to ensure goals are met

The steps for integrating ESG into FM include the following:

Awareness: Facilities managers must grasp the significance of ESG issues and their impact on their organisation. Understanding the challenges E, S, and G encompass is crucial to their role in addressing these issues. Sustainability is not a widely embraced concept in many countries in Africa.

Knowledge: Once facilities managers understand ESG issues, they are empowered to examine their company's ESG goals and aspirations more deeply. FM professionals should proactively explore the ESG frameworks their organisation uses and identify those offering FM possibilities.

Resources: Integrating ESG into FM processes may require additional resources from within or outside the company, such as technology, training, or personnel. In Africa, there is a skills shortage in the FM industry and a need to use technology suitable for the continent. Facilities managers should recognise the importance of resource management, identify the necessary resources, prioritise ESGs, and devise a plan to acquire them or collaborate with existing groups.

Develop: Once facilities managers have the necessary information and tools, they can start making an ESG strategy for their organisation's FM and Real Estate plan. Setting goals and aims for ESG factors, such as reducing energy consumption by 20% in the next year, improving employee satisfaction by 10%, ensuring 100% of waste is recycled, figuring out who the important people are, and making an execution plan that includes communication may all be part of this. ESG should move from compliance with regulations to having them in decision-making.

Data management: For ESG to be appropriately integrated into FM processes, facilities managers must have access to the correct data and promote openness and honesty. This could include information about how much energy is used, how waste is handled, or how workers are treated. Facility managers should set up ways to collect this information, check its correctness and reliability, and then review it. Limited data and improper reporting standards can be a problem for African companies.

Application: Once a plan has been made to include ESG in FM processes, it's time to implement it on a large scale by setting ESG goals that can be checked and followed. This could mean putting new tools in place, training staff, or altering how things are done. Implementing ESG in Africa requires access to funds and infrastructure, engagement with all stakeholders concerned with the ESG project, good governance, and transparency.

Monitor: The facilities managers should check and review their ESG performance daily to ensure they meet their goals and aims. This could mean doing regular checks, getting peer feedback, or monitoring progress with success measures, such as key performance indicators (KPIs) or sustainability ratings.[20]

7.10 Management of Water

Water management must manage the water supplies and wastewater entering and exiting the facility and recognise the high costs and risks.[21] Water is scarce and needs to be maintained and regarded as such. The cost of treating water and water supply is little. Organisations should reduce water consumption and harvest rainwater where possible, and end user behaviour should determine the extent of consumption. Campaigns can help, and active measures should be taken to help conserve water. Site design can make use of rainwater run-off. Potable water is likely safe for human consumption when it gets to the facility. Most plants have a treated water source, but the facilities manager is responsible for maintaining the water's quality after it reaches the facility.

Facilities managers must consider water management to follow the legislation, make cost-effective decisions, and show savings. Generally, there are three approaches to water management:

1 Reducing losses
2 Reducing the amount of water used by equipment or processes
3 Reusing water that would usually be thrown away

7.10.1 Water Management Plan

Creating a water management plan is analytical and sequential, including much more than just preparing a report and conducting a cost-benefit analysis. A water management strategy must consider both the technical, such as installing water-saving plumbing fittings and the human, such as changing staff members' fixed operational procedures and water-using attitudes. A water management plan seeks to prevent water wastage and for effective water management. It should be part of the maintenance plan and can be combined with plans for energy management, operations and maintenance, compliance with regulations, and emergency plans.

7.10.1.1 Plan Outline

Get a team together: Put together a dedicated water management team with professionals and staff, and ensure a team head manages and carries out the plan. Employees from across the organisation should be on the team, including someone who knows how to comply with regulations and facilities or a building manager who knows about the building's major mechanical systems and infrastructure. The team could include a few of the following:

- Representatives from the facilities' management
- The director of the physical plant or the chief operating engineer
- A representative from the maintenance department
- Representatives of departments known to use significant volumes of water
- Design or water management consultants
- Qualified plumbing and mechanical, landscape, or other water management contractors

Develop a water management policy: This policy should provide guidelines for establishing and accomplishing water management goals. The policy should be included in the organisation's performance and sustainability report to be tracked and used to help make water quality and efficiency more critical while being accountable.

Assess water usage: Knowing how much water a building uses is essential for water management plans. A water survey provides a comprehensive report about how the building uses water. The water management team can use it to set a standard for tracking growth and determining the programme's success. It also helps the water management team set goals and determine the most critical projects based on how much money they will save and how well they will pay for themselves. To evaluate a building's water use and find water quality problems, there is a need to do the following:

- Find information that is easy to get about water usage
- Determine what the building's main water uses are.
- Understand the way water moves through its systems.
- Determine how much it costs to operate

A standard of water use data found on water bills and metres can be used to determine water usage. A water use study can be done by listing the building's most water-consuming features, systems, and equipment. This can be done by looking at old data, visiting the building to list water-using equipment, and checking water use when possible. It can also be done by setting up a building water balance. It should show how much water different end users use compared to the whole facility's water use (i.e., baseline). It is a great way to identify, assess, and prioritise ways to make the facility more water efficient. All kinds of buildings are concerned with water quality and safety. In addition not knowing how to use tools and methods can be hazardous for water safety (if it is in the plan).

Set goals and communicate them: Once the water management team knows how the building uses water, the next step in the planning process is to get building owners, facilities management staff, senior management, and any other vital decision-makers jointly to make a list of water management policies and goals. Objectives should be set with employees from across the organisation to obtain a variety of perspectives and meet stakeholders' needs. The goals will guide the water management programme and help ensure things keep improving. Once policies and objectives for water management are made, they should be communicated to the whole organisation with the help of top management or the building owners. Having support from the top gives the project authority and lets employees know that saving water is a top concern. A way for feedback to provide comments should be set up for input, generating ideas, and reporting problems.

Formulate an action plan: The water management team can make a thorough action plan based on the water balance and any significant improvement areas during the assessment process. This includes turning water management possibilities into specific projects or changes to maintenance and then putting those projects in order of importance. The action plan should list the projects and methods that can be used at the location to meet the water management goals. To develop an action plan, there is the need to create a list of actions or projects, determine their costs and possible savings, identify funding sources, determine simple payback and net present value, prioritise projects, and document goals in a thorough action plan.

Implement the action plan: The water management team should develop a focused strategy for implementing the action plan. This will help the project succeed and allow the water management team to reach its goals. This could mean getting support for specific ideas and ways of doing things. When implementing the action plan, there is a need to make standard operating procedures (SOPs), checklists, and other materials to help the maintenance staff turn practices into their regular responsibilities. These can include actions for water efficiency, water safety, emergency preparedness, and waste management. Documentation is needed to monitor a completed task, preventative maintenance, or ongoing monitoring.

Evaluate progress: The water management team should do official reviews of water use and quality data and regularly implement the implementation to ensure they meet the water management goals. The organisation can use this review to evaluate its success, set new goals, and continually improve. The water management team can also use the study to show and promote how well the water management programme works, and this can help the programme, future projects, and efforts for a long time.

Recognise achievements: to attract and retain support for a building's water management programme, the team might consider giving awards for water management actions and successes. This includes acknowledging the work of those who helped reach the goals for water management and promotion both inside and outside the programme about its success.[22]

The influence of top management: The organisation's top managers must continue to support the water management plan. This means they must commit to following through with the written policy on water management and be ready to provide the funds, staff, and other resources needed to carry out the water management plan.

7.10.2 Water Management Options

Facilities managers can handle almost all of a building's water use in many different ways. Some methods only change how much water is used in a building. On the other hand, various methods use and maintain fittings and equipment differently. Changing or replacing tools and equipment may be best for long-term savings. Sometimes, just making one choice can save cost. Sometimes, there is a need to use many options. A detailed water management plan should consider all the different water management options. It should be clear from this strategy that changes to a building's water system would affect other systems, like the heating system. Some water management actions include reusing and recycling water, repairing leaking plumbing fittings and heating and cooling systems, wise water use for landscaping, rainwater collections and restrictions on water use.

7.10.2.1 Conservation Actions

The amount of water used indoors differs greatly from location to location. The manager carrying out the water conservation plan should correctly audit each important water-using process, device, and position. The best water management techniques will then be selected and evaluated.

7.10.2.2 Consumer Participation

Consumer participation involves changing end users' behaviour towards water consumption. It focuses on raising end users' awareness of the need to implement water-saving measures. Most of these measures must be initiated by the water management team but are implemented by the end users. An appropriate awareness campaign is critical to ensure the consumers' involvement in sharing the cost of utility bills with staff. Other methods include staff awareness programmes like environmental fairs, suggestion awards, annual resource conservation recognition, and suggestion awards.

7.10.2.3 Promotion of Water Management

The organisation's top management should communicate the policy to all staff. They should also communicate goals and fully support the water-saving plan. When users are informed about water savings, they should be given suggestions and be rewarded for doing so. Water savings marketing could be from promoting poster competitions, publishing water savings success stories, writing an article for the organisation, having the results in the organisation's yearly report, making a detailed plan for public relations, and using local newspapers, radio stations, and TV interviews.

7.11 Waste Management

Africa made 125 million tonnes of municipal solid waste (MSW) in 2012. Of that amount, 81 million tonnes, or 65%, came from Sub-Saharan Africa. This production is predicted to rise to 244 million tonnes annually by 2025. Still, about half of the MSW made in Africa is dumped in cities and towns. On streets, open areas, rivers, and stormwater drains, only 55% (68 million tonnes) of waste is picked up every week. It takes anywhere from less than 20% to over 90% of MSW in Sub-Saharan Africa, with 44% being the average. The average rate of collecting MSW on the continent should rise to 69% by 2025.

In contrast to city centres, which usually have great waste removal services, suburbs and peri-urban areas generally have bad waste collection services. Things worsen in rural areas, where waste pickup services are only sometimes available. Because of this, most African countries need to improve how they collect MSW, which lets waste get into the environment, especially in river and sea areas.

7.11.1 Waste Collection Coverage

Over 90% of the waste made in Africa is still dumped on land, often in places that must be monitored or controlled. This is true even though 80% to 90% of the MSW made in Africa can be recycled. Only 4% of the waste made in Africa is collected. Busy but poor informal reclaimers who must learn that the waste can be used as a secondary resource usually do this.[23]

Facilities managers are in charge of managing waste in many buildings. They need to find the waste, devise a way to reduce it, and monitor the type of waste produced so that they can make changes to the buildings they are in charge of. This would also help sort and dispose of waste to cut costs and follow regulations.

Once the waste has been identified, a waste audit should be conducted to determine how the building is used and the changed activities that could influence sustainability. This can help the organisation determine which processes are wasteful and need the most attention. Additionally, it would help determine the areas that need growth. The next step is getting support from the organisation's top management by assembling a team with everyone interested in waste management.

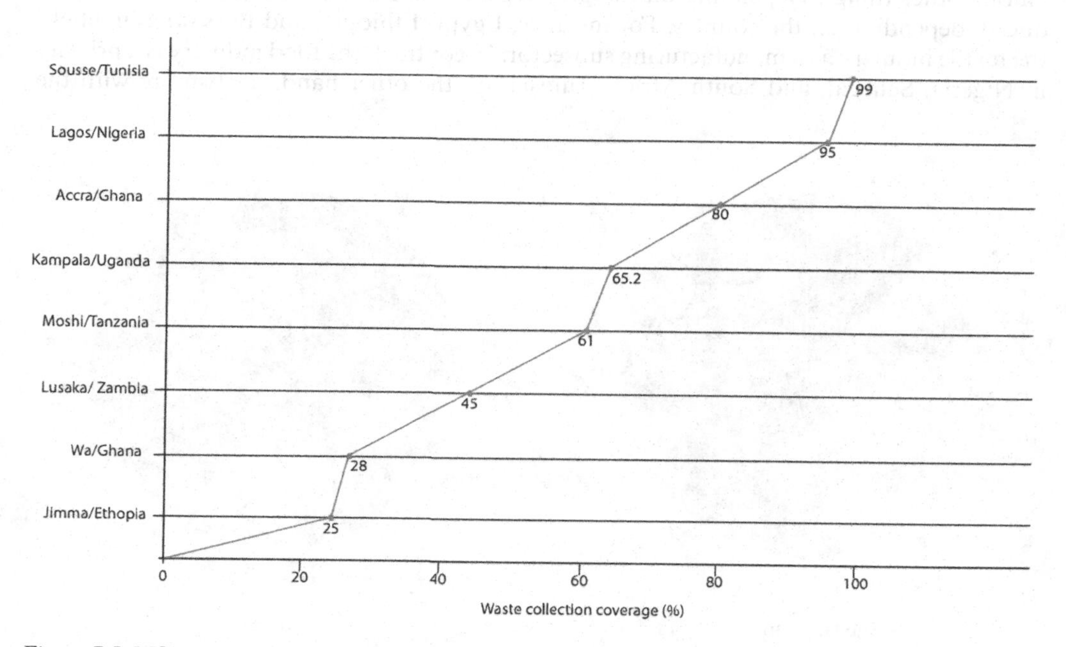

Figure 7.5 MSW collection coverage for cities in Africa.

Source: Godfrey, L., Ahmed, M.T., Gebremedhin, K.G., Katima, J.H., Oelofse, S., Osibanjo, O., & Yonli, A.H. (2019). Solid waste management in Africa: Governance failure or development opportunity. *Regional development in Africa, 235*(10), 5772

The facilities manager is also responsible for making a waste management plan that lists the problems, needs, and goals for better waste management. They need to make a policy that requires employees to keep records of how they handle waste, ensure that the policy considers health and safety, and develop a plan to ensure that the activities are done well. The facilities manager should also monitor waste management tasks using measures and goals. Being aware of the different types of waste in the organisation can also help with this, both for safety and efficiency reasons. Facilities managers should deal with the changes in the process and train employees to deal with waste so that the method works.

7.11.2 Types of Waste

Home wastes are the wastes produced by the residents and the types of waste usually made during daily activities in residential locations. Paper, cans, bottles, textiles, glass, metals, e-waste, and hazardous wastes like paint and aerosol spray are a few examples. Household garbage is one of the primary sources of waste. For instance, in Uganda, home wastes make up between 52% and 80% of the total weight of waste, followed by waste from marketplaces, businesses, industries, and other sources. Food waste is the bulk of waste generated; paper, plastics, and ceramics comprise a far smaller portion. This also happens in other countries in Africa, such as Kenya. On average, moist, organic, biodegradable waste makes up 57% of MSW in Africa. Waste plastic is a growing source of worry for Africa; it accounts for 13% of the continent's MSW, with the bulk of it dumped on land (Figure 7.6).

***Industrial waste*:** The industrial sector frequently produces a large quantity of waste, which may be from power plants or factories that make cars, textiles, or building materials, among other things. Depending on the primary materials used, different wastes are produced, depending on the country. For instance, Egypt, Ethiopia, and Botswana highlight the textile industry as a manufacturing subsector. In contrast, the food industry is a priority in Nigeria, Senegal, and South Africa. Tunisia, on the other hand, deals more with the

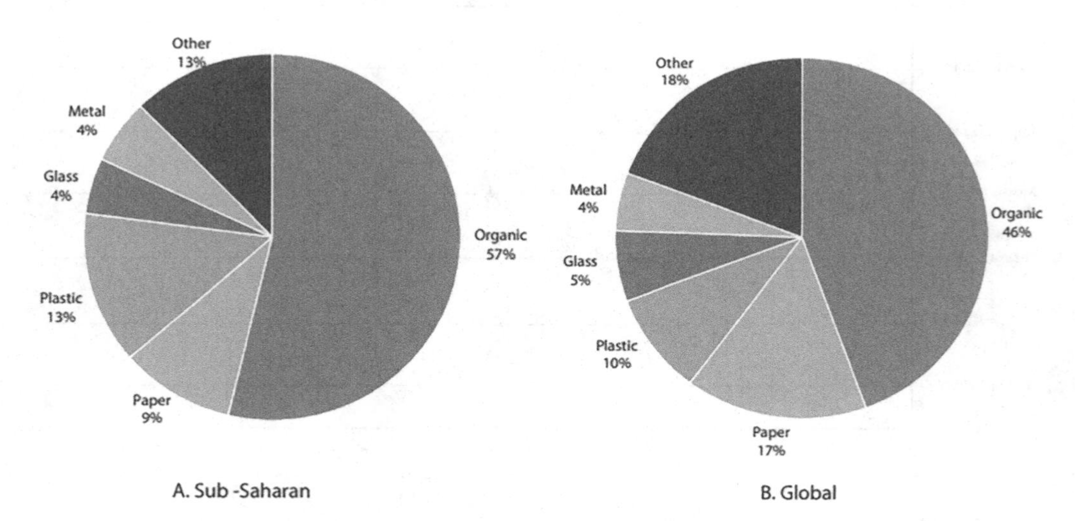

Figure 7.6 MSW composition in Sub-Saharan Africa and globally.

Source: Godfrey, L., Ahmed, M.T., Gebremedhin, K.G., Katima, J.H., Oelofse, S., Osibanjo, O., & Yonli, A.H. (2019). Solid waste management in Africa: Governance failure or development opportunity. *Regional development in Africa*, *235*(10), 5772

garment sector. Institutions like schools create institutional waste. Construction and demolition wastes, including waste from wastewater treatment facilities, are also here.

Manufacturing, medical and agricultural waste: Wastes produced during production include scrap metal from the manufacturing sector. Medical waste, sometimes called hospital waste, can be dangerous and must be handled carefully. Agricultural wastes can be used as raw materials for the production of biogas and bio-compost, both of which are plant fertilisers.

7.11.3 Problems With Waste Management Practices

Africa's ways of handling waste must be changed because they lead to big problems. The way people behave within cities is one of these problems. Their culture and social backgrounds shape how they behave. Other issues include needing more funds or human resources, cities growing too quickly, poor roadways, poor infrastructure, and requiring the right tools and equipment. Other significant issues with Africa's waste management are inaccessible roads, inadequate education and communication routes between stakeholders, regulations that are not implemented, ineffective methods for waste collection fees, the problem of not having any planned systems for processing waste and having short-term contracts with the private sector.

Most waste management plans are based on the idea of an "advanced economy," which means that different technical, financial, and administrative frameworks are expected, mainly for primary collection. These models are inappropriate in Africa because various social strata live elsewhere. They fail because they are not possible.[24]

7.11.4 Waste Management Legislation in Selected Countries in Africa

In South Africa, the Environmental Conservation Act (Act 73) of 1989 was the first time waste was legally defined, and guidelines were put in place about dealing with waste. Still, the main focus was on approving, supervising, and managing dumps. The goal was to reduce the damage that many waste places (both regulated and uncontrolled) did to the environment by not being managed well. People thought the Minister of Environmental Affairs and Tourism might make regulations to help people reduce, reuse, and recycle waste materials, but this tool was not used. There were few legislations or regulations about waste between 1989 and 2007. The White Paper on Integrated Pollution and Waste Management and the first National Waste Management Strategy laid the groundwork for integrated pollution and waste management. Between 2008 and 2017, regulations like the National Environmental Management: Waste Amendment Act (Act 26) 2014 were made because of the National Environmental Management (NEM): Waste Act (Act 59) of 2008.

The recycling and waste sector in South Africa has gone through four stages in the past 30 years. The first stage, which began in 1989, was known as "The Age of Landfilling." When the second step, "The Emergence of Recycling," began in 2001, the Polokwane Declaration was made public. This paper set the groundwork for the legislation that banned plastic bags that can only be used once. On the other hand, the government waste numbers show that only 10% of the country's waste has been collected. When the NEM Act became law in 2008, the third stage, "The Flood of Regulation," started. Because of this, there are now many more legislations about dealing with waste. The fourth step, "The Push towards EPR," started when the Integrated Industrial Waste Tyre Management Plan (IIWTMP) was implemented in 2012. A plan called Extended Producer Responsibility (EPR) was put in place by the IIWTMP to help tyre makers handle their liabilities for used tyres. A Producer Responsibility Organisation set up this system.

In Nigeria, the central, state, and local governments work together to make up the federalist system of government. Because of this, the state's ruling order decides who is in charge of SWM. In 1988, the Federal government created the Federal Environmental Protection Agency (FEPA) to handle SWM at all levels of government, among other things. FEPA's responsibilities were given to the new Federal Ministry of Environment when things changed in 1999. People are more aware of environmental problems because of this group, which is also in charge of following Nigeria's environmental laws. A law from 1999 in the Federal Republic of Nigeria says that waste control is the responsibility of the local government at the state and local government levels.

Regarding Nigeria, state governments either clean up after local governments, or help them do it. For instance, the Rivers State Environmental Sanitation Authority (RSESA) runs SWM in Port Harcourt. On the other hand, Abuja's Abuja Environmental Protection Board (AEPB) runs SWM in that city. Cleanup in Lagos State is the responsibility of the Lagos Waste Management Authority (LAWMA), the Lagos State Environmental Protection Agency (LASEPA), the local government, and the Ministry of Environment and Physical Planning.

The United Nations Economic and Social Council studied Ghana, Egypt, Kenya, and Zambia to evaluate waste management practices.[25] Egypt's environment deteriorates because people throw away waste incorrectly, which harms the environment. Law 4 of 1994 regulates the environment and is run by the Ministry of State for Environmental Affairs, the Egyptian Environmental Affairs Agency (EEAA). The country has been able to come up with an acceptable method to handle hazardous waste, but it needs to be enforced.

Landfills in Ghana handle biomedical waste. Controlling and managing e-waste is more challenging because recycling is difficult. The regulations are the Pesticides Control and Management Act of 1996, the Environmental Sanitation Policy of Ghana, the Local Government Act of 1994, and the Guidelines for Biomedical Waste of 2000. The life cycle method for handling hazardous waste in an integrated manner has been attained.

There is no official data on hazardous waste in Kenya. The most common way to deal with waste is to burn it, and some hazardous waste ends up in landfills. The National Environmental Management Authority (NEMA) is in charge, and laws like Local Government Act 265 and Public Health Act Cap 242 are in place. The country has collaborated with UNEP and UN-Habitat to assist cities like Nairobi and Kisumu with solid waste management strategies.[26]

7.11.5 Solid Waste Constituents

Biodegradable organic materials include most of the waste produced in much of Africa. Most waste is biodegradable because of human consumption, which has a lot of kitchen and compound waste. Hence, an effective collection system is needed to stop disease outbreaks and other harmful environmental effects. In addition, the e-waste produced by the global trend of electronic gadgets is growing toxic in Africa. Hence, a workable collecting system is needed to stop adverse environmental effects.

7.11.6 Waste Collection and Transportation

There are three critical stages of waste collection in Africa. Stages include informal, primary, and secondary levels. Most household-to-community collection point collection occurs during the informal and primary periods. Institutional groups like urban councils and private operators are in control during the secondary stage. Usually, waste is moved from transfer stations to landfills or other places for final disposal. Private contractors that go door to door to collect waste directly from households have also been used in waste collection. This is usual in cities in East Africa, where private waste collectors have been hired to remove waste from shopping malls and the areas around them.

However, the council, a public-sector organisation, still creates waste collection agreements that the local market and hospitals rely on. The "bring to court" strategy is another common waste collection technique in African countries. This strategy involves the collecting truck operating according to a set plan on particular days of the week. It gets there to signal for people or organisations to bring their waste to be disposed of. Depending on the socioeconomic status of the persons, such as low-income and high-income groups, the frequency with which wastes are dumped differs. The quantity of waste collected by government workers changes, and waste is physically loaded onto trucks. In addition, the solid waste collection rate has grown in most African countries because the private sector began operating there rather than relying only on municipal governments.

7.11.7 Recycling

According to a UN report in 2018, recycling is growing across Africa, driven more by societal needs and public sector planning than poverty, unemployment, and other factors. About 70–80% of the MSW produced in Africa is recyclable. Yet, just 4% of MSW is now recycled. Municipalities and commercial groups suffer when informal waste pickers recover valuable materials from the waste.[27]

7.11.7.1 Recyclable Components

There are three main ways to recycle: Reuse, resource recovery, and composting.

Reuse
Reuse may be achieved in two ways. Firstly, by using products more than once to extend their useful lives, and secondly, by turning waste into valuable items like toys, artwork, decorations, sandals, and carry bags. Rather than throwing away unused household things, donate them to someone who can use them.

Resource recovery
Recycling involves crushing or grinding recyclable materials like cans, glass, paper, and plastic before using them to make new products at manufacturing facilities. Glass is gathered to be recycled into new glass products, such as whole bottles, broken glass, or cullet (crushed glass), which are then cleaned and reused. Cans and other rusted metal items are among the steel and aluminium scraps gathered and sold to steel mills for recycling. Depending on the kind of plastic, collected plastics are recycled into items like hollow fibre, garbage collecting bags, barrier liners, and plastic irrigation pipes. Owing to the small weight of plastic, labour costs for collecting, sorting, and processing, transportation costs, and finally, power and water use costs for washing and processing recovered materials all significantly influence the economics of plastic recycling.

A resource recovery system
Separation at the source is the basis of an Integrated Waste Management System (IWMS) combining recycling. The challenge is to concentrate enough recyclable materials such that it is profitable to sell them to a final customer, even if that customer is some distance from the site of separation. Many resource recovery special programmes must be changed

because they rely on removing components from mixed waste. Usually, such efforts involve recovering at dumps. It has been challenging to produce enough revenue from sales to pay project participants even the legal minimum wage. Recycling is appreciated in an IWMS because it saves money on landfill management and landfill airspace by avoiding the dumping of recyclables and the money gained from the sale of materials. Creating a new process to increase the worth of recyclables and make the best use of waste management cost savings from separation at the source is possible. Such a system may be made of source-separation collection and transfer.

Buyback centres

From the collecting point, the recyclables can be transferred to a buyback centre or a recycling business for distribution to an end user. When distances allow it, people may also send their recyclables directly to the buyback centre in exchange for payment. Local organisations may turn recyclables into cash at the centre. All of these choices are practical and financially possible when they are separated at the source. Moreover, the municipality can still pick up recyclables from drop-off locations, transport them to a buyback centre, or sell them directly to a recycling company or end user. A city may make the cost of this transportation external by allowing a contractor that delivers recyclables for access to this waste.

The government may permit recycling collection in areas with more people. The municipality may stop collecting actual waste in this situation. A current expenditure can, therefore, be compensated by a future revenue stream or, at least, cost reductions. Large amounts of recyclables will remain disposed of in landfills until source separation is made. Waste drivers will continue to exist as long as this is the case. The construction of safety gear, proper sanitary facilities at landfills, and the supply of health and safety awareness training to salvagers are important.

Effective resource recovery starts with separation at the source. However, resource recovery might be carried out in various unique ways. Anyone can bring materials directly to a buyback facility. As source separation grows, the available quantities may call for machinery construction to process the materials, such as breaking the glass or grinding the plastic. This may give enough raw resources to firms that can use the material. For the tourist industry or other markets, small-scale community-based initiatives that use waste as a raw material may be satisfactory. Municipalities can also enter into direct buyback agreements with consumers and business associations on Plastics. It is still a solution that developing countries have yet to implement. However, the essentials are still highly relevant to the discussions that follow.

Composting

Although it may be possible for the community to make compost on a commercial scale, it should be encouraged that all organic waste be composted at home and used in the backyard garden to improve the soil and grow vegetables and fruits.[28]

7.12 Building Energy Management Systems

Computer-based systems called building energy management systems (BEMS) monitor and control a building's mechanical and electrical parts, like the power systems, lights, ventilation, and heating and cooling systems. They connect the building services plant to a central computer to control the temperature, humidity, and electricity. Energy management tools can lower annual energy expenditure and make it work better. Energy management systems can make a building's machinery and lighting systems last longer and save funds on maintenance costs by reducing equipment use that is not needed. BEMS are comprised of several main parts, including a sensing infrastructure for monitoring energy use and external factors, data processing software for managing sensed data and putting energy-saving plans into action, a section for user contact, and an actuator infrastructure for adjusting the environment. The BEMS programme helps employees get the most out of the building by monitoring control features and alerts. BEMS are essential for monitoring how much energy is used, especially in large, complex structures and places with multiple buildings.

Both digital and analogue input sources send temperature and humidity readings to BEMS. One of the inputs could be a show of whether the equipment works. Changes in comfort are caused by messages sent from the central computer to site equipment like fans, pumps, and valves, communicating how to set their controls or turn on their devices. BEMS can monitor and control essential systems, like lights. Outstation software sets local hubs immediately to connect these inputs and outputs to the central monitoring master station. The user can then change the temperature, pressure, humidity, and the schedule for when things turn on themselves (Figure 7.7). Trained personnel can use the BEMS to change settings to save energy without affecting comfort or services. BEMS are used to monitor and control the building. A local area network (LAN) connects the remote tools. The software usually has a graphical user interface (GUI) or human-machine interface (HMI) comprising pictures of the machines managed (Figure 7.7). When an event logger is turned on, it shows the conditions inside the building and what is going on there right now. The main job of BEMS is to control the hot mix locally and run the heaters and pumps from afar to keep the room at the right temperature. BEMS manage chillers, cooling, and air distribution systems in buildings with air conditioning. For example, it can turn fans on and off or

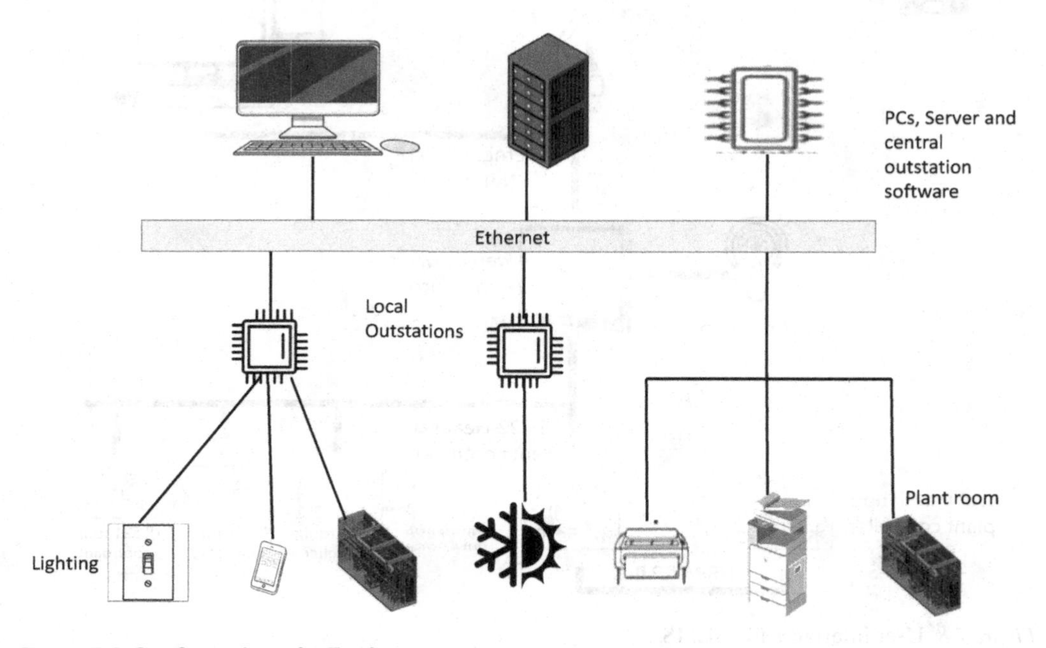

Figure 7.7 Configuration of BEMS.

Source: Author

open and close vents. BEMS can keep track of energy meters and control lights or any other energy-using devices.

BEMS allow data to be collected and controlled from multiple locations, and Wi-Fi, phone networks, and satellite systems can be used for distant monitoring. In the outstations, they also have distributed intelligence. They can connect to smart devices like laptops and cell phones to set off alarms that let the on-call workers know about problems or events in the building (Figure 7.8). By giving practical feedback and a complete control plan, BEMS can improve building management and performance.

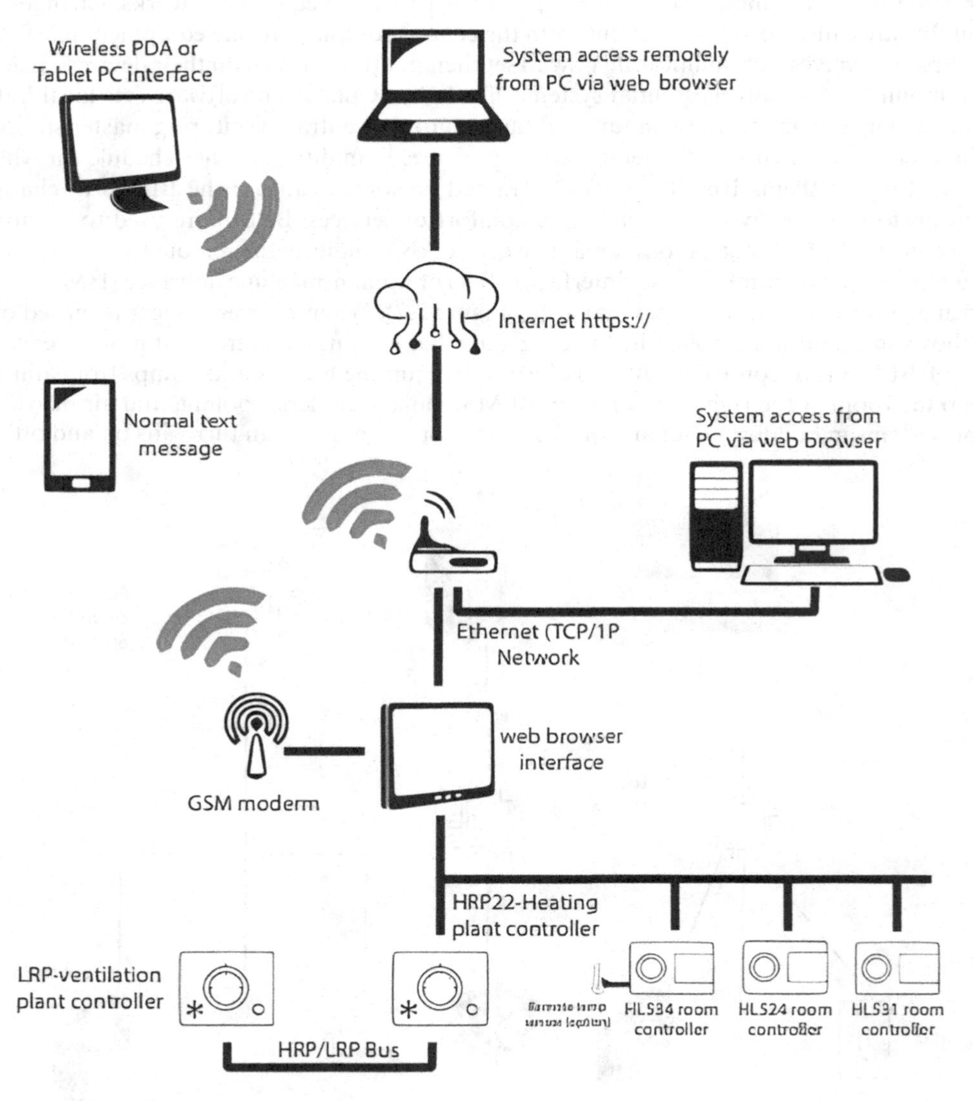

Figure 7.8 User interfaces for BEMS.

Source: Adapted from Sayed, K., & Gabbar, H.A. (2018). Building energy management systems (BEMS). In *Energy conservation in residential, commercial, and industrial facilities*, (pp. 15–81)

Large organisations use BEMS to control buildings spread out over large areas. Intelligent outstations are built into modern systems and can be examined locally in a plant room to find problems. For BEMS to work well, they need to be well-designed and described, come with clear instructions, and have a virtual user interface. For BEMS, it is essential to have the right front-ends. Web platforms, the internet, and telecommunication tools are just a few ways to access current BEMS (Figure 7.8). Putting in BEMS can save 10–20% of the energy used in a building. BEMS cannot pay for poorly built, maintained, or managed systems. Building operators are more likely to use BEMS to make the building more efficient if there are easy ways for them to get to it and use it in a way that fits their duties. At least once a month, BEMS options should be certified to make sure they meet the needs of the building.

Attention should be paid to the following when managing BEMS:

- Ensure that all of the system's cables, links, cabinets, and screens are in good shape
- Ensure all equipment that isn't needed is switched off
- Ensure the buildings are at the right temperature for those who use them
- Write down and keep track of alarms and say how they should monitored, recorded, directed, and handled
- Ensure that a BEMS has safety features that start working on their own to protect people, property, or equipment
- Ensure that the BEMS gives you information
- Keeping controls in good shape affects how well a building works
- Identify the best way to do maintenance at the start of the project
- Write down essential changes to the BEMS, such as any adjustments to set points and control strategies, software updates, network additions, problems found, or maintenance done
- Ensure the tasks are in the correct order
- Before starting a system, knowing the users' energy management goals and spending rules is essential
- Check the sensors' quality and ensure they are in the right places
- Look at the actuators' control outputs and ensure that the things being controlled work across their whole range of operation
- Check that the equipment works as written
- Ensure that the sources are functional and working properly
- Check and fix any problems with the switching devices
- Check that the controllers' battery packs have enough power and that the controllers immediately restart when the power goes out
- To maximise BEMS purchases to reduce the building's energy use and demand charges. This can be done in many ways, such as by changing goals or using special software projects or energy-saving techniques

With BEMS, it might be possible to lower operating costs and improve comfort. BEMS can monitor the state of plants, the environment, and energy use so that the person in charge can get real-time information on how the building is running; that often leads to problems that wouldn't have been reported otherwise, like a plant on all the time or high energy use. When energy counters connect to BEMS, they show real-time trends in energy consumption. They also record the building's energy performance over time, which can be logged and examined in some ways, including visually and quantitatively. BEMS may improve management information by tracking performance trends and enhancing planning and spending that could make the users more aware of how to save energy to save as much money as possible; finding out if current buildings and tools can work together is essential. It is possible to set up heating, ventilation, and lighting systems in different areas based on their use so that BEMS can work well in a current building.

The benefits of a BEMS installation include the following:

- Lets users check the performance of controls and easily make changes
- Improves comfort for users of buildings because it can monitor and control HVAC systems to make sure the correct temperatures and air quality for people who use buildings
- Controls that save energy and will lower energy bills.
- Helps buildings follow regulations about energy efficiency and the environment, like LEED and GBCSA
- Ability to record and store info for managing energy
- Provides quick updates on the state of plants
- Buildings with BEMS may be more valuable because users want features that save energy
- System alarms are generated automatically when equipment fails, or things fail to function
- Records both planned and unplanned maintenance requirements
- Assists buildings to save on their carbon emissions and make the future more sustainable

BEMS may be used as a tool to enhance building performance once it has been installed and entirely correctly commissioned. Even the best-designed and commissioned control plan will change to meet the users' and the facility's needs. Without compromising on comfort levels, a skilled BEMS operator may frequently examine the settings of BEMS to gradually lower room set points, operation hours, and energy usage. It sometimes takes one or two whole heating seasons to fine-tune the building controls to their ideal settings. This optimisation procedure is crucial when BEMS control is used for big multi-building complexes and buildings distributed over a large region. The BEMS operator can monitor energy and operations without going inside the buildings. This central BEMS bureau model is popular among FM providers and big estates since it is very cost-effective. Because optimisation is ongoing, it is important to keep track of any system modifications. According to occupant concerns, too many buildings have set points incorrectly designed many years ago and have high operation hours. Facilities running everything at maximum efficiency 24 hours a day are still common. These facilities might save significant money, energy, and carbon emissions with little optimisation. For the service provider to maintain their integrity in the system's operation, the person who maintains BEMS must get training.

BEMS must be designed to work correctly and have an easy-to-operate user interface. Continuous evaluation and adjustment should be made as performance improves over time. It should be maintained so that operating it is simple.

The installation of BEMS is independent. The buildings must be tested to determine larger systems needed for THE size, form, and reception of signals of the BEMS. This will specify what needs monitoring and managing, the connections, hardware, and cabling requirements and their advantages. In addition, it will show the location and capacity required for it. A thorough life cycle costing estimate based on discounted cash flow should be included in the financial motivation for a BEMS. Savings obtained from BEMs could result from enhanced maintenance, increased dependability, and decreased energy use.

Careful planning and design of adequate controls are necessary to create a good building—the client's brief for an acceptable control system that demands comfort and energy efficiency. The essential energy characteristics must be outlined in the designer's specs so that contractors understand what the control system must do. The best way for customers to achieve low-carbon buildings is to include a goal for a low-carbon building in the client brief. Initial comments immediately relate to the resulting control system's design, selection, installation, and operation. A low-carbon building may only be possible if the design team is given clear instructions.

BEMS may have outstations that can be accessed using mobile phones. Another issue with outstation is that buildings may be prone to cyber-attacks in some organisations when some of the functionalities are located in the cloud. In addition, it is important to look at the potential for system growth at each outstation. If all of the points on the original are occupied, adding one point can cause a high cost. Older systems might need to be replaced, which can be an issue when the manufacturers do not manufacture them again. BEMS connected to meters for energy tracking might help find savings.[29]

7.13 Conclusion

This chapter focused on sustainable development (SD), and its three pillars emphasising social, environmental, and economic dimensions. Sustainable FM helps to achieve the UN's 17 SDGs.

It discussed environmental management, which guides growth to make the most of opportunities, avoid dangers, reduce issues, and prepare people for problems they cannot prevent by making them more adaptable and resilient. The ISO 14001 standard, an environmental management system, describes implementing an efficient environmental management system in a company. It aims to assist companies in maintaining their financial success while paying attention to their environmental obligations and effects. It can also help you grow sustainably and reduce the adverse impact of that growth on the environment. Organisations can use the guidelines in ISO 14001 to create and implement an EMS that includes the company's environmental policy, the environmental aspects of its operations, legal and other requirements, a clear set of goals and targets for environmental improvement, and a set of environmental management programmes. To follow ISO 14001, a system must have a clear structure for assigning responsibility for environmental management, training, knowledge, and competency courses for all facility workers and ways for the EMS to be communicated inside and outside the facility. It also needs to have a way to keep track of environmental management paperwork, a way to ensure that paperwork is handled correctly, and steps for making sure that operating controls of environmental effects and emergency preparedness and response are followed.

The chapter covered sustainable procurement, which improves efficiency and effectiveness and enhances the partnership between employees and suppliers. It can help with risk management, protect a company's image, save cost, create value, make it more future-proof and transparent, and set a brand apart. It further looked at net zero carbon emissions. The move to low or no carbon emissions is becoming a big issue. Energy efficiency is the most essential part of building with net zero carbon emissions. A building with net zero carbon pollution uses more green energy sources to help it reach its goal. Before showing off green energy systems or offsets, lowering the energy a building needs is essential. This will keep these systems' prices and installation to a minimum.

Energy performance rating systems were explored. The most famous green building systems in South Africa are Green Star South Africa and EDGE. Two local rating systems exist in Egypt: TARSHEED and the Green Pyramid Rating System. Some projects in Nigeria and Morocco already have LEED green building certification.

In addition, we have computer-based systems called BEMS that monitor and control a building's mechanical and electrical parts, like the power systems, lights, ventilation, and heating systems. Water management must manage the water supplies and wastewater entering and exiting the facility and recognise the high costs and risks. Facilities managers are in charge of managing waste in many buildings. They need to find the waste, devise a way to cut it down and monitor the type of waste that is made so that they can make changes to the buildings in charge for which they are responsible. This would also help sort and dispose of waste to cut costs and follow regulations.

7.14 Guidelines

- Involve the facilities manager from the design stage
- Create an energy use and emissions baseline
  Start with simple goals
- Align goals with SDGs
- Embrace sustainable facilities management
  Use environmental management systems
- Have a clear set of goals and targets for environmental improvement
- Have a set of environmental management programmes
- To follow ISO 14001, a system must have a clear structure for assigning responsibility for environmental management, training, knowledge, and competency courses for all facility workers
- Have ways for the EMS to communicate inside and outside the facility
- Keep track of environmental management paperwork
- Handle paperwork correctly for EMS
- Follow steps for operating controls of environmental effects and emergency preparedness and response
- Practise green purchasing
- Use technology to assist in green purchasing
- Reduce pollution
- Go for energy performance ratings
- Use BEMS
- Have data on energy consumption for the BEMS
- BEMS must be interoperable by using communication protocols
- BEMS need a good internet connection
- Ensure that the EMS gives information
- Keep controls of BEMS in good shape as they affect how well a building works
- Identify the best way to do maintenance at the start of the project
- Write down essential changes to the BEMS, such as any adjustments to set points and control strategies, software updates, network additions, problems found, or maintenance done
- Ensure the tasks are in the correct order in use of BEMS

- For BEMS to reduce the building's energy use and demand charges, change goals or use special software projects or energy-saving techniques
- Use water management to manage water
- Have a goal and policy for the use of water
- Assess water usage
- Have a water plan and implement it
- Involve top management in water management
- Embrace water conservation practices
- Promote a water conservation plan in the organisation
- Embrace waste management practices such as recycling
- Comply with waste management regulation
- Stakeholders education and training on waste management
- Facilities managers can move towards net zero carbon emissions
- Employing techniques for ongoing development for net zero
- Occupant awareness of net zero
- Preventive maintenance for net zero
- Understand embodied carbon for net zero
- Construct an ongoing energy monitoring system to ensure that systems retain increased energy efficiency for net zero
- Convert systems that burn fuel to electric ones for net zero
- Use certificates for renewable energy to compensate for emissions for net zero

7.15 Questions

1. What is sustainable development?
2. What are the sustainable aspects of FM over the life cycle of the building, from the design, construction, and operations phases?
3. What are the reasons for environmental management?
4. How would you implement an environmental management system in your organisation?
5. How would you implement or improve green purchasing in your organisation?
6. How is the environmental performance of buildings determined in your country?
7. Provide examples of the best-performing green buildings in your city or country
8. What is zero carbon? How can you implement the concept in your organisation?
9. What could be the possible challenges of implementing zero carbon in your organisation?
10. What is energy efficiency?
11. How do you practice energy efficiency in your workplace?
12. How does ESG relate to facilities management?
13. Prepare a water management plan for your organisation. If an existing plan is in place, discuss ways to improve it
14. How do you practice waste management in your workplace?
15. What are the government legislations that guide waste management practices in your workplace?
16. How would you implement or improve recycling practices in your workplace?
17. Discuss the applicability of building energy management systems to your workplace
18. What should you pay attention to when managing a building energy management system?

Notes

1 Adewunmi, Y., Omirin, M., & Koleoso, H. (2012). Developing a sustainable approach to corporate FM in Nigeria. *Facilities*, *30*(9/10), 350–373.
2 Lok, K.L., Opoku, A., Smith, A., Vanderpool, I., & Cheung, K.L. (2023). Sustainable facility management in UN developmentg. In *IOP Conference Series: Earth and Environmental Science* (Vol. 1176, No. 1, p. 012022). IOP Publishing.
3 Barrow, C. (2006). *Environmental management for sustainable development*. Routledge.
4 Morrow, D., & Rondinelli, D. (2002). Adopting corporate environmental management systems: Motivations and results of ISO 14001 and EMAS certification. *European management journal*, *20*(2), 159–171.
5 Marjanovic, L., Mclennan, P., & Hamid, A.J.B.S. (2014). An investigation of the value proposition for environmental management systems implementation in facility management operations in Singapore. *International Journal of Facility Management*, *5*(2).
6 IFMA. (2016) *Sustainable procurement for FM*. Accessed February, 2024 at FMJ January/February 2016: Sustainable Procurement for FM (ifma.org).
7 Aaltonen, A., Määttänen, E., Kyrö, R., & Sarasoja, A.L. (2013). Facilities management driving green building certification: A case from Finland. *Facilities*, *31*(7/8), 328–342.
8 Amasuomo, T. T., Atanda, J., & Baird, G. (2017). Development of a building performance assessment and design tool for residential buildings in Nigeria. *Procedia engineering*, *180*, 221-230.
9 Moussa, R.R. (2019). The reasons for not implementing Green Pyramid Rating System in Egyptian buildings. *Ain Shams Engineering Journal*, *10*(4), 917–927.
10 NEC Addis Ababa, Ethiopia | U.S. Green Building Council (usgbc.org).
11 Vodafone SV-C2 building | U.S. Green Building Council (usgbc.org).
12 Ridge Hospital | U.S. Green Building Council (usgbc.org).
13 GBCSA. (2023). https://gbcsa.org.za/resources-listings/case-studies/.
14 Munene, L.N. (2019). Reducing carbon emissions: Strathmore University contributions towards sustainable development in Kenya. *African Journal of Business Ethics*, *13*(1).
15 IQ Next. (2023). Role of Facility Management in the Race to Net Zero!
16 McDonough Bolyard Peck (MBP). (2023). A road map to carbon neutrality for facilities managers. MBP.
17 Mansour, Y., Ghandour, S., Krisanda, S., Tarabeih, K., & Goubran, S. (2023). *American University Cairo carbon footprint report*. documents.aucegypt.edu/Docs/about_sustainability/CFP 2023 Report.pdf.
18 Arup. (2021). Net zero buildings: Where do we stand? (IFMA Knowledge Bank).
19 Adewunmi, Y., Alister, A., Phooko, B., & Nokukhanya, T. (2019). Energy efficiency practices in facilities management in Johannesburg. *Journal of Facilities Management*, *17*(4), 331–343.
20 Saunders, J. (2023). What is ESG? The 7 Steps Model. IFMA.
21 Reid, R.N. (2003). *Water quality systems: Guide for facility managers*. CRC Press.
22 EPA. (2023). *Water management planning*. EPA.
23 Godfrey, L., Ahmed, M.T., Gebremedhin, K.G., Katima, J.H., Oelofse, S., Osibanjo, O., & Yonli, A.H. (2019). Solid waste management in Africa: Governance failure or development opportunity. *Regional development in Africa*, *235*(10), 5772.
24 Abigo, A. (2016). *Facilities management: Enhancing solid waste management practices for urban marketplaces in Nigeria*. PhD thesis, University of Brighton, UK.
25 Godfrey, L., & Oelofse, S. (2017). Historical review of waste management and recycling in South Africa. *Resources*, *6*(4), 57.
26 Akpan, V.E., & Olukanni, D.O. (2020). Hazardous waste management: An African overview. *Recycling*, *5*(3), 15.
27 Bello, I.A. et al. (2016). Solid waste management in Africa: A review. *International Journal of Waste Resources*, *6*, 1–4. https://doi.org/10.4172/2252-5211.100021.
28 Liebenberg, C.J. (2007,). Waste recycling in developing countries in Africa: Barriers to improving reclamation rates. In *Proceedings of the Eleventh International Waste Management and Landfill Symposium, Cagliari, Italy* (pp. 1–5).
29 Sayed, K., & Gabbar, H.A. (2018). Building energy management systems (BEMS). In *Energy conservation in residential, commercial, and industrial facilities*, (pp. 15-81).

Bibliography

Aaltonen, A., Määttänen, E., Kyrö, R., & Sarasoja, A.L. (2013). Facilities management driving green building certification: A case from Finland. *Facilities, 31*(7/8), 328–342.

Abigo, A. (2016). *Facilities management: Enhancing solid waste management practices for urban marketplaces in Nigeria*. PhD thesis, University of Brighton, UK.

Adewunmi, Y., Omirin, M., & Koleoso, H. (2012). Developing a sustainable approach to corporate FM in Nigeria. *Facilities, 30*(9/10), 350–373.

Adewunmi, Y., Alister, A., Phooko, B., & Nokukhanya, T. (2019). Energy efficiency practices in facilities management in Johannesburg. *Journal of Facilities Management, 17*(4), 331–343.

Akpan, V.E., & Olukanni, D.O. (2020). Hazardous waste management: An African overview. *Recycling, 5*(3), 15.

Alpin. (2021). *Alpin annual sustainability report*. Accessed June 2023 at Alpin-Annual-Sustainability-Report-2021.pdf (alpinme.com)

Amasuomo, T.T., Atanda, J., & Baird, G. (2017). Development of a building performance assessment and design tool for residential buildings in Nigeria. *Procedia engineering, 180*, 221–230.

Arup. (2021). Net zero buildings: Where do we stand? (IFMA Knowledge Bank).

Barrow, C. (2006). *Environmental management for sustainable development*. Routledge.

Bello, I.A. et al. (2016). Solid waste management in Africa: A review. *International Journal of Waste Resources, 6*, 1–4. https://doi.org/10.4172/2252-5211.100021

BSI. (2015). *ISO 14001:2015: Your implementation guide*. Accessed August 2023 at ISO-14001-implementation-guide-2016.pdf (bsigroup.com)

Cervigni, R., Rogers, J.A., & Dvorak, I. (Eds.) (2013). *Assessing low-carbon development in Nigeria: An analysis of four sectors*. World Bank. doi:10.1596/978-0-8213-9973-6.

Daoud, A.O., Othman, A., Robinson, H., & Bayati, A. (2018). Towards a green materials procurement: Investigating the Egyptian green pyramid rating system. In *3rd International Green Heritage Conference*.

EPA. (2023). *Water management planning*. EPA.

GBCSA. (2023). https://gbcsa.org.za/resources-listings/case-studies/

Godfrey, L., & Oelofse, S. (2017). Historical review of waste management and recycling in South Africa. *Resources, 6*(4), 57.

Godfrey, L., Ahmed, M.T., Gebremedhin, K.G., Katima, J.H., Oelofse, S., Osibanjo, O., & Yonli, A.H. (2019). Solid waste management in Africa: Governance failure or development opportunity. *Regional development in Africa, 235*(10), 5772.

IFMA. (2016). *Sustainable procurement for FM*. Accessed February, 2024 at FMJ January/February 2016: Sustainable Procurement for FM (ifma.org)

IFMA. (2024). *ESG and facility management*. Accessed August 2024 at ESG and Facility Management (ifma.org)

IQ Next. (2023). Role of Facility Management in the Race to Net Zero!

Liebenberg, C.J. (2007). Waste recycling in developing countries in Africa: Barriers to improving reclamation rates. In *Proceedings of the Eleventh International Waste Management and Landfill Symposium, Cagliari, Italy* (pp. 1–5).

Lok, K.L., Opoku, A., Smith, A., Vanderpool, I., & Cheung, K.L. (2023). Sustainable facility management in UN development goals. In *IOP Conference Series: Earth and Environmental Science* (Vol. 1176, No. 1, p. 012022). IOP Publishing.

Mansour, Y., Ghandour, S., Krisanda, S., Tarabeih, K., & Goubran, S. (2023). *American University Cairo carbon footprint report*. https://aucegypt0.sharepoint.com/sites/documents/sustainability/Forms/AllItems.aspx?id=%2Fsites%2Fdocuments%2Fsustainability%2FCFP%202023%20Report%2Epdf&parent=%2Fsites%2Fdocuments%2Fsustainability&p=true&ga=1

McDonough Bolyard Peck (MBP). (2023). A road map to carbon neutrality for facilities managers. MBP.

Morrow, D., & Rondinelli, D. (2002). Adopting corporate environmental management systems: Motivations and results of ISO 14001 and EMAS certification. *European management journal, 20*(2), 159–171.

Moussa, R.R. (2019). The reasons for not implementing Green Pyramid Rating System in Egyptian buildings. *Ain Shams Engineering Journal, 10*(4), 917–927.

Munene, L.N. (2019). Reducing carbon emissions: Strathmore University contributions towards sustainable development in Kenya. *African Journal of Business Ethics*, *13*(1).
Reid, R.N. (2003). *Water quality systems: Guide for facility managers*. CRC Press.
Saunders, J. (2023). What is ESG? The 7 Steps Model. IFMA.
Sayed, K., & Gabbar, H.A. (2018). Building energy management systems (BEMS). In *Energy conservation in residential, commercial, and industrial facilities*, (pp. 15–81).

8 Intelligent Buildings

8.0 Introduction

Research has shown that the number and type of intelligent buildings affect sustainability scores and that there needs to be an optimisation between the user's needs, the building's functionality, and the core intelligence function of the technology against the parameters enforced by building rating systems. A positive correlation was observed between the variables. The buildings reported a reduction of 41% in energy consumption, 39% in water consumption, 36% in carbon dioxide emissions, as well as significant economic and social benefits. An intelligent building is defined as a dynamic and responsive infrastructure that integrates disparate building systems such as lighting, heating, ventilation, and air conditioning (HVAC), security, facilities management, and so on, to manage resources effectively, provide high-performance benefits, and optimise processes, comfort, energy costs, and environmental benefits.[1]

Intelligent buildings are not common in developing countries due to factors such as instrumentation and control, which deal with the measurement and control of design and the implementation of systems that are incorporated to achieve the concept of intelligent buildings. Other factors include connectivity, interoperability or ability of IT systems to use information, the use of data management and analytics in systems, privacy and security of systems, IT professional support, top management support, viable funding strategy from the organisation, stakeholders.[2] There is a need for guidelines that will assist managers with implementing intelligent building concepts. This chapter focuses on intelligent buildings.

8.1 The Meaning of Intelligent Buildings

Intelligent Buildings (IBs) are challenging to define, and no one term is generally recognised. Understanding what is being discussed when this word is used is essential, but having a standard definition of IB is not possible. Separate IB concepts might be in place in different countries, locations, and disciplines.[3]

According to Wigginton and Harris (2013), "there are almost 30 different meanings of the term 'intelligence' when relative to structures." According to another study on Intelligent Building Automation in Construction, the most popular definition of an intelligent building was "any facility that offers a responsive, effective, and supportive environment within which the organisation may fulfil its business objectives". However, the Intelligent Building Institute (IBI) of the United States has a generally accepted standard for IBs. This may be summed up as a building "that creates an environment that is both cost-effective and productive via optimisation of its four fundamental components, including its structures, systems, services, and management, as well as the interactions between them."[4]

DOI: 10.1201/9781032656663-10

There are definitions of intelligent buildings focused on FM; for example, performance-based, services-based, and system-based definitions are the three categories of IB definitions. Performance-based definitions focus on a building's functionality and user needs rather than the available technologies or systems. Based in the UK, the European Intelligent Building Group's performance-based concept of an IB is "a structure made to provide users with the most effective environment while also effectively using and managing resources and minimising the life costs of hardware and facilities" (Wang, 2009).

IBs are also described in terms of services and the calibre of services offered by buildings. An example of a definition based on services is given by the Japanese Intelligent Building Institute (JIBI): "[An] IB is a building having the service functions of communication, office automation, and building automation and is suitable for intelligent activities. Users' services are highlighted."

Furthermore, IBs are described by system-based IB definitions that specifically discuss the technologies and technology systems that IBs should include. The Chinese IB Design Standard (GB/T50314-2000) states that IBs provide communication network systems, building automation, and office automation. Including structure, system, service, and management correctly makes the building efficient, safe, and comfortable for the people using it.

Intelligent buildings can change from different levels of automatic control to a management system for the building. These systems include network management (IT and communications), security and emergency management (access control, safety, and video surveillance), and workforce management. Building infrastructure management includes managing lifts and escalators, parking, and intelligent water management. Energy management includes things like controlling HVAC and lighting systems.[5]

8.2 What Makes a Building Intelligent?

For intelligent buildings to work, they need technological tools, especially information technology (IT) tools. But just because these systems use technology doesn't mean they are intelligent. The building's technology should also be set up correctly and related to its services. The features of the systems should be designed to meet user needs and do what they're supposed to do. To ensure the technology systems work as planned, it is also essential to set and handle them correctly, including how they connect and work. The application software should be changed and used along with the system hardware and software for the facility's automation, control, efficiency, and management. A building can have technology systems, but they will only be intelligent if they work correctly. Instead, the technology systems might make things easier for users and managers. Multi-industrial system engineering is a part of intelligent building systems (IBS). It needs the right combination of building services, design, structure, information technology, facilities management, and environment. There is a strong link between IBs and economic and cultural factors (Wang, 2009).

8.3 Intelligent Buildings in Africa

8.3.1 Drivers of Intelligent Building Construction in Africa

There may be political incentives because the government manages the environment. For example, South Africa's Minister of Environmental Affairs signed the Paris Agreement on April 22, 2016, which made people more aware of climate change and more determined to deal with its impacts. This deal clarifies that South Africa and the other parties will be held accountable.[6]

First, every country needs housing, educational institutions, jobs, and health care. This is a social drive. As the number of people living in many African cities grows, improving living conditions in homes is essential. Technological growth and digital competitiveness are two examples of how education leads to innovation.

Technology is the second major force that is changing things. It's changing how people work, how often they move to and from work, where they work, how they make links and keep them, and how things are made or sold. Robots, the Internet of Things (IoT), intelligent sensors, smart devices, artificial intelligence (AI), and other technologies all impact change. These technologies are also the main ones that make smart cities, new business opportunities, and automatic processes possible. Some other relevant technologies include big data, drones, 3D printing, innovative materials, and automated building tools – these new ideas power smart homes in Africa. Third, there are legal factors like the Paris Agreement, a major international agreement that will force the world to implement climate change plans. The fourth part is about economic drivers related to the benefits of energy-efficient buildings and green architecture that go beyond fighting climate change. These include creating jobs, protecting resources, being resilient over the long term, and improving social aspects, like quality of life and energy security, that are important for economic success.[7]

Lastly, environmental factors affect how the built environment changes to deal with climate change issues in buildings. For example, buildings must stay cool in hot weather and deal with worsening wet weather. Flood risk areas should also be avoided when building. Buildings should use as few fossil fuels as possible and leave as little carbon impact as possible.[8]

8.3.2 Associations

The Continental Automated Buildings Association (CABA) is an international professional body with a presence in African countries such as Nigeria, Egypt, and Botswana. Its members include companies offering services to the connected home and intelligent building industries, software developers, building developers, owners, facility managers, governmental and research organisations, B2B groups, conference organisers, and media businesses.

On the academic side, no African institutions offer intelligent building courses. Also, more research on intelligent buildings needs to be done in Africa. Some of the research publications from Africa can be seen in Table 8.1.

8.4 Facilities Management and Intelligent Buildings

A building's condition can be brought close to its original state with facilities management. However, a gap continues throughout the building's lifespan because user expectations change as organisations do. This might result from information technology improvements that place more demands on the organisation and require a change of practical abilities. Technological and operational gaps can occur from the beginning of design, partly due to the time lag after conception, during which time alterations may occur. However, the maintenance gap only happens once the building is commissioned and starts to be used. However,

Table 8.1 Publications on Intelligent Buildings in Africa

Author	*Country*	*Findings*
Khashaba (2014)	Egypt	Identified essential elements for intelligent public buildings.
El-Motasem et al. (2020)	Egypt	Egypt needs more regulations on how to build an intelligent building, and there isn't enough study on SBPs, their design, management system, life cycle costs, and risks.
Eseosa and Temitope (2019)	Nigeria	Looked at how BMS can be used to make buildings' electrical and mechanical systems use less energy and how it can save money on costs in operations and repair.
Iwuagwu and Iwuagwu (2014)	Nigeria	The benefits of intelligent buildings include cost, environment, and greenhouse gas effects. Challenges include poor maintenance culture, lack of technical know-how, awareness, and erratic power supply.
Osunsanmi et al. (2020)	South Africa	The primary challenges with maintaining intelligent buildings are the cost, the occupants' behaviour, and how the buildings are maintained.

Source: Author

the facilities manager may significantly contribute at the planning stage by ensuring that the new building considers any changes made to earlier buildings. The building, the systems, and the people must all function and interact correctly, and the facilities manager is essential to this.[9]

Facilities management is necessary for intelligent buildings to identify needs, defend investments, and gain benefits. Facilities managers want intelligent buildings to manage single services, oversee facility performance, respond quickly to changing situations, and offer crucial management data.[10] There is a link between FM and IB information.[11] Facilities management often sees intelligent buildings as tools for managing buildings by putting communication and automation systems in the offices and other parts of the building. Powerful microprocessors and network technologies allow building management to get much information. Connecting, integrating, or opening facilities management and building automation systems is necessary to create a single software design by expanding the features of a standard management tool to include FM and building control networks. Buying, running, and keeping a lot of systems is expensive. Data duplication and coordinating systems from different vendors can cause problems.

Putting FM and IB together has been called "Integrated Facilities Management." IBS are similar to IB and FM in that they all use system integration, flexibility, adaptability, services, user efficiency, security, sustainability, and ICT. The IB and FM concepts bring together the control of a building and the actions of its users. On the other hand, FM systems need to be more flexible regarding space. To do FM duties, though, you need to be able to change how things are done. There are different kinds of flexibility, such as spatial flexibility and adaptation. What the building can do to meet the needs of its users in terms of tracking and controlling HVAC, energy use, and lights is called its "talent." IB systems often include mobility through building technology and active setups like automatic doors.

Intelligent building management and facilities management go hand in hand. Extensive IB hardware facilities and operations are often part of what FM experts mean when discussing facilities management. On the other hand, intelligent building experts often include important FM parts when describing what intelligent buildings are made of. These two terms cover a wide range of topics and points of view. Modern building control and management services are provided by IB systems that are very advanced and hard to understand. The IB method is the best way to track what each building facility needs to do. Another thing that makes an intelligent building more appealing is that it can effectively combine FM features with IB systems. Because IB systems are high-tech tools, FM can profit from them (Wang, 2009).

8.5 The Benefits and Challenges of Intelligent Buildings in Africa

Intelligent buildings are suitable for the environment because they follow codes for sustainable design, which helps them lower their carbon footprint and control how much energy they use when they don't need to. These projects are also good for the environment because they support using cleaner technologies like biogas. They make work more efficient and effective because they use a lot of different kinds of intelligence, mostly sensory (watching the environment, managing the visual aspects of the environment, making communication, network connectivity) and digital, electronic, and innovative technology.

Benefits include extending the life of the building and enhancing operations while lowering maintenance and management costs and improving facility dependability, efficiency, and performance. Additionally, they can increase occupancy rates and market rentals, giving facilities managers and renters access to security systems that communicate with other building systems.[12]

8.5.1 Benefits of Intelligent Buildings

Reduction in Building Expenses: With the Building Management System (BMS), building owners can expect significant long-term financial savings. A Building Automation System (BAS) primarily helps reduce utility bills by using different monitoring and control tools for electromechanical equipment. Employing these features will often make enough money to pay to install and use the BAS quickly.[13]

Improved Comfort and Productivity: Controlling the temperature inside a building better will give building owners more power over its comfort. The building will be warm and cooled more effectively and efficiently. The air circulation will also improve, affecting the facility's efficiency significantly. The buildings learn from occupants' behaviour. There has been increased productivity since using this long-term solution, and customer satisfaction has increased and complaints have decreased due to the increased comfort levels.

Environment: The main advantage of information technology's rapid progress has been the creation of systems that can measure, assess, and react to change. This has inspired progress in the architecture of our physical world, notably buildings.[14]

Greenhouse Rating: A molecule in the atmosphere known as a greenhouse gas (also abbreviated as GHG) can both absorb and emit thermal infrared radiation. This mechanism mainly causes the greenhouse effect. Buildings will become more environmentally friendly due to using a BAS because it reduces energy use and greenhouse gas production.

BMS and Energy Conservation Potential: The Energy Performance of Building Directive (EPBD) considers energy efficiency as "the actual consumption, calculated or expected quantities of energy necessary to cover different needs linked to the standardised usage of a building." Energy efficiency has become very important to building owners and investors, who see it as a way to make money. However, energy efficiency is quickly becoming a part of real estate management, facilities management, and operations planning, since there is more awareness of problems with energy use, and cost-effective technologies are improving.[15] Energy conservation is also becoming increasingly popular in the home construction industry. It has been said that a well-designed BMS can save up to 75% of the original circuit load in lighting energy, which is about 5% of all the energy used by residential and commercial properties. Furthermore, producing hot water or air cooling can save up to 10% of the energy used, equal to up to 7% of all the energy used by residential and commercial properties.[16]

Time-saving: Automating routine duties can save a lot of time. Intelligent buildings reduce staff downtime by reducing the time needed for maintenance.

Safety: Intelligent buildings can detect gas, water, and fire leaks. They can also self-diagnose and warn people when something goes faulty or performance decreases.

Health and Care: People's health is the most important thing; the proper controls for temperature, light intensity, and air conditioning reflect that.

Assistive Demotics: These can improve the quality of life for disabled and elderly individuals by providing a safe and comfortable place. This is good for the environment because it tries to eliminate CO2 pollution.

8.5.2 Challenges of Using Intelligent Buildings

BMS have been popular in developed countries for over 50 years. However, they haven't been used much in Africa for several reasons. Intelligent buildings can be challenging because they need significant capital in African countries (Ogunleye, 2012).

Some of these challenges include:
Technical know how
Since intelligent buildings are still a relatively new technology in many African countries, there is a critical shortage of qualified workers who can install the systems in intelligent buildings, particularly in older buildings. Additionally, because there are so few intelligent building specialists, it is expensive to hire their skills. Combined with the fact that the equipment needs to be imported from outside rather than made locally, the cost of the systems in intelligent building installation can be too much for most developers.

Awareness
There is a need to be more knowledgeable about using intelligent buildings and their advantages, including financial savings, energy efficiency, and environmental friendliness. Even among specialists, an intelligent building system is still not a common characteristic since modern computing arrangements in buildings are still in their infancy. The arrangement of intelligent technologies faces significant obstacles since computing is essential to their integration.

Power supply
Because the amount of electricity produced needs to be increased, a larger percentage of the population often uses solar panels or diesel generators to provide their electricity. Since a dedicated, uninterrupted power source would be necessary for the perfect

operation of intelligent buildings, this inadequate/interrupted power supply makes installing systems in intelligent buildings costly. Intelligent buildings employ fossil fuels to run generators. Burning fossil fuels releases greenhouse gases and may expose people to health risks.

Power reliability due to load shedding or power failure to ensure continuous functionality of the technology systems can be a problem. During power cuts, the systems lose set points, and this causes much discomfort for the occupants during low or high temperatures. When there are power cuts, the system takes time to reset. Sometimes, a draught can be too much for occupants, making the place too hot or cold, especially during a hot summer or cold winter. It is challenging to get the right ambience as the occupants have different temperature preferences.

Maintenance and upgrades

Commercial buildings in Africa have a tradition of poor maintenance. The intelligent building modules are expensive and require maintenance throughout their lifespan to fulfil their installation function. Additionally, equipment must be replaced due to the intelligent building's ongoing improvement. Most developers are discouraged by this since they want to avoid investing money, which means this machinery will soon be outdated.

Economy

The state of the country's economy is another problem that most developers care about. Developers of commercial buildings have had difficulty getting the funding they need for their projects. And even when they do, not everyone is willing to pay to lease or buy. Because of the economy, only a few developers put the Internet of Things in their buildings because they see it as a luxury.

High cost of installation

The initial cost of installing and upgrading systems is high. Financially, it can mean it takes longer to see the return on investment, resulting in senior management not being willing to invest in intelligent building systems.

Cybersecurity

Cybersecurity can be hacked. However, all gadgets linked to a single control network (Cloud) can show who the wrong users are.

Technological limitations

This can come from the small life span of the electronic equipment.

Gaps in analysing occupants' behaviour

Some organisations need more processes to control occupants' behaviour, such as misuse of equipment, which can cause systems to malfunction and not capture the required information. Users manually operating the system can disturb the system.

Without integrated facilities management, IB involves installing and using advanced and integrated building technology systems. When support services are not integrated, for example, security management not being part of the facilities management department, it can result in a delay due to procurement problems. This problem also makes planning for occupants difficult and makes measuring a building's efficiency in terms of energy consumption and sustainability hard.

The age of some buildings makes it difficult and expensive to modernise such facilities.

Intelligent building benefits	Intelligent building challenges
Reduced building cost	Technical know how
Improved comfort and productivity	Awareness
Environment	Power supply
Conservation	Maintenance
Green house rating	Economy
Time saving	Poor maintenance culture
Safety	High cost of installation
Health and care	Cybersecurity
Asssistive demotics	Technology

Figure 8.1 Intelligent buildings benefits and challenges.

Source: Author

8.6 Differences Between Smart Buildings, Green Buildings, and Intelligent Buildings

There are differences in the functional and technical components and the benefits of conventional, smart, green, and intelligent buildings. According to Buckman et al. (2014):

> Smart buildings integrate and account for intelligence, enterprise, control, materials and construction as an entire building system, with adaptability, not reactivity, at the core to meet the drivers for building progression: energy and efficiency, longevity, comfort and satisfaction. The increased amount of information available from this broader range of sources will allow these systems to become adaptable and enable a Smart Building to prepare itself for context and change over all timescales.

The goal of green buildings is to reduce the different adverse effects on the environment and resource use while increasing the positive impact over the life cycle of a building. Green building responsibilities include planning, designing, building, and running buildings with many environmental factors in mind, like making the best use of energy, water, and materials, improving the quality of the living environment, and lowering adverse effects on the environment. "Green building" is based on high performance, meaning energy efficiency can't be traded for lower comfort or indoor air quality.[17]

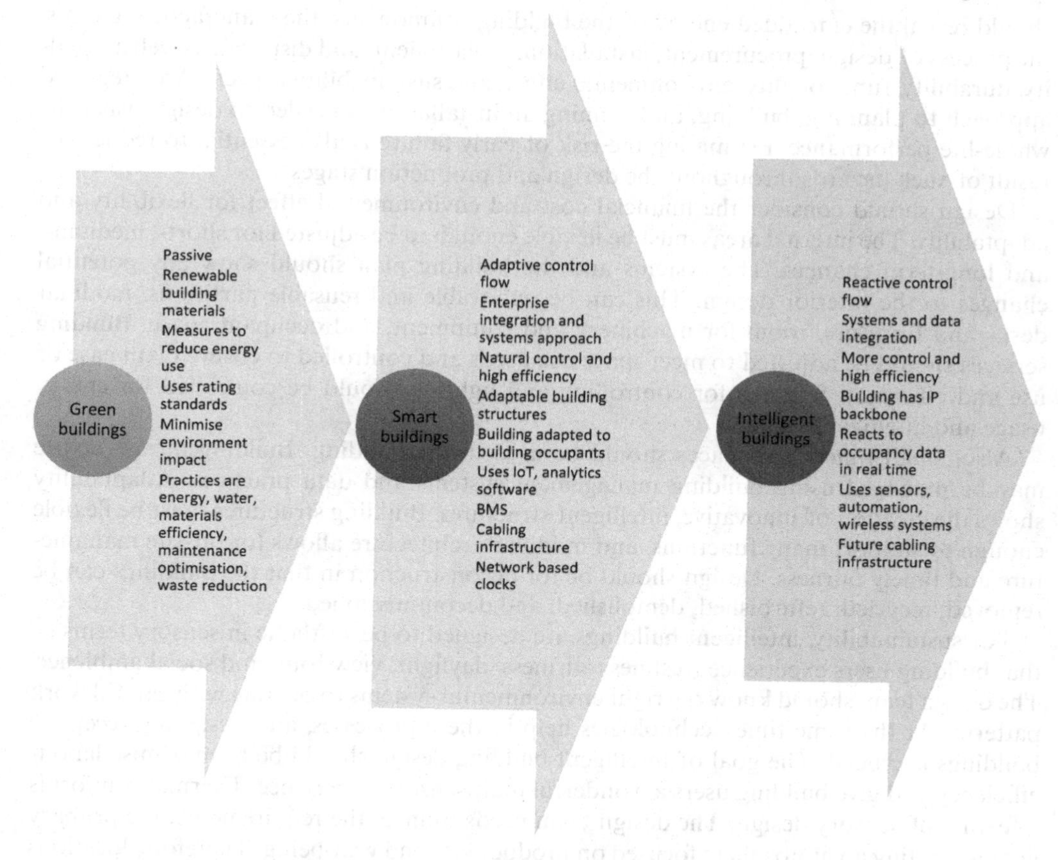

Figure 8.2 Differences between green, smart, and intelligent buildings.

Source: Author

Intelligent buildings need to be energy efficient. Intelligent buildings consider intelligence, enterprise, control, resources, and construction as a whole building life cycle to meet the building goals, such as energy and efficiency, longevity, comfort, and satisfaction. As more and better information comes in from different sources, these systems will be able to react. An intelligent building can get ready for context and change in all periods. Control, enterprise, building materials, and construction make systems that can think for themselves. Intelligent buildings focus on systems that can use information to improve buildings dynamically.[18]

8.7 The Structure of Intelligent Buildings

8.7.1 Design

The design of the building is also part of the structure consideration of intelligent buildings. The performance of a building over its whole lifespan, flexibility and adaptation, and sustainability should all be considered in the design of intelligent buildings. The design should reveal the embedded energy of the building components, the manufacturing costs, the process of design, procurement, installation, replacement, and disposal, as well as quality, durability, functionality, environmental effect, and sustainability aspects. An integrated approach to planning, building, and running an installation is needed to design using the whole-life performance. Estimating the risk of early failure is also essential to reduce the result of such hazards throughout the design and production stages.

Design should consider the financial cost and environmental effect for flexibility and adaptability. The internal areas must be flexible enough to be adjusted for short-, medium-, and long-term changes. The systems and the building plan should show any potential changes to the interior design. This can be removable and reusable partitions, modular desks and furniture, room for machinery and equipment, and occupant space. Building services should be adjusted to meet space use needs and controlled to ensure plant ease of use and efficiency. Systems for controlling and lighting should be controlled for energy usage and intelligent systems.

Also, small functional spaces should be used in the building. Building infrastructure may be managed using building management systems and data practices. Adaptability shows the creation of innovative, intelligent structures. Building structures must be flexible enough to support many functions, and modular architecture allows for off-site manufacture and timely fairness. Design should be for deconstruction in that the buildings can be removed, recycled, refurbished, demolished, and decommissioned.

For sustainability, intelligent buildings are designed to be aesthetic in sensory terms so that building users experience freshness, airiness, daylight, views out, and social ambience. The design team should know the right environmental systems to operate with careful work patterns. At the same time, technologies help in these processes, and designing receptive buildings is crucial. The goal of intelligent building design should be to maximise labour efficiency and give building users a wonderful multisensory experience. Thermal comfort is a feature of sensory design. The design team needs to meet the requirements; the primary goal is creating an atmosphere focused on productivity and well-being. Therefore, locations suitable for living and working should be included in intelligent buildings. This can only be done with creative thinking and sensitive design conceptualisation. The customer must specify the appeal of the spaces while making a new building or changing an existing one. The integrated team will assume these needs while seeing sensory aesthetics, function, sustainability, and value.[19,20]

The structure of intelligent buildings should consider modelling principles. Some of the necessary modelling principles are cost identification, reliability analysis, reliability-centred maintenance studies, replacement frequency analysis, energy use, and maintenance needs. The design includes involvement from suppliers and building industry experts and would lower maintenance needs and life cycle costs.

8.7.2 Structure

Intelligent buildings mainly consist of ten "Quality Environment Modules" (QEMs), which are as follows:

- Health and energy saving
- Making the best use of space and being flexible
- Cost-effective operation and maintenance with a focus on efficiency
- Comfort for people

- Work efficiency
- Measures for safety and security, such as fire, earthquake, tragedy, and damage to buildings
- Culture
- The concept of high technology
- The construction process and arrangement
- Health and sanitation

The ten modules mentioned above include the definition's first level, the fundamental level. In the second level, other demanded facilities or critical elements can be added later based on request.

A building must be designed intelligently. Different buildings can serve different purposes, such as residential, industrial, commercial (office or retail), educational, public service, religious, and transportation. Each building's different parts can be organised into categories based on importance (P1 is the highest priority, P8 is the lowest). Table 8.2 shows how modules can be put into four different kinds of buildings.

The initial set of eight key modules' individual critical modules can be given many variables known as "secondary" and "sub-factors." Figure 8.3 shows the evaluation criteria for intelligent buildings that affect and direct a building's life cycle obtained from the quality environment module.

Building management systems (BMS), building automation systems (BAS), sensors, smart materials, intelligent skins or interactive facades, and passive design methods are essential for intelligent buildings. This level of criterion selection will help the project team

Table 8.2 Second Level of Intelligent Building Criteria

Raised floor	Firefighting	Day lighting	Training
False Ceiling	Electrical services	Indoor touring guidance	Shared meeting and conference Services
Curtain wall	Plumbing and drainage	Public Address	Restaurants
HVAC	Maintenance Management	PABX	Entertainment areas
Roof and floor loading	Property management Asset	Office automation	Building directory
Floor height	and facility auditing	Parking and Public	Interior design Transportation
Raise space	Security control	Voice mail after hour operation	Emergency escape High speed
Fixture and furnishing	Indoor air quality		data communication
Vertical transportation	Cleaning	Energy saving and conservation	Satellite conferencing.
Building automation	Artificial Lighting	Trend logging and analysis	Internet Gateway
Fire detection	Structural Monitoring	Domestic hot water supply	Gas supply

Source: Omar, O. (2018). Intelligent building, definitions, factors and evaluation criteria of selection. *Alexandria Engineering Journal*, *57*(4), 2903–2910

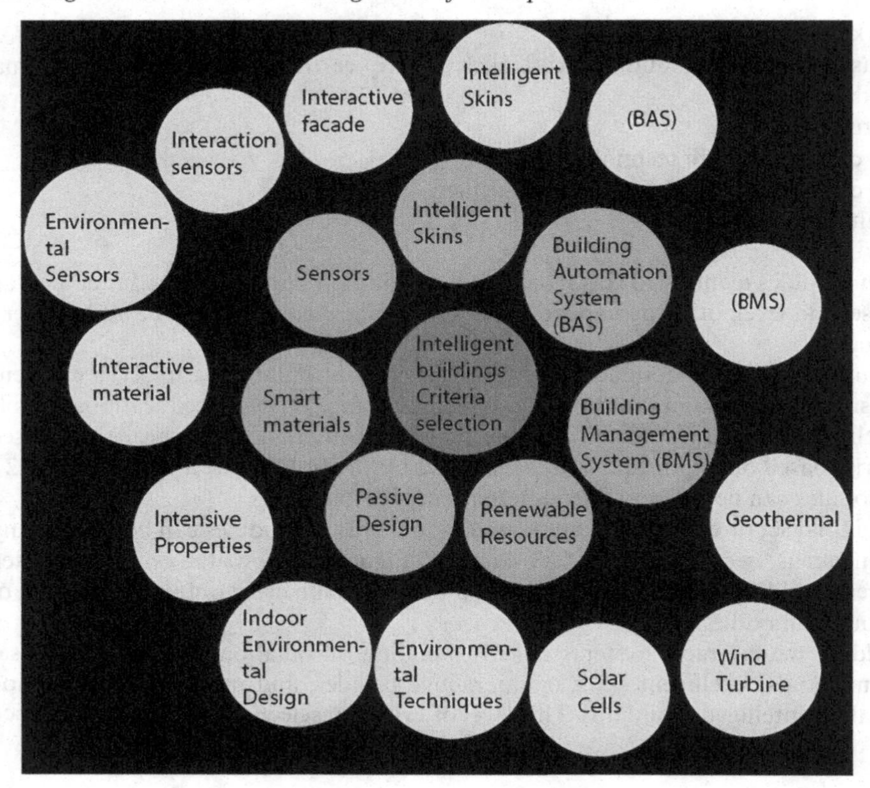

Figure 8.3 Intelligent buildings criteria selection – fundamental level and secondary level.

Source: Adapted from Omar, O. (2018). Intelligent building, definitions, factors and evaluation criteria of selection. *Alexandria Engineering Journal*, *57*(4), 2903–2910

reach its primary goal of efficient energy management by making good use of its resources and learning how to reduce energy consumption while still meeting high standards.

At the secondary level, most main parts are split into two or more parts to provide more ways to help reach the primary goal of reducing energy consumption. These include interactive facades, intelligent skins, BAS, BMS, environmental sensors, interaction sensors, interactive materials, intensive properties, indoor environmental control, environmental techniques, solar cells, wind turbines, and geothermal.

The core level, which has 64 parameters, includes the third level of evaluation criterion selection. This level ensures that the primary and secondary levels relate to the progress of intelligent building design's final goal of lowering energy consumption while making clean energy and CO2 emissions (Omar, 2018).

8.7.3 Work Productivity

8.7.3.1 Occupant Productivity

Productivity is linked to how well individuals reach their goals and is more easily defined and measured in offices or other commercial buildings. The performance of employees in offices is influenced by many psychological, social, environmental, and organisational factors. The behavioural standards that affect employees' social lives in the workplace, such as social contacts and collaborations, degree of privacy, and security, are discussed as personal and social aspects. Productivity is affected by the characteristics of an organisation's culture, structure, and connections.

The built environment's physical dimensions and Indoor Environment Quality (IEQ), office design, layout, warmth, colour, privacy concerns, thermal, visual, acoustic comfort, Indoor Air Quality (IAQ), view, and biophilia are examples of environmental influences on productivity. An intelligent building's energy management system can control occupants' IAQ and thermal and visual comfort during the building's operational stage. Some IEQ variables, such as interior plan and layout, biophilia and view, appearance and feel, colours, forms, texture, and artistic and cultural touches, should be considered throughout the design, building, or renovation stages. Energy management systems have little control over IEQ aspects like privacy concerns or acoustic comfort. Still, for people using common spaces like open-plan workplaces or libraries, these are the main sources of discomfort and a large source of lost productivity. The indoor/outdoor noise source(s) must either be removed from the indoor/outdoor environment or protected from the occupants to ensure acoustic comfort. Acoustic comfort is often measured concerning sound pressure levels, reverberation, or the magnitude of an acoustical quantity in decibels (dB).

8.7.3.2 Productivity Standards and Guidelines

Many building standards that include guidelines for making and operating green, sustainable, or energy-efficient buildings stress how important it is for people to be productive and for the world to be healthy. On top of that, these standards include suggestions and instructions for making things acceptable. The Green Building Council of South Africa (GBSCA) rates many things that have to do with designing and operating a building in a way that is good for the environment. These include management, energy, transportation, water, materials, land use and ecology, emissions, and new ideas. Quality of the physical environment can help people be more productive at work in many ways. The Green Pyramid Rating System (GPRS) in Egypt looks at things like the site's sustainability, how easy it is to get to, the environment, how much energy it uses, the materials and resources used, the IAQ, management, new ideas, and worth. The Leadership in Energy and Environmental Design (LEED) and the Building Research Establishment Environmental Assessment Method (BREEM) are used as guidelines all over Africa. Buildings that meet the worldwide WELL standard are more likely to make individuals satisfied, healthy, and productive.

8.7.3.3 Productivity and Indoor Environmental Quality

An intelligent building management system can control lighting, daylighting, and IAQ to make people more productive. Controlling the temperature, wind, and humidity inside can make people feel more comfortable. The amount and quality of light, sunshine, and glare all affect and explain visual comfort. The ventilation system can also control IAQ to let in the fresh air and lessen levels of internal air pollutants like CO2 and Total Volatile Organic Compounds (VOCs).

8.7.3.4 Productivity Assessment

Occupant productivity assessment and data collection are grouped into subjective assessment methods. Some examples are:

- Questionnaires and surveys about the comfort and performance of occupants
- Indirect assessment methods, such as observing and analysing performance
- Physical assessment methods

During the operations part, the building is evaluated and assessed again. This is called post-occupancy evaluation (POE). It can be done using one or more of the following methods – subjective methods, such as surveys and interviews, indirect methods, such as examining the number of office workers who are ill or absentee, or objective and direct methods, such as using sensors to measure the physical characteristics of the indoor environment.

Gathering and analysing users' feedback on indoor environment factors and other performance-related problems related to biased evaluation methods is essential. The different ways to ask and get replies are through web polls, emails, smartphone apps, and questionnaires. Many people have used the Building Use Studies Occupant Survey (BUS), the Building Assessment Survey and Evaluation (BASE), and The Leesman Index to determine how productive and satisfied people are. A multiple-point scale is often used to make these kinds of poll questions. For example, the Leesman rating method is based on an 11-minute poll that asks office workers about their activities, characteristics, and the facilities they offer. The operation areas are then scored based on how well they meet LMI standards. The BUS survey method is one of the most common survey methods. It has 45 questions about IAQ, heat comfort, personal control over the environment, reported output, and visual and audio comfort.

Indirect assessment techniques are based on converting results from people or organisations into occupant productivity. This goal can be reached in many different ways, whether at the individual or company level. Individual factors include showing up or not showing up for work, having physical or medical problems, expenditure on medical care, working speedily and efficiently, and meeting goals. Methods for direct assessment are based on measuring environmental variables like air temperature, air velocity, ventilation rate, daylight, level of illumination, glare, background noise, and the amount of air pollutants present. These variables are measured using environmental sensors and wireless sensor networks. Wearable tech should be used to connect these methods to physical characteristics, like body mass index, skin temperature, heart rate, metabolic rate, health, and weariness.

Experimental or observational approaches are used in physical and environmental direct evaluation methods to examine comfort and productivity. Most observing methods include passive occupant presence, preference, and adaptable behaviour. Motion detectors are often used as occupancy monitors because they can tell when someone comes in or leaves and when there is extra space. Sometimes, motion monitors are added to CO2 sensors to make them more accurate. Video cameras can track occupancy trends correctly as long as cost and privacy are not significant issues. Wearable technology and smartphones are two other ways to discover changes in usage trends. Multiple-input multiple-output (MIMO), wireless body reflection, and eye-tracking technologies are some virtual reality tools used in the observing method approach. Monitoring and tracking tools, such as touch or built-in sensors, can examine changes in energy-saving behaviours, how they interact with the building envelope, and how they use features like thermostats, fans, windows, and doors.[21]

8.8 Intelligent Building Systems

Buildings with independent control systems, like lighting, HVAC, security, and other new management and integrating technologies can be considered intelligent. The main parts of these devices are sensors and controls.

8.8.1 Building Control Systems

Intelligent buildings are comprised of independent parts that can be designed to work with more advanced systems. The following sections discuss some of these systems.[22]

8.8.1.1 Lighting Systems

Intelligent lighting systems can do many things, like turn lights on and off, change the amount of light and shade, and let you control the lights through your computer. Automated shading systems shift the angle of blinds based on outside factors like temperature and sunlight. Other examples include lighting with occupancy sensors to detect when someone is there, lighting with daylight-sensing photo sensors to adjust the lighting inside based on the amount of natural light and more (Gadakari & Mushatat, 2020).

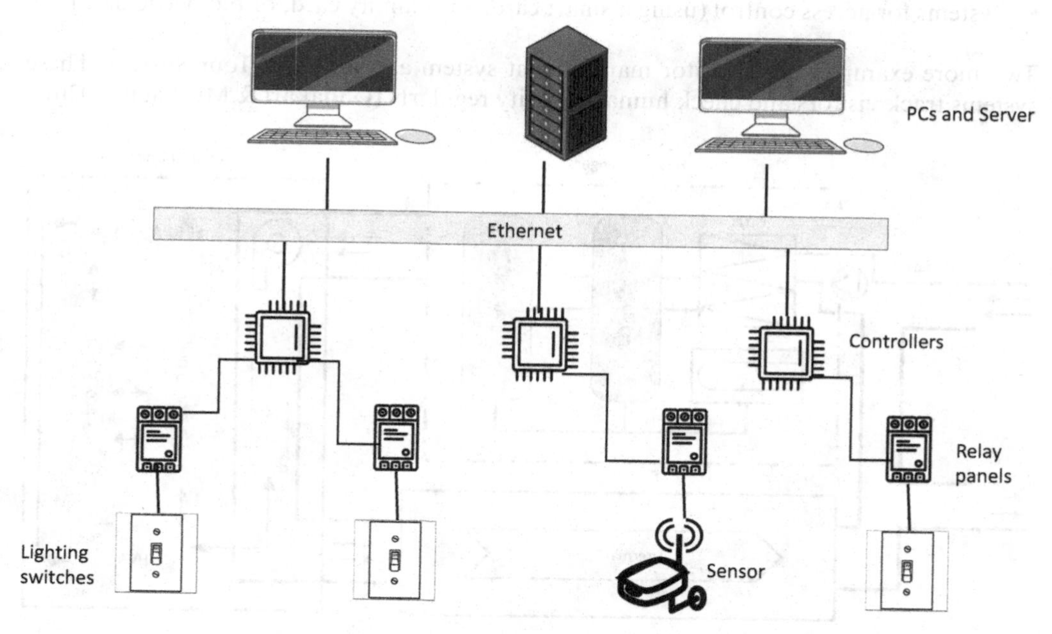

Figure 8.4 A typical lighting control system illustrates the efficient use of intelligent devices to control lights. These processors manage relay panels strategically distributed throughout a building. The computers and the system server are interconnected via an Ethernet network connection, facilitating seamless sharing of plans and updates. The controller, equipped with a user interface panel, simplifies the setup and monitoring of the lighting control system.

Source: Author

8.8.1.2 HVAC Control Systems

Intelligent HVAC systems can improve indoor air quality by checking the temperature and humidity or changing how often air changes depending on how many people are in the room. Sensors for temperature, moisture, and CO2 are used in these systems. It also has switches, transformers for transferring electricity from one point to another, control boards for regulating lights, and valves to control water flow. One can also conserve energy by setting up zones that only control HVAC in rooms that are being used or that turn off automatically when natural air is present. Temperature and CO2 sensors can control the louvres of doors and windows. In parking lots, carbon monoxide sensors are connected to pulse fans that move air and remove hazardous chemicals that can't be seen (Gadakari & Mushatat, 2020).

8.8.1.3 Security Systems

There are many things that clever security systems can do, which can be split into three groups:

- Systems for tracking (like smart carpeting that keeps track of people and their movements)
- Systems for surveillance (like cameras that can identify motion and faces)
- Systems for access control (using a smart card, a proximity card, or biometric data)

Two more examples are a visitor management system and a Guard Tour System. These systems track visitors and check human security regularly (Gadakari & Mushatat, 2020).

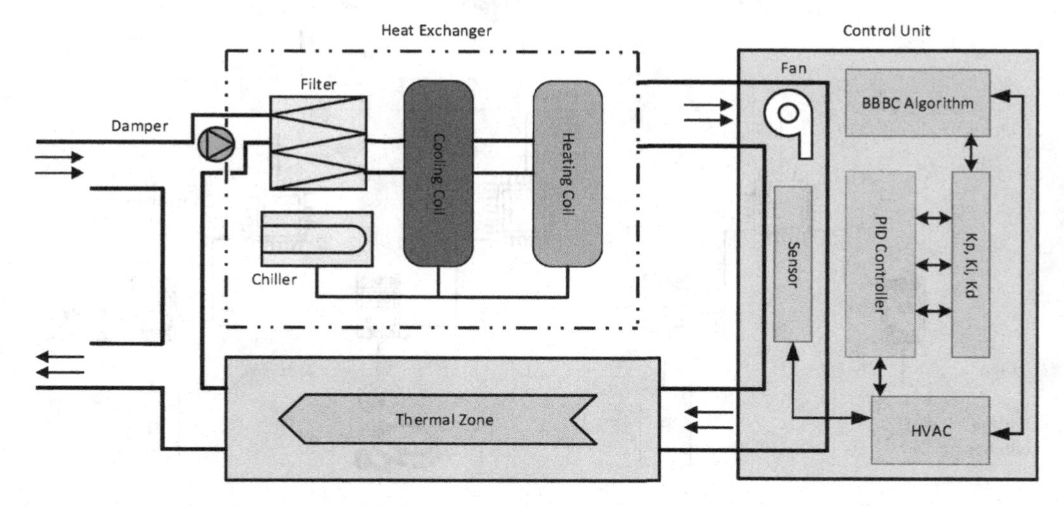

Figure 8.5 HVAC system and control unit (Almabrok et al., 2018). This system has a variable-speed fan that can change the airflow and water flow rates from the chiller to the heat exchanger. Based on the system output signal, the proportional–integral–derivative (PID) controller adjusts the motor speed to maintain a set point.

Source: Almabrok, A., Psarakis, M., & Dounis, A. (2018). Fast tuning of the PID controller in an HVAC system using the big bang-big crunch algorithm and FPGA technology. *Algorithms*, *11*(10), 146

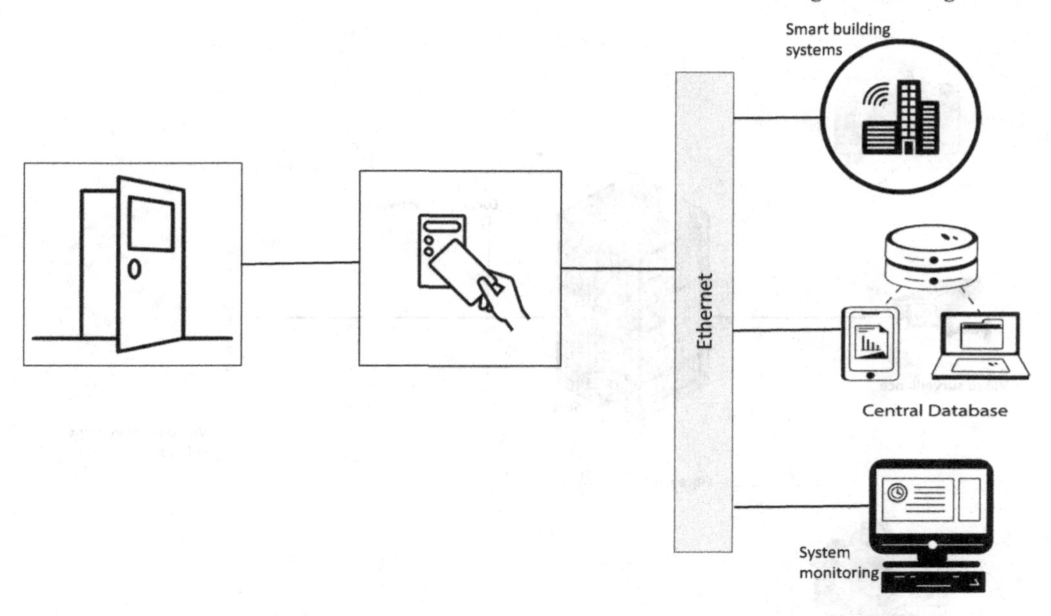

Figure 8.6 The access control system shows how information is spread to the local control panel when a user arrives at a door and submits an access card containing encoded credential information to a card reader. In a system with a centralised database, the host computer receives the data from the control panel, information about the door's position, and the time the card was presented. After information verification, the host computer compares the access level to the location of the door and the time of day. After confirming, the host computer sends an order to unlock the control panel's door, giving the user entrance to the building.

Source: Author

8.8.1.4 Video Surveillance Systems

Closed-circuit television (CCTV) or video surveillance systems are part of a facility's total security and life safety plan. The strategy may include operational and physical security measures and extra security or life safety technologies, such as access control and intrusion detection. Legal factors, including an individual's right to privacy and belief in security, must be considered while installing video surveillance systems. Systems for video surveillance have been determined by analogue technology for many years (Sinopoli, 2009).

8.8.1.5 Fire Alarm Safety Systems

A strong trend in intelligent fire systems is based on their ability to identify and protect against fire reliably. Integrated fire alarm systems connect fire alarm monitors to the fire alarm panel (see Figure 8.8), which lets the fire department discover the fire, swiftly shut down the building, set off escape systems, and control other safety systems. Life safety systems come in many forms, some of which can identify gas leaks, earthquakes, and floods.

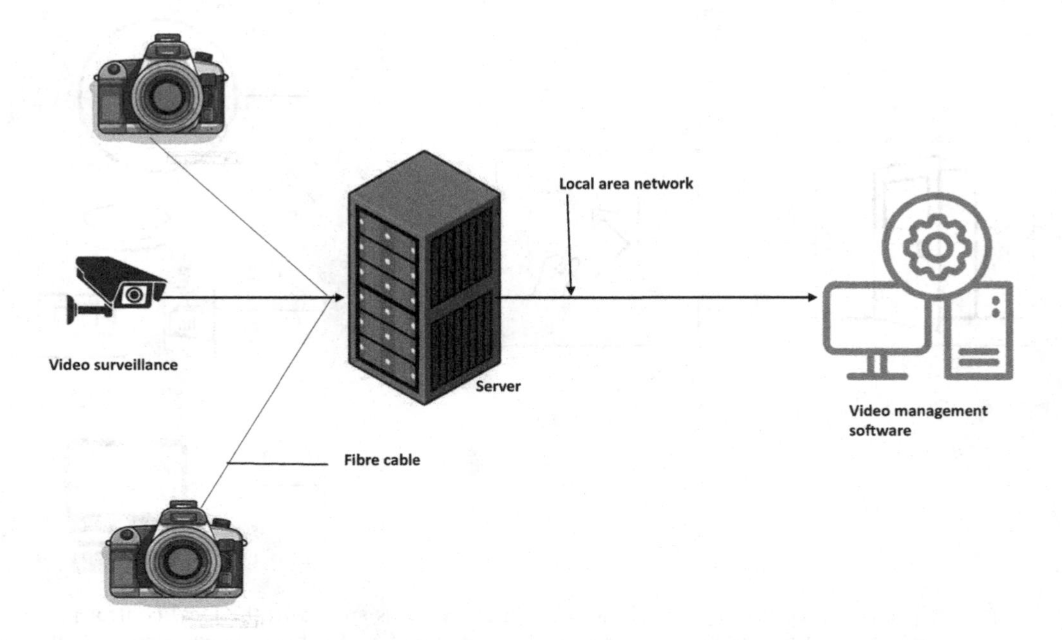

Figure 8.7 Video surveillance that links to a video server links to video management software on a data network server. Analogue or digital cameras link to the server in this setup, which also links to the network. A video surveillance system focuses on the video server, an ordinary data network hardware. The server's network connectivity offers remote or network-based recording, storing, viewing, and system administration options.

Source: Author

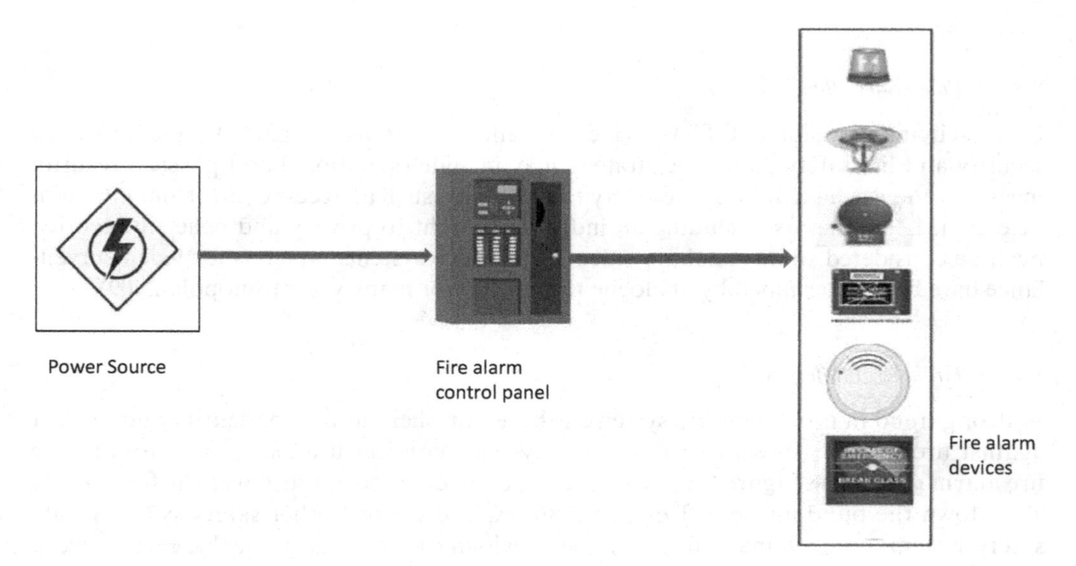

Figure 8.8 Fire alarm control panel.

Source: Author

8.8.1.6 Feedback and Display Systems

Building occupants should be taught about building data, such as operating and environmental data, signage, and interfaces, using feedback and display systems. Most of the information on digital signage is related to operational metrics gathered in real-time while events are happening, room assignments, control options, and other information (Gadakari & Mushatat, 2020).

8.8.2 Integrative Building Systems

Subsystem interactions are vital to intelligent buildings. To make a building intelligent and responsive, the core services of each component must be controlled and monitored by a shared network that makes the whole system homogeneous and interoperable. The independent subsystems should be interconnected with integrative building systems (Gadakari & Mushatat, 2020).

The Building Management System (BMS), Energy Management System (EMS), and Facilities Management System (FMS) are all parts of the BAS. The BMS manages and observes the coordinated central operation of all building systems to increase efficiency. A good BMS should be a technologically advanced, open system that works with products from different companies and meets industry standards. It should also be easy to put in buildings that are already there without much work. It should be made so that in an emergency, critical services can still be given separately (Gadakari & Mushatat, 2020).

A good BMS should be a technologically advanced, open system that works with products from different companies and meets industry standards. It should also be easy to put in buildings already there with little work. It should be made so that in an emergency, critical services can still be given separately.

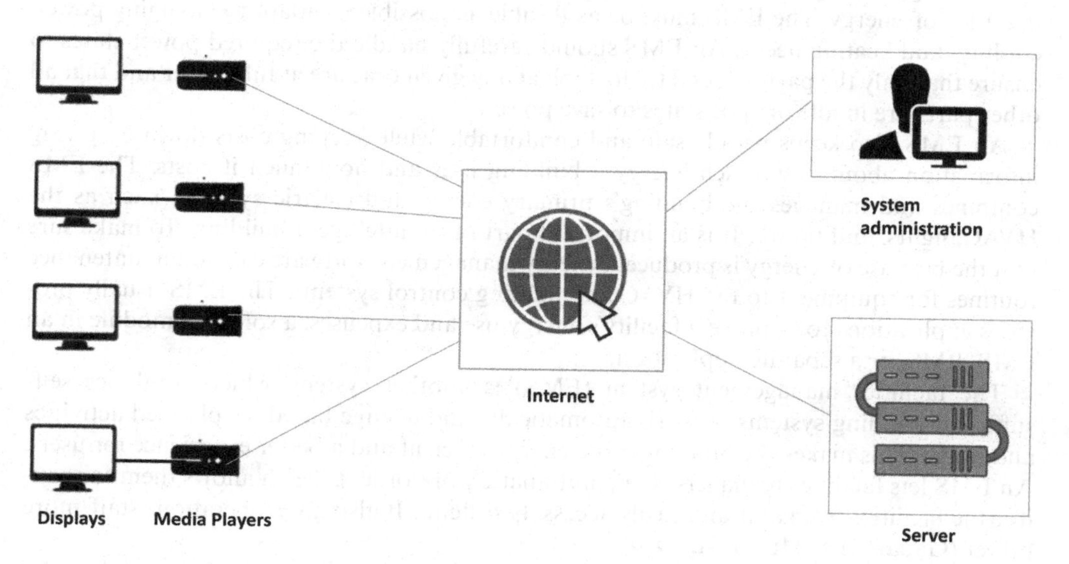

Figure 8.9 Digital signage systems.
Source: Author

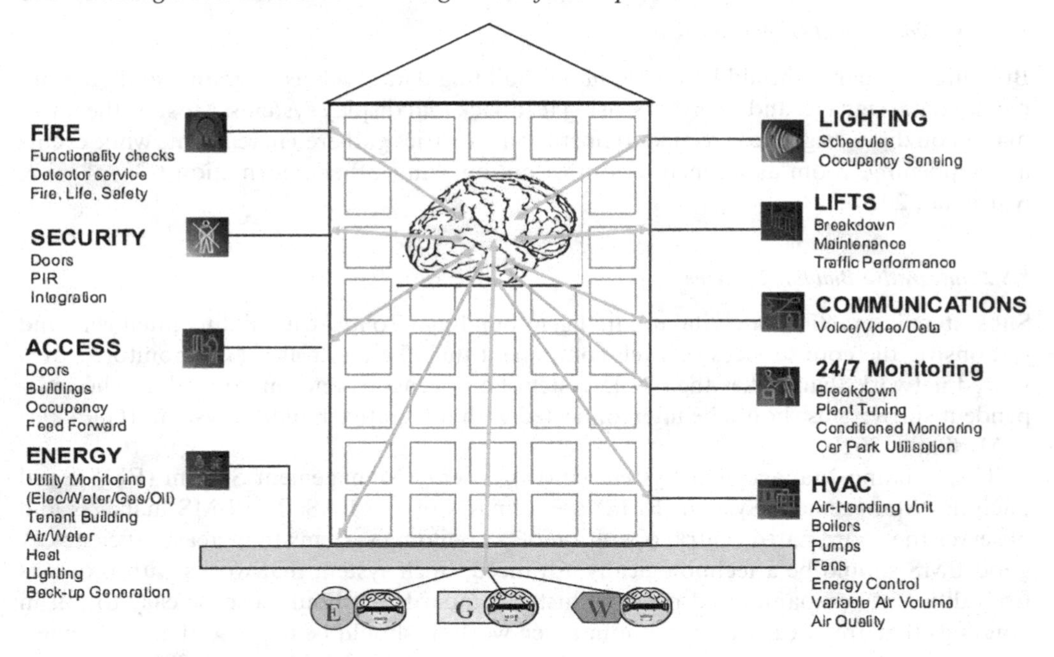

Figure 8.10 Major elements of an intelligent building.
Source: Continental Automated Buildings Association (www.caba.org)

An energy management system (EMS) manages the self-sufficient building systems that use a lot of energy. The EMS must be as flexible as possible to adapt to changing power, cooling, and heating needs. An EMS should carefully handle the required power states to ensure that only the parts needed to do a job at any given time are at full power and that all other parts are in idle or split states to save power.[23]

An EMS also keeps people safe and comfortable while keeping costs down by giving information about how much energy a building uses and how much it costs. The EMS combines and manages the building's primary energy and electric systems, such as the HVAC, lights, and power. It is an important part of an intelligent building. To make sure that the best use of energy is produced, energy management software can add maintenance routines for equipment to the HVAC and lighting control systems. The EMS usually provides applications to improve a facility's energy use and expenses, a software module in an FMS, BMS, or a separate application.

The facilities management system (FMS) is another system, which combines self-sufficient building systems to work automatically and change based on planned activities and usage. This makes the building more energy efficient and a better experience for users. An FMS lets facilities managers set up automatic work orders, which allows them to monitor the facilities from afar and easily access its systems. It also gives operations staff more power (Gadakari & Mushatat, 2020).

An FMS is also a tool that helps manage service requests, inventory, sales, and assets, focusing on facilities management tasks. Companies that specialise in applications for

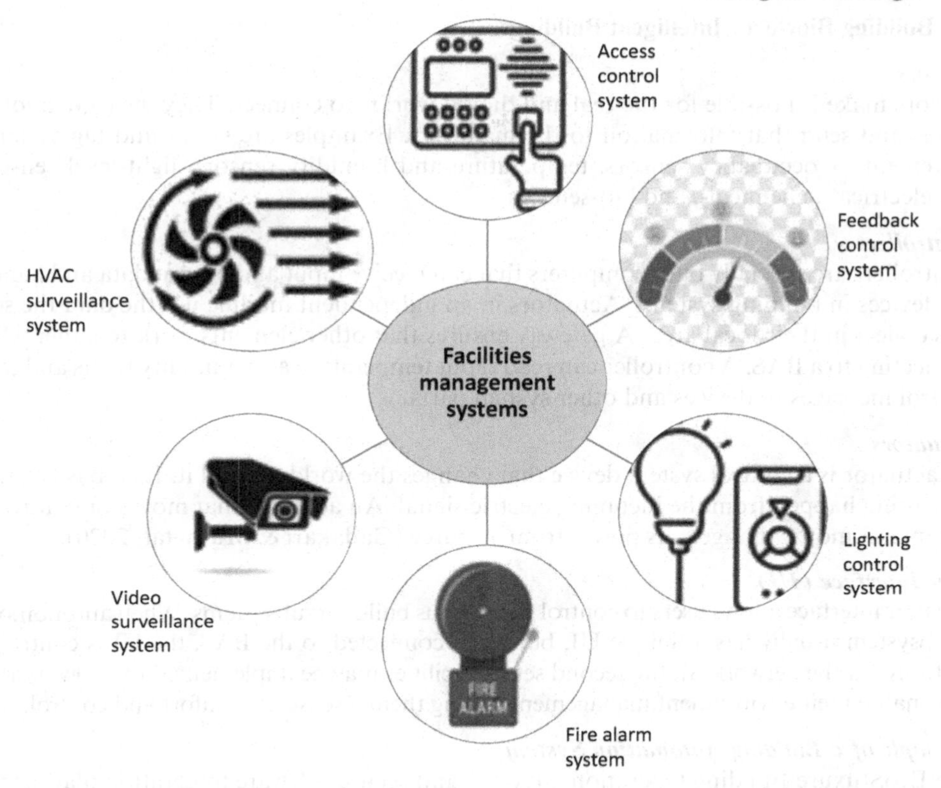

Figure 8.11 Building management systems.

Source: Author

facilities management or offer a full range of business process services, such as purchasing, accounting, and human resources, often sell these systems (Gadakari & Mushatat, 2020).

8.8.3 Integrated Communication Systems (ICS)

These include IP-enabled devices, communication methods, local area networks, sharing networks, telephones, and IT help. Transmission systems used for building control include BACnet, LonWorks, Modbus, PROFIBUS, and European Installation Bus (EIB). An Industrial Control System (ICS) is a network that connects and shares information between building technologies. It should be able to handle data and information on a solid network and adapt to new situations. BACnet is a data transmission system for BA and control networks based on the LonWorks protocol and ANSI/EIA 709.1 Control Networking Standard. Modbus is a widely used system in industrial automation, allowing devices like PLCs, HMIs, and meters to communicate using Ethernet and common serial standards. PROFIBUS is an open standard for automation and field networks since 1989. EIB is a fieldbus designed to improve electricity setups in homes and buildings by removing control information transfer from main wires (Wang, 2009).

8.9 Building Blocks of Intelligent Buildings

Sensors

Sensors make it possible for the real and digital worlds to connect. They measure a lot of things and send that information to the managers. Examples are status and tag systems, movement or occupancy sensors, temperature and humidity sensors, light-level sensors, and electrical-, chemical-, and bio-sensors.

Controllers

Controllers are specially built computers that can receive input and output data and handle the devices in building systems. Actuators in an independent module use the data the sensors collect in their local area. A gateway ensures that other elements work together when connecting to a BAS. A controller can read input temperature and humidity levels and send control messages to devices and other system parts.

Actuators

An actuator is a control system device that changes the world around it. Its job is to make something happen from the incoming electric signal. An actuator that moves or controls a system is a motor that gets its power from a source (Gadakari & Mushatat, 2020).

User Interface (UI)

The user interface allows users to control the various building subsystems. When autonomous, a subsystem usually has a unique UI, but when connected to the BAS, the UI is controlled centrally via the network. In the second setup, facilities may be implemented to allow users to personalise their environment management, giving them a sense of comfort and control.

Example of a Building Automation System

The EcoStuxure Building Operation, an open and secure software integration platform, is an example of a BAS from Schneider, a provider of systems in South Africa. Collaboration across complex third-party designs is made possible by it. Industry-recognised open protocols are supported, including BACnet, Modbus, LON, and Zigbee. The SmartX hardware and software components of the system are also BTL-certified. Over an Ethernet IP, EcoStruxure Building links hardware, software, and services securely. Edge control, open architecture, plug-and-play, IoT-enabled applications, analytics, and services are provided. It supports the building, data centre, industry, and infrastructure sectors and encompasses power, IT, buildings, machines, plants, and the grid.

The system's advantages include making buildings future-ready and guaranteeing new technological advancements that will be simpler to accept and implement. It further assists with digital transformation, giving them an advantage in the current digital economy. Digitalisation and big data enable organisations to exchange information and analyse workplace management, energy, lighting, HVAC, and security systems. It is simple to use since it provides visibility across large estates and allows access to building data anytime, anywhere, through a laptop, tablet, or smartphone. It can optimise comfort and productivity, boost engineering efficiency by up to 30%, and raise building value.

Living space sensors and cabling management are part of the IP field controllers and accessories. It is scalable and flexible, and it can:

- Allow data transmission from connected equipment
- Speed up the diagnosis and resolution of problems
- Connect to mobile phones for added convenience
- Improve the market-leading cybersecurity[24]

Figure 8.12 Connected Products: SmartX IP Controllers and SmartX Living Space Sensors.

Source: Schneider Electric. (2017). *Unlocking value, unleashing productivity*. Accessed October 2022 at 12589_SE_ESX_Building_Mgt_bro_v9.2.pdf (walshadv.com)

Case Studies

Case Study 1

Eastgate Centre, Harare, Zimbabwe

The Eastgate Centre, mainly composed of concrete, includes a ventilation system that functions similarly to the self-cooling mounds of African termites. The structure draws influence from native Zimbabwean stonework in addition to biomimicry. The facility maintains a consistent temperature throughout the year without using traditional air conditioning or heating, thus lowering energy use. The structure consumes 10% less energy than a typical building of the same size. Natural ventilation has allowed Eastgate's owners to save $3.5 million, which has trickled down to the tenants, who now pay 20% less rent than those who live in nearby buildings.

Figure 8.13 Dar Headquarters, Egypt.

Source: El-Motasem, S., Eid, A.F., & Khodeir, L.M. (2020). Identification of key challenges of smart buildings projects in Egypt. *The Academic Research Community Publication*, *4*(1), 34–43

Case Study 2

Dar Headquarters, Egypt

Dar's office facility is in Egypt's Smart Village in Giza. It has 42,300 m2 of space inside and cost $100 million to build and costs about $70,000 a month to run.

The features of the building were pointed out as follows:[25]

- LEED Gold was achieved for the building, and LEED Gold is now the goal for running and upkeep
- 1000 SM photovoltaic panels on the roof collect solar energy and meet about 5% of the building's energy needs
- A big glass entryway that lets in natural light through walls inside and outside. This courtyard is meant to keep the outside temperature from affecting the temperature inside the building
- An 18 m2 area of the atrium roof's movable skylight was set aside for the passive smoke exit system
- When sustainable methods are used together, they cut energy use by about 30%
- Saving water for both inside the building and for watering plants
- Daylight monitors to control the lights in the office
- When there is enough natural light, the lights turn off immediately
- A facilities management system was put in place during testing and launching
- BMS (building management system) and CCTV to keep the building safe and handle and run its systems
- Electronic access control – the building's monthly costs make up 0.07% of its total cost

Case Study 3

The subject of case study 3, which is an anonymous case study, is a company in Johannesburg which runs trains and occupies a space on the ground floor of an office building. The area is constantly occupied because the activity runs around the clock.

For example, concern was raised about the atrium's layout and air conditioning system. The atrium's atmosphere was unsuitable for 24/7 operation due to the structural architecture of the structure, which was the root of the problem. A group of engineers created a long-term and sustainable solution in 2016. There were two ways to approach the project. One was to offer a short- to medium-term solution to minimise business interruptions, particularly in the early morning hours when wintertime lows of 2 degrees were common.

Short-term solutions:
The maintenance service provider put into practice the following measures to deal with faults and be proactive. In the case of a temperature fluctuation, text the technician. To adapt set points to ambient temperature, use the BMS. In this method, when the temperature decreased before the occupants complained, a text message would be sent to the technician immediately. When the temperature changes, the BMS system detects it, and the technician remotely modifies the air conditioning to match the new temperature.

These were the advantages of this solution:

- Response times improved, which temporarily decreased labour and the number of consumer complaints
- Because of the more favourable surroundings, the residents felt more comfortable
- The user accessed the information digitally before experiencing it, meaning they could quickly report the fault to the helpdesk

Table 8.3 Long-Term Solutions

Findings	*Solution*
The structure is a hot well (like a greenhouse). High heat gains during summer and high heat loss during winter. HVAC system trying to cool or heat a glass box, which is a total waste of energy	Install infill glass with an air gap to reduce the amount of radiant heat entering the building during summer and the amount of internal heat lost during winter.
The sidewall air distribution system must be corrected for the structure's air conditioning. Supplying air in cooling mode, a dumping effect is created as cold air drops. Supplying air in heating mode, the air rises without meeting people as hot air rises.	Provide a displacement system of introducing air from the ceiling level (just above the screen top level) downwards. The air supply quantities for this method can be reduced so that dumping and draughts are not created.
Finding	*Solution*
The existing heat pump chiller system model cools water during cooling mode and heats water during heating mode. The cold/hot water is then circulated to the 2 air Handling Units. The time delay between these 2 Units is +/–30 minutes as the water needs to be heated from 6°C to 43°C or cooled from 43°C to 6°C, which is one of the complaints that it takes too long to attain comfort temperatures.	Installation of new two split package systems, one in each of the existing AHU plant rooms, with the condensing sections mounted outside and screened off. If one system is faulty and breaks down, 50% capacity is still available from the other system serving the atrium.

Source: Anonymous

Challenges:

- During power outages, the BMS lost the set points, which made the occupants extremely uncomfortable during cold or hot temperatures
- The technician's phone had to be on at all times to receive the SMS message, occasionally resulting in delayed answers
- Users manually controlled the system while it was powered on, causing problems; in some cases, they turned the system off
- The remedy did not resolve the issue with the building's air conditioning and architecture

Figure 8.14 shows the long-term solution.

The implemented solution:

- Temperature control
- Insulated roof and façade with heat-reduction glass film
- Ducting and air supply were installed across the area. They were installed just above the screen top level and suspended from the columns and trusses.
- The chiller plant was replaced with two modern energy-efficient HVAC units.
- Floor grilles were installed on the floor to provide heating from the bottom upwards to prevent loss of heat caused by the dome
- Provided insulation under the slab of the atrium floor to prevent heat loss to the basement

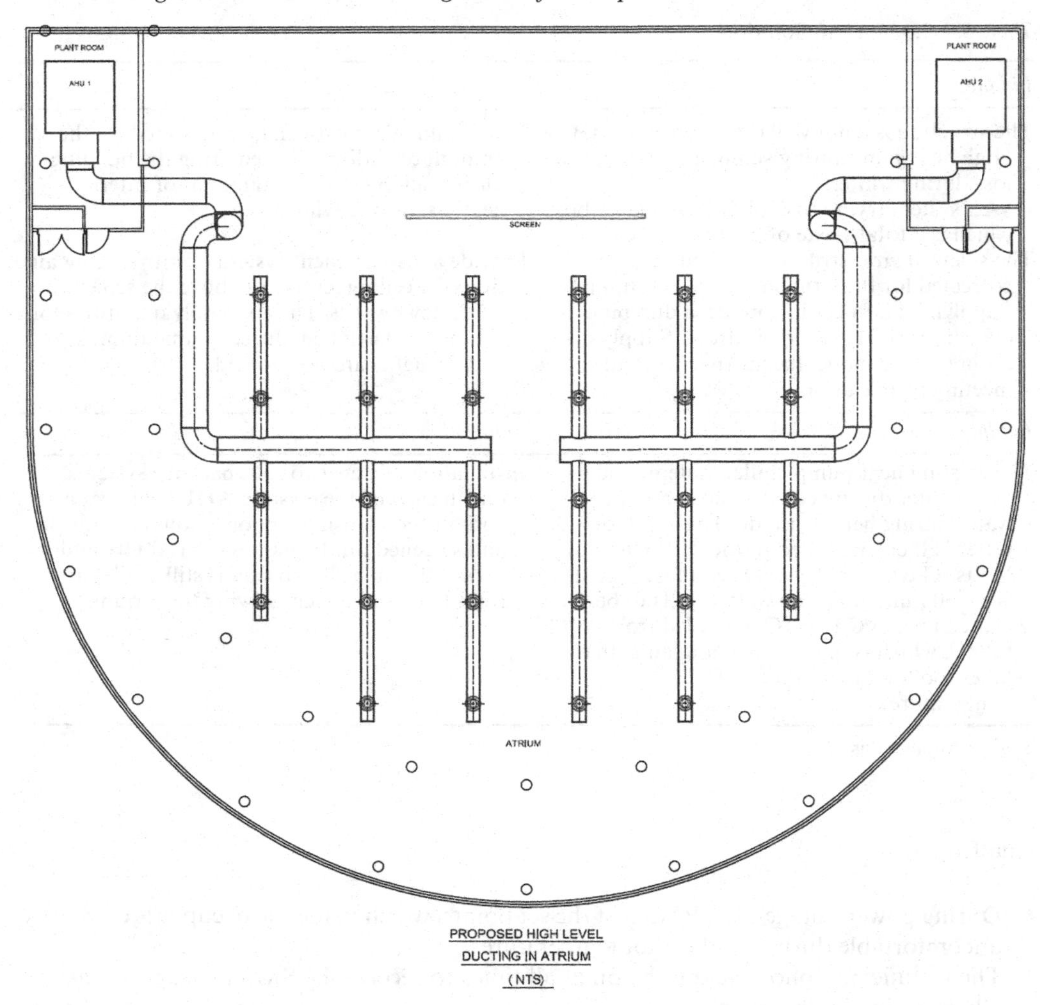

Figure 8.14 Proposed air distribution system.

Source: Anonymous

Humidity Control

- Provide indoor plants as this will provide additional humidity
- Temperature control
- Duty managers to take charge of the air conditioning control room
- Install thermometers in the atrium area to monitor the temperature

Case Study 4

This is a corporation in Johannesburg, South Africa, manages properties and uses the Schneider Building Management System. This system combines energy management, energy monitoring, and HVAC. This system can recognise energy power outlets and assign workspaces based on demand and user needs. The strategy resulted in a 30% decrease in utility bills. Because the system includes an autonomous data point that transmits warnings when unusual behaviour is identified, it gives us information that helps us consume less energy and water. Because the system can manage illumination, the lights and air conditioners automatically switch off when the room is empty, preventing water leaks before they create waste and lowering energy costs. Tenants enjoy a convenient and cosy atmosphere thanks to data control and monitoring.

Smart Building Solutions utilised in the organisation:
Class A (5 Sites) – Schneider BMS

- Monitor power incoming, before, and after the UPS (Voltage ranges)
- Environmental monitoring (Temp/Humidity/Water/Smoke)
- It comes with a SIM card for monitoring
- The technology, which uses the cloud to monitor all the sites, delivers alerts and can be accessed using a web browser
- The PowerTap Link C device and the site modem are connected by an RJ45 connection (which would be fitted with an MTN SIM card)
- PowerTag Link C transforms a switchboard into a linked device. The restaurant's router connects to the Internet, obtains data from the switchboard, and transmits it to a secure Schneider Electric cloud server
- PowerTap Energy is a wireless energy sensor that monitors circuits and loads. There are several varieties, and it sends data to PowerTap Link C
- The PowerTap Link C receives temperature measurement information from the PowerTap Ambient in T°C. Depending on the installation, a relay antenna may be needed for improved communication

Case Study 5

The Edge, Amsterdam

This building is the most intelligent and sustainable in the world, with a BREEAM rating of over 98%, and is owned by Deloitte. It was opened in 2014, developed by OVG Real Estate, and designed by PLP Architecture. It uses sustainability principles, technology, the Internet of Things, flexibility, and user experience. The type is an office and covers a gross floor area of 51,000 square metres. The building has 28,000 sensors linked to a network that coordinates the workplace and people. It gathers information on people's preferences and uses that to organise activities in the space. The building produces more energy than it consumes due to its design and solar and geothermal energy use. The building has helped its users be productive and has reduced costs and negative environmental influences.[26]

Lessons learned

- That using smart technology in buildings has helped to achieve sustainability and collaboration and innovation among users
- Cost savings can be achieved through using value engineering of the building concrete facade
- Many of the technologies were new so those that were not ready to take the risk were not ready to use them
- Some of the technologies did not meet building codes
- Consideration of users' needs at the beginning of design is important
- Involve the regulatory authorities right from the design stage in using new technology

Table 8.4 The Structure of the Edge Building

Design	*Structure*	*Work productivity*	*Building systems*
Modelling The design was based on a model on the same side of the table, where all professionals had the same goal. 3D computer models were used to get the ideal building shape **Flexibility** Each user is connected to the building using an app to book spaces depending on the work that will be done. Users can personalise the indoor air quality. Use of an app to book coffee and lockers. Each worker is responsible for his activity for the day and conditions the building to accommodate the activity. **Passive features and sustainability** Passive features for sustainability include parking for bicycles and electric charging points for cars Windows that opened into the atrium or with an external view Use of additional nearby rooftops to place solar panels for energy Each façade is detailed in line with its goal and orientation Ventilation systems circulate air through rooms pushing Rainwater harvesting for irrigation of landscape and flushing toilets Use of the earth as a battery to store energy Change management in that the building can be converted to other uses in future The use of power over Ethernet of computer networks allows reduced energy consumption. Using licence plate recognition to access parking improves traffic flow and reduces pollution from traffic.	Solar panels to generate energy Lighting is powered by Ethernet cables with sensors for temperature, light, and air. Light panels are linked to an app. The atrium in the building for collaboration and lighting in the space and faces the north The use of solar panels for energy Windows that opened into the atrium or with an external view Use of facade	BREEAM Certification at 98% Use of post-occupancy evaluation and collection data on the use of the building that is used for maintenance.	BIM is integrated with IoT and uses digital cartography BIM is used for safety The systems are connected to all aspects of the building infrastructure, such as HVAC, security, lighting, temperature, access control, and fire safety. Automatic signalling with visitor management Use of coded light and level control by users' phones. Use of licence plate recognition to access parking Use of a single platform for integration and systems

Source: Author

- Work closely with suppliers on their innovation for marketing and use of their new technology
- The use of data brought about an increase in the capacity of those who manage it
- The success has brought about the integration of the real estate, facilities management and technology departments as they are relevant to each other, and it helps buildings to be more responsive to predictive maintenance
- Since many professionals are involved, it helps with cooperation and risk sharing among the professionals[27]

8.10 Conclusion

Intelligent buildings need the right combination of building services, design, structure, information technology, facilities management, and environment. Factors that drive or limit the use of intelligent buildings are economic, legal, environmental, organisational and technological factors. Facilities management is necessary for intelligent buildings to identify needs, defend investments, and gain benefits. Intelligent building management and facilities management go hand in hand. Extensive IB hardware facilities and operations are often part of what FM experts mean when discussing facilities management. On the other hand, intelligent building experts usually include important FM parts when describing what intelligent buildings are made of. The chapter touched on the benefits and challenges of intelligent buildings. The differences between smart, green and intelligent buildings are that smart buildings integrate and account for intelligence, enterprise, control, materials and construction as an entire building system, with adaptability, not reactivity, at the core to meet the drivers for building progression while the goal of green buildings is to reduce the different adverse effects on the environment and resource use while increasing the positive impact over the life cycle of a building. Intelligent buildings consider intelligence, enterprise, control, resources, and construction as a whole building life cycle to meet the building goals, such as energy and efficiency, longevity, comfort, and satisfaction.

Regarding the structure of intelligent buildings, managers should ensure the design reflects the performance of a building over its whole lifespan, flexibility and adaptation, and sustainability, which should all be considered in the design of intelligent buildings. The structure should meet health, energy savings, cost savings, space management, comfort, work efficiency, safety and security, culture, technology and cleanliness requirements. Intelligent building systems should be used in the buildings such as building control systems, integrative building systems and integrative communication systems.

8.11 Guidelines

- Switch to a proactive maintenance approach
- Gain top management support in implementing IB
- Gain stakeholders' support in implementing the concept
- Use friendly communication channels to support users' participation
- Managers should consider the cost implications of running intelligent buildings
- Managers should get the right training on technology for IBs
- Ensure there is local competence in the management of imported technology
- Ensure the buildings are connected to functional alternative energy sources
- Ensure there is knowledge of regulatory requirements for IBs
- In the use of equipment, ensure there is compliance with manufacturer's guidelines

- Have competence on life cycle analysis of buildings
- Network with other professionals to gain knowledge of IB management
- Ensure that the design takes into consideration life cycle analysis, flexible, adaptable and sustainable
- The structure should meet requirements for health, energy savings, cost savings, space management, comfort, work efficiency, safety and security, culture, technology and cleanliness
- Intelligent building systems should be used in the buildings, such as building control systems, integrative building systems and integrative communication systems
- Ensure that the technology is good in terms of connectivity and interoperability and takes care of cybersecurity

8.12 Questions

1 What are intelligent buildings?
2 What is the link between facilities management and intelligent buildings?
3 What are the differences between green, smart, and intelligent buildings?
4 Discuss the development of intelligent buildings in Africa
5 Discuss the development of intelligent buildings in your country
6 What features of intelligent buildings are in your work buildings?
7 What will be the benefits and challenges of implementing intelligent buildings in your workplace?
8 Discuss any three components of intelligent building systems

Notes

1 Gadakari, T., Hadjri, K., & Mushatat, S. (2016). Relationship between building intelligence and sustainability. In *Proceedings of the Institution of Civil Engineers - Engineering sustainability* (Vol. 170, No. 6, pp. 294–307). Thomas Telford.
2 Owusu-Manu, D.G., Ghansah, F.A., Ayarkwa, J., Edwards, D.J., & Hosseini, R. (2022). Factors influencing the decision to adopt Smart Building Technology (SBT) in developing countries. *African Journal of Science, Technology, Innovation and Development*, *14*(3), 790–800.
3 Wang, S. (2009). *Intelligent buildings and building automation.* Routledge.
4 Omar, O. (2018). Intelligent building, definitions, factors and evaluation criteria of selection. *Alexandria.*
Engineering Journal, *57*(4), 2903–2910.
5 Shen, L.S., Durba, A., Di Sui, P.E., David Williams, P.E., & Wriedt, N. (2021). Intelligent Buildings Literature. Center for Energy and Environment, USA. Accessed September 2022 at IB Literature Review (ashb.com).
6 Modise, A. (2016). South Africa joins Nations of the World in ratifying the Paris Agreement on Climate Change. Department of Environmental Affairs.
7 Cilliers, Z., & Euston-Brown, M. (2018). *Aiming for zero-carbon new buildings in South African metros.*
8 Roberts, S. (2008). Effects of climate change on the built environment. *Energy Policy*, *36*(12), 4552–4557.
9 Clements-Croome, T.D.J. (1997). What do we mean by intelligent buildings? *Automation in Construction*, *6*(5), 395–400.
10 Kelly, C. (1996). Limitations to the use of military resources for foreign disaster assistance. *Disaster Prevention and Management: An International Journal*, *5*(1), 22–29.
11 Himanen, M. (2003). *The intelligence of intelligent buildings: The feasibility of the intelligent building concept in office buildings.* VTT Technical Research Centre of Finland.
12 Ogunleye, B.M. (2012). Management of intelligent building in Nigeria.

13 Domingues, P., Carreira, P., Vieira, R., & Kastner, W. (2016). Building automation systems: Concepts and technology review. *Computer Standards & Interfaces*, *45*, 1–12.
14 Iwuagwu, U., & Iwuagwu, M. (2014). Adopting intelligent buildings in Nigeria: The hope and fears. In *2nd international conference on emerging trends in engineering and technology* (Vol. 10, No. 6, pp. 160–163).
15 Shaikh, P.H., Nor, N.B.M., Nallagownden, P., & Elamvazuthi, I. (2018). Intelligent multi-objective optimization for building energy and comfort management. *Journal of King Saud University-Engineering Sciences*, *30*(2), 195–204.
16 Papadopoulou, E.V. (2012). Energy efficiency and energy saving. In *Energy management in buildings using photovoltaics* (pp. 11–20). Springer.
17 Zhang, L., Wu, J., & Liu, H. (2018). Turning green into gold: A review on the economics of green buildings. *Journal of Cleaner PProduction*, *172*, 2234–2245.
18 Buckman, A.H., Mayfield, M., & Beck, S.B. (2014). What is a smart building? *Smart and Sustainable Built Environment*.
19 Clements-Croome, D., & Croome, D.J. (Eds.) (2004). *Intelligent buildings: Design, management and operation*. Thomas Telford.
20 Clements-Croome, D. (2011). Sustainable intelligent buildings for people: A review. *Intelligent Buildings International*, *3*(2), 67–86.
21 Mofidi, F., & Akbari, H. (2020). Intelligent buildings: An overview. *Energy and Buildings*, *223*, 110192.
22 Gadakari, T., & Mushatat, S. (2020). The development and validation of an ontology of intelligent buildings. *Global Dwelling: Approaches to Sustainability, Design and Participation*, *193*, 101.
23 Sinopoli, J. M. (2009). *Smart buildings systems for architects, owners and builders*. Butterworth-Heinemann.
24 Schneider Electric. (2017). *Unlocking value, unleashing productivity*. Accessed October 2022 at 12589_SE_ESX_Building_Mgt_bro_v9.2.pdf (walshadv.com).
25 El-Motasem, S., Eid, A.F., & Khodeir, L.M. (2020). Identification of key challenges of smart buildings projects in Egypt. *The Academic Research Community Publication*, *4*(1), 34–43.
26 The World Economic Forum. (2017). Is this the world's smartest, greenest office building? Accessed January 2024, https://www.weforum.org/agenda/2017/03/smart-building-amsterdam-the-edge-sustainability/.
27 The World Economic Forum, The Edge, Case study by Boston Consulting Group as part of the future of Construction Project at the World Economic Project, World Economic Forum Switzerland.

Bibliography

Almabrok, A., Psarakis, M., & Dounis, A. (2018). Fast tuning of the PID controller in an HVAC system using the big bang-big crunch algorithm and FPGA technology. *Algorithms*, *11*(10), 146.

Buckman, A.H., Mayfield, M., & Beck, S.B. (2014). What is a smart building? *Smart and Sustainable Built Environment*.

Cilliers, Z., & Euston-Brown, M. (2018). *Aiming for zero-carbon new buildings in South African metros*.

Clements-Croome, D. (2011). Sustainable intelligent buildings for people: A review. *Intelligent Buildings International*, *3*(2), 67–86.

Clements-Croome, D., & Croome, D.J. (Eds.) (2004). *Intelligent buildings: Design, management and operation*. Thomas Telford.

Clements-Croome, D.J. (1997). What do we mean by intelligent buildings? *Automation in Construction*, *6*(5), 395–400.

Domingues, P., Carreira, P., Vieira, R., & Kastner, W. (2016). Building automation systems: Concepts and technology review. *Computer Standards & Interfaces*, *45*, 1–12.

El-Motasem, S., Eid, A.F., & Khodeir, L.M. (2020). Identification of key challenges of smart buildings projects in Egypt. *The Academic Research Community Publication*, *4*(1), 34–43.

Eseosa, O., & Temitope, F. (2019). Review of smart based building management system. *World J Innov Res*, *7*(2), 14–23.

Gadakari, T., Hadjri, K., & Mushatat, S. (2016). Relationship between building intelligence and sustainability. In *Proceedings of the Institution of Civil Engineers - Engineering sustainability* (Vol. 170, No. 6, pp. 294–307). Thomas Telford.

Gadakari, T., & Mushatat, S. (2020). The development and validation of an ontology of intelligent buildings. *Global Dwelling: Approaches to Sustainability, Design and Participation*, *193*, 101.

Himanen, M. (2003). *The intelligence of intelligent buildings: The feasibility of the intelligent building concept in office buildings*. VTT Technical Research Centre of Finland.

Holmes, C.K. (2019). *South Africa's readiness of the smart built environment towards 2035*. MBA thesis, Nelson Mandela University, South Africa.

Iwuagwu, U., & Iwuagwu, M. (2014). Adopting intelligent buildings in Nigeria: The hope and fears. In *2nd international conference on emerging trends in engineering and technology* (Vol. 10, No. 6, pp. 160–163).

Kelly, C. (1996). Limitations to the use of military resources for foreign disaster assistance. *Disaster Prevention and Management: An International Journal*, *5*(1), 22–29.

Khashaba, S. (2014). *The use of intelligent buildings to achieve sustainability through an architectural proposal for public buildings in Cairo*. World Sustainable Buildings. Madrid: GBCe, 216–225.

Modise, A. (2016). South Africa joins Nations of the World in ratifying the Paris Agreement on Climate Change. Department of Environmental Affairs.

Mofidi, F., & Akbari, H. (2020). Intelligent buildings: An overview. *Energy and Buildings*, *223*, 110192.

Ogunleye, B.M. (2012). *Management of intelligent building in Nigeria*.

Omar, O. (2018). Intelligent building, definitions, factors and evaluation criteria of selection. *Alexandria Engineering Journal*, *57*(4), 2903–2910.

Osunsanmi, T.O., Aigbavboa, C.O., Oke, A., & Onyia, M.E. (2020). Making a case for smart buildings in preventing corona-virus: Focus on maintenance management challenges. *International Journal of Construction Management*, 1–10.

Owusu-Manu, D.G., Ghansah, F.A., Ayarkwa, J., Edwards, D.J., & Hosseini, R. (2022). Factors influencing the decision to adopt Smart Building Technology (SBT) in developing countries. *African Journal of Science, Technology, Innovation and Development*, *14*(3), 790–800.

Papadopoulou, E.V. (2012). Energy efficiency and energy saving. In *Energy management in buildings using photovoltaics* (pp. 11–20). Springer.

Roberts, S. (2008). Effects of climate change on the built environment. *Energy Policy*, *36*(12), 4552–4557.

Schneider Electric. (2017). Unlocking value, unleashing productivity. Accessed October 2022 at 12589_SE_ESX_Building_Mgt_bro_v9.2.pdf (walshadv.com).

Shaikh, P.H., Nor, N.B.M., Nallagownden, P., & Elamvazuthi, I. (2018). Intelligent multi-objective optimization for building energy and comfort management. *Journal of King Saud University-Engineering Sciences*, *30*(2), 195–204.

Shen, L.S., Durba, A., Di Sui, P.E., David Williams, P.E., & Wriedt, N. (2021). *Intelligent buildings literature*. Center for Energy and Environment, USA. Accessed September 2022 at IB Literature Review (ashb.com)

Sinopoli, J.M. (2009). *Smart buildings systems for architects, owners and builders*. Butterworth-Heinemann.

Umair, M., Afzal, B., Khan, A., Rehman, A.U., Sekercioglu, Y.A., & Shah, G.A. (2018). Self-configurable hybrid energy management system for smart buildings. In *2018 15th International Conference on Control, Automation, Robotics and Vision (ICARCV)* (pp. 1241–1246). IEEE.

Wang, S. (2009). *Intelligent buildings and building automation*. Routledge.

Wigginton, M., & Harris, J. (2013). *Intelligent skins*. Routledge.

Zhang, L., Wu, J., & Liu, H. (2018). Turning green into gold: A review on the economics of green buildings. *Journal of Cleaner Production*, *172*, 2234–2245.

Index

Note: page numbers in *italics* refer to information in figures; page numbers in **bold** refer to information in tables.